P9-ANZ-977

Algebraic Topology
Homology and Cohomology

Algebraic Topology
Homology and Cohomology

Andrew H. Wallace

University of Pennsylvania

W. A. Benjamin, Inc.
1970
New York

Algebraic Topology: Homology and Cohomology

Standard Book Number 8053–9482–6
Library of Congress Catalog Card Number: 79–108005
Manufactured in the United States of America
12345 M 43210

W. A. Benjamin, Inc.

New York, New York 10016

Preface

This text is intended as a two-semester course in algebraic topology for first or second year graduate students. It is sufficiently self-contained to be taken as a first course in algebraic topology, although in my own teaching I have usually preferred to precede this material by a more elementary treatment, such as is given in my *Introduction to Algebraic Topology* (Pergamon Press, 1957, henceforth referred to as *Introduction*). Prerequisites for this course are introductory courses in general topology and algebra; but in fact these could be running concurrently since, at the beginning of the study of algebraic topology, all the student needs to know are the definitions of topological spaces, continuous maps, groups, and homomorphisms. The use of any deeper properties does not occur till later.

Basically, the aim of algebraic topology is to attach topologically invariant algebraic structures to topological spaces. Here the term "invariant" means that homeomorphic spaces will have isomorphic structures attached to them. The idea behind this, of course, is that if the algebraic structures corresponding to two spaces are not isomorphic then it follows that these spaces are not homeomorphic. In other words, topologically invariant structures are tools for distinguishing topological spaces from one another.

In this text several algebraic invariants are constructed and studied: the fundamental group, singular and Čech homology groups, and various kinds of cohomology groups. The latter are shown to have a ring structure that considerably increases their strength as topological invariants; that is, it happens that spaces that cannot be distinguished from each other by looking at their homology or cohomology groups can be distinguished by constructing their cohomology rings.

As in *Introduction*, the fundamental philosophy adopted here is that topology is a form of geometry. Accordingly, geometrical motivations and interpretations are given as much emphasis as possible. On the other hand, as soon as the student goes beyond the construction of the singular homology groups it becomes clear that there are certain underlying algebraic patterns running through the whole subject. So it becomes desirable to leave the geometry aside at certain moments (for example, in Chapter 3) to study these patterns as pure algebra. Perhaps the more thoroughly logical approach would be to present all the algebra first, so that the homology and cohomology theory could be started in the same way as, for example, analytical geometry is usually begun, namely, with all the necessary algebraic tools already available. Such an approach appears to be unsuitable for beginners, however, since the algebra involved here has no readily available motivation other than in the homology and cohomology theory to which it is to be applied, and without motivation the definitions would almost certainly appear rather artificial.

Briefly, then, the arrangement of the course is as follows. In Chapter 1, singular homology theory is described. In a sense, this is a review of the material of Chapters 5–8 of *Introduction*, but it differs in two respects from the treatment given there. In the first place, the exposition here moves a bit faster; in the second (more important) place, many of the arguments are carried out in a more algebraic way than in *Introduction*. This more algebraic approach leads very naturally to the introduction of the axiomatic method of studying homology that is briefly described in Chapter 2. And of course, it also motivates the notion of algebraic complexes, which are described in Chapter 3, and which provide the algebraic foundation of all that follows. Chapter 2 also contains an account of the homology theory of simplicial complexes, with methods of computation. Cohomology makes its first appearance in Chapter 3, where it is treated in the singular and simplicial cases. Chapter 4 describes the ring structure on cohomology groups. Chapters 5, 6, and 7 describe the Čech theories.

There are two appendixes. Appendix A contains a discussion of the fundamental group. This is not relegated to an appendix because of any inferiority or lack of importance, but simply because it is an independent topic, which does not fall naturally into the sequence of subjects described in Chapters 1–7. Appendix B is a reminder of the various ideas and theorems of general topology which are needed throughout the text.

The student is urged to treat the exercises as an integral part of the text. They give practice in the style of thinking involved in algebraic topology; besides, many of them are results that are used elsewhere in the exposition.

Philadelphia, Pennsylvania, July 1969 ANDREW WALLACE

Contents

I

Singular Homology Theory

The motivation for singular homology lies in the fact that certain topological spaces can be distinguished from each other (i.e., can be shown to be nonhomeomorphic) because every closed curve on one forms the boundary of a surface in that space, while this property is not satisfied by the other space. For example, it is intuitively clear that every closed curve on a sphere is a boundary, while on the torus in Fig. 1 the curve α is not a boundary. In fact, if the torus is cut along α we have a piece of surface whose boundary is not α but two copies of α. More generally, the bounding or nonbounding properties of r-dimensional surfaces in a space can sometimes be used to distinguish that space from nonhomeomorphic spaces.

To make this more precise and suitable for algebraic treatment, the notion of r-dimensional surface in a space E is replaced by the notion of a linear combination of singular simplexes on E. Geometrically, this is thought of as a surface built up from simplexes embedded in E, and indeed this geometrical picture motivates all the main theorems of singular homology theory, which is outlined in this chapter.

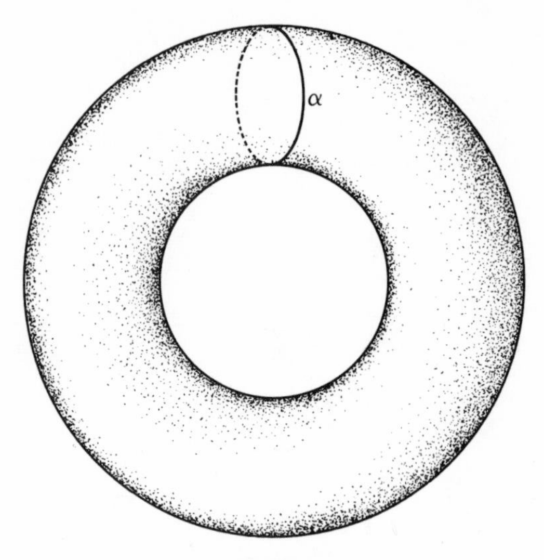

FIGURE 1

1-1. EUCLIDEAN SIMPLEXES

Definition 1-1. For $i = 0, 1, \ldots, r$ let $y_i = (y_i^1, y_i^2, \ldots, y_i^N)$ be a point of Euclidean N space. These $r + 1$ points are called *linearly independent* if the matrix

$$\begin{pmatrix} y_0^1 y_0^2 & \cdots & y_0^N 1 \\ y_1^1 y_1^2 & \cdots & y_1^N 1 \\ \vdots & & \vdots \\ y_r^1 y_r^2 & \cdots & y_r^N 1 \end{pmatrix}$$

is of rank $r + 1$.

It should be checked—and this is left as an exercise—that if $y_0, y_1, \ldots, y_r$ are linearly independent, then the smallest linear subspace of N space containing all these points is of dimension r. Conversely, in an r-dimensional linear subspace it is always possible to find $r + 1$ linearly independent points.

Definition 1-2. Let $y_0, y_1, \ldots, y_r$ be linearly independent points in Euclidean N space E_N. Let S be the set of points of E_N whose coordinates are given by the formulas

$$y^i = \sum_{j=0}^{r} \lambda_j y_j^{\ i}$$

where the λ_j runs through all sets of real numbers satisfying the conditions

$$\lambda_j \geqq 0, j = 0, i, \ldots, r$$

$$\sum_{j=1}^{r} \lambda_j = 1$$

Then S is the *Euclidean r simplex* with vertexes $y_0, y_1, \ldots, y_r$. Usually S will be denoted by $[y_0 y_1 \cdots y_r]$.

Exercises. 1-1. A convex set S in a Euclidean space is one such that if two points belong to it then the segment joining them is contained in it. Prove that, in Definition 1-2, S is the smallest convex set containing $y_0, y_1, \ldots, y_r$.

1-2. Prove that the λ_j in Definition 1-2 are uniquely determined by the point y. If the same point y is also given by

$$y^i = \sum_{j=1} \mu_j y_j^i$$

$$\mu_j \geqq 0$$

$$\sum \mu_j = 1$$

then $\mu_j = \lambda_j$ for each j. (To prove this use the linear independence condition on the y_i.)

1-3. Prove that $[y_0 y_2 \cdots y_r]$ is a point, line segment, triangle, or tetrahedron when r = 0, 1, 2, 3, respectively.

Definition 1-3. Continuing with the notation of Definition 1-2, the λ_j are called the *barycentric coordinates* of the point y in S. This definition makes sense since, by Exercise 1-2, they are uniquely determined by y.

Definition 1-4. Let x_i be the point $(0, \ldots, 0, 1, 0, \ldots, 0)$ with 1 in the $(i + 1)$ place and all other coordinates zero, in Euclidean $(r + 1)$ space. The r simplex Δ_r with these as vertexes (clearly they are linearly independent) is called the *standard Euclidean r simplex.*

Note that in this case the barycentric coordinates of a point coincide with the Euclidean coordinates.

As a notational convention, $x_0, x_1, \ldots, x_r$ always denote the vertexes of Δ_r; different letters are used for the vertexes of any other simplex. This convention will cause no ambiguity provided that, for each n, Euclidean n space is identified with the hyperplane $x^{n+1} = 0$ in $(n + 1)$ space.

Associated with a Euclidean simplex are subsets, also simplexes, which are called *faces.* They are defined as follows.

Definition 1-5. Let S be a Euclidean r simplex and let $p + 1$ of its vertexes be picked. For convenience, let the vertexes be numbered so that those selected are $y_0, y_1, \ldots, y_p$. Then the Euclidean p simplex with vertexes $y_0, \ldots, y_p$ is called a *p-dimensional* face of S. The simplex with vertexes $y_{p+1}, \ldots, y_r$, the remaining vertexes of S, is called the *face opposite* the face $[y_0 y_1 \cdots y_p]$.

Note that a simplex of dimension r has faces of all dimensions from 0 to r according to the number of vertexes chosen as $y_0, \ldots, y_p$ in Definition 1-5. In particular, there are $r + 1$ faces of dimension 0, namely, the vertexes, and one r-dimensional face, namely, the simplex S itself. Opposite to each vertex there is an $(r - 1)$-dimensional face. These $(r - 1)$-dimensional faces usually are thought of as faces in the elementary geometrical sense, that is, the triangular faces of a tetrahedron for $r = 3$, or the sides of a triangle for $r = 2$. Here, however, a tetrahedron, for example, has, in the more general sense, faces of dimension 0 (vertexes), of dimension 1 (edges), of dimension 2 (faces in the ordinary sense), and of dimension 3 (the whole tetrahedron). Also in the case of a tetrahedron, opposite one-dimensional faces are opposite edges in the usual geometrical sense, namely, nonintersecting edges.

Exercise. 1-4. Show that a simplex S is the point-set union of all segments joining a point of one face to a point of the opposite face.

1-2. LINEAR MAPS

Definition 1-6. Let S and T be Euclidean simplexes of dimensions p and q respectively, not necessarily in the same Euclidean space. Let $y_0, y_1, \ldots, y_p$ be the vertexes of S and let $z_0, z_1, \ldots, z_p$ be vertexes of T, not necessarily all of them and not necessarily all distinct. Let f map the point y of S, where

$$y^i = \sum \lambda_j y_j{}^i$$
$$\lambda_j \geqq 0 \qquad \sum \lambda_j = 1$$

on the point z of T, where

$$z^i = \sum \lambda_j z_j{}^i$$

Then f is called the *linear map of S into T* which maps the vertexes $y_0, y_1, \ldots, y_p$ of S into the vertexes $z_0, z_1, \ldots, z_p$ of T.

Definition 1-6 clearly defines a map of S into the Euclidean space containing T. The fact that the image of f is actually contained in T should be checked as an exercise. The following exercises present some of the essential properties of linear maps.

Exercises. **1-5.** In the notation of Definition 1-6, the image of f is a face of T. In fact, it is the face whose vertexes are the distinct vertexes of T contained in the set $z_0, z_1, \ldots, z_p$.

Note that this result emphasizes the fact that f is determined as soon as its values on the vertexes of S are assigned.

1-6. In the notation of Definition 1-6, the barycentric coordinates of $z = f(y)$ are linear functions of those of y. In addition, the Euclidean coordinates of z are linear functions of those of y. Show also that if f is a map from one Euclidean space to another which expresses the coordinates in the second space as linear functions of those in the first, then f induces a linear map on any simplex in the first space.

This result implies that the linear map f is continuous if S and T are given the induced topologies as subspaces of their respective Euclidean spaces. In addition, of course, it justifies the name "linear map."

1-7. If, in particular, $p = q$ and $z_0, z_1, \ldots, z_p$ are all the vertexes of T, then f is one-to-one and is, in fact, a homeomorphism. Also f^{-1} is the linear map of T into S which maps the vertexes $z_0, z_1, \ldots, z_p$ of T on the vertexes $y_0, y_1, \ldots, y_p$ respectively of S.

1-8. If S, T, U are Euclidean simplexes and $f: S \to T$ and $g: T \to U$ are linear maps, then gf is a linear map of S into U.

1-9. Let $y_0, y_1, \ldots, y_p$ be the vertexes of S and let S_0 be the face opposite y_0. Let f be a linear map of S_0 into a simplex T and let z be a vertex of T. For each $y \in S_0$ let g map the segment $y_0 y$ linearly on the segment $zf(y)$, that is, a point dividing the first segment in a given ratio is mapped on the point dividing the second in the same ratio. Prove that g is a linear map of S into T.

1-3. SINGULAR SIMPLEXES AND CHAINS

Definition 1-7. A *singular simplex of dimension p, or singular p simplex* on a topological space E is a continuous map σ of the standard Euclidean p simplex Δ_p into E.

Note that the singular simplex is a map and not a set in E, although much of singular homology theory can be motivated intuitively by thinking of σ as a simplex embedded in E.

Definition 1-8. A *singular p chain on a topological space E over an additive Abelian group $\mathcal{G}$* is a formal linear combination $\sum a_i \sigma_i$ of singular p simplexes σ_i on E with coefficients a_i in G; only a finite number of terms are nonzero. Two such chains $\alpha = \sum a_i \sigma_i$, $\beta = \sum b_i \sigma_i$ are added by adding corresponding coefficients, so,

$$\alpha + \beta = \sum (a_i + b_i)\sigma_i$$

The p chains over G thus form an additive Abelian group $C_p(E; \mathcal{G})$.

This definition may also be formulated by saying that $C_p(E; \mathscr{G})$ is the group of all functions on the set of singular p simplexes in E, taking values in $\mathscr{G}$, and equal to zero except on finitely many singular simplexes. Then the addition in $C_p(E; \mathscr{G})$ is defined by adding functional values. Corresponding to a singular simplex σ there is a function of this type, for instance f_σ, which takes the value 1 on σ and zero on all other singular simplexes, and it is easy to see that $C_p(E; \mathscr{G})$, as defined in this second formulation, is generated by the f_σ. The first definition is obtained from the second by substituting σ for f_σ.

The notion of singular p chain over $\mathscr{G}$ as defined here is a little more general than the corresponding notion defined in *Introduction*, where the group $\mathscr{G}$ is always taken to be the additive group Z of integers. In this special case the reference to the group is usually dropped so that $C_p(E)$ stands for $C_p(E; Z)$.

Intuitively, a p chain on E can be thought of as a kind of p-dimensional surface embedded in E and built up from simplexes. If the coefficients are positive integers, this gives the additional intuitive appeal that pieces of this surface can be regarded as being counted with various multiplicities. Otherwise the idea is more abstract and algebraic, although a geometric significance is given for negative coefficients too (cf. Definition 2-7).

Let E and F be two topological spaces and f a continuous map of E into F. Intuitively surfaces embedded in E should be carried over into F by the map f. The following definition precisely expresses this idea.

Definition 1-9. Let $f: E \to F$ be a continuous map and σ a singular p simplex on E. The composition $f\sigma$ is a continuous map of Δ_p into F, so it is a singular simplex on F denoted by $f_1(\sigma)$. If $\alpha = \sum a_i \sigma_i$ is in $C_p(E; \mathscr{G})$, define the operation f_1 by

$$f_1(\alpha) = \sum a_i f_1(\sigma_i)$$

Thus f_1 becomes a homomorphism of $C_p(E; \mathscr{G})$ into $C_p(F; \mathscr{G})$. It is called the *induced homomorphism on chains associated* with f.

The notational convention introduced here will always be used for this notion; that is, a subscript 1 will be attached to the name of a continuous map to obtain the corresponding induced homomorphism on chain groups.

The properties of these induced homomorphisms are given by the following theorem, whose proof is left as an exercise.

Theorem 1-1. (1) *If f: $E \to E$ is the identity map, then f_1 is the identity homomorphism.*

(2) *If f: $E \to F$ and g: $F \to H$ are continuous maps, then*

$$(gf)_1 = g_1 f_1$$

1-4. THE BOUNDARY OPERATOR

The next step is to introduce an algebraic operation on chains which will represent in a natural way the geometric operation of taking the boundary of a surface. The operation will be defined on singular simplexes and then extended to chains so as to be a homomorphism. A simple way of formulating this definition is to define first the boundary of a singular simplex in Euclidean space defined by a linear map. In this situation an explicit formula can be written down. Then induced homomorphisms on chains are used to define the boundary of any singular simplex.

Definition 1-10. Let $[y_0 y_1 \cdots y_p]$ be a p-dimensional Euclidean simplex in some Euclidean space E (or in some subspace of E) and, as usual, let $[x_0 x_1 \cdots x_p] = \Delta_p$ be the standard Euclidean p simplex. Then the symbol $(y_0 y_1 \cdots y_p)$ denotes the linear map $\Delta_p \to [y_0 y_1 \cdots y_p]$ mapping x_i on y_i for each i. Thus $(y_0 y_1 \cdots y_p)$ is a singular simplex on E or on any part of E containing $[y_0 y_1 \cdots y_p]$.

Note that the *set* $[y_0 y_1 \cdots y_p]$ is independent of the order of the vertexes, but the *map* $(y_0 y_1 \cdots y_p)$ does depend on this order. The order $x_0, x_1, \ldots, x_p$ of the vertexes of Δ_p is supposed to be fixed once and for all. Note also that the symbol $(x_0 x_1 \cdots x_p)$ is simply the identity map of Δ_p on itself. This means that if σ is any singular simplex on any topological space, then

$$\sigma = \sigma(x_0 x_1 \cdots x_p)$$

By Definition 1-9 this means

$$\sigma = \sigma_1((x_0 x_1 \cdots x_p))$$

where σ_1 is the induced homomorphism on chains over any coefficient group, corresponding to the continuous map σ.

Definition 1-11. The *boundary operator* d acting on the singular simplex $(x_0 x_1 \cdots x_p)$ is defined by the formula

$$d(x_0 x_1 \cdots x_p) = \sum_{i=0}^{p} (-1)^i (x_0 x_1 \cdots \hat{x}_i \cdots x_p)$$

where the circumflex denotes the omission of the marked vertex.

The right-hand side of the formula in this definition will be shown to be a $(p - 1)$ chain on Δ_p consisting of a linear combination of singular simplexes, one on each $(p-1)$ face of Δ_p, and with coefficients ± 1, so that each of these faces is, so to speak, counted once. Thus the algebraic formula represents the intuitive idea of the boundary of the simplex. The arrangement of the plus and minus signs is chosen just so that the boundary itself will have zero boundary (cf. Theorem 1-4), in agreement with the geometric notion that the boundary of a simplex should be a closed surface.

The operation just defined will be extended now to singular simplexes on any space, and here the essential feature of the extension is that it is constructed to commute with induced homomorphisms corresponding to continuous maps.

Definition 1-12. Let σ be a singular simplex on a topological space E (i.e., a continuous map of Δ_p into E) and let σ_1 be the corresponding induced homomorphism on chain groups over any coefficient group. Then the *boundary of* σ, $d\sigma$, is defined by the formula

$$d\sigma = \sigma_1(d(x_0 x_1 \cdots x_p))$$

By Definition 1-11 the right-hand side is defined here. Also, in view of the remark following that definition, the operation d has now been defined so as to commute with σ_1. Theorem 1-2 shows that, in fact, it commutes with any induced homomorphism.

Theorem 1-2. *Let $f\colon E \to F$ be a continuous map and f_1 the corresponding induced homomorphism on chain groups. Then*

$$f_1 d = d f_1$$

Proof. Since the chain groups are generated by singular simplexes, it is sufficient to show that $f_1 d$ and df_1 have the same effect on a singular simplex σ on E.

$$\begin{aligned} f_1(d\sigma) &= f_1\sigma_1(d(x_0 x_1 \cdots x_p)) && \text{by Definition 1-12} \\ &= (f\sigma)_1(d(x_0 x_1 \cdots x_p)) && \text{by Theorem 1-1} \\ &= d(f\sigma) && \text{by Definition 1-12 applied to } f\sigma \\ &= df_1(\sigma) && \text{by Definition 1-9} \end{aligned}$$

∎

At this point a very natural question arises. $d(x_0 x_1 \cdots x_p)$ has been already defined by an explicit formula, whereas $d(y_0 y_1 \cdots y_p)$, where $[y_0 y_1 \cdots y_p]$ is any Euclidean simplex, must be defined by using Definition 1-12. The question is whether there is a similar explicit algebraic formula for $d(y_0 y_1 \cdots y_p)$. The affirmative answer is given by the following theorem.

Theorem 1-3. *For any Euclidean simplex* $[y_0 y_1 \cdots y_p]$,

$$d(y_0 y_1 \cdots y_p) = \sum (-1)^i (y_0 y_1 \cdots \hat{y}_i \cdots y_p)$$

Proof. Take $\sigma = (y_0 y_1 \cdots y_p)$ Then, by Definition 1-12,

$$\begin{aligned} d(y_0 y_1 \cdots y_p) &= d\sigma \\ &= \sigma_1 d(x_0 x_1, \ldots, x_p) \\ &= \sigma_1 (\sum (-1)^i (x_0 x_1 \cdots \hat{x}_i \cdots x_p)) \end{aligned}$$

Now $\sigma_1(x_0 x_1 \cdots \hat{x}_i \cdots x_p)$ is simply (Definition 1-9) the composition of the linear maps $(y_0 y_1 \cdots y_p)$ and $(x_0 x_1 \cdots \hat{x}_i \cdots x_p)$; it is easy to see that this is $(y_0 y_1 \cdots \hat{y}_i \cdots y_p)$ as required by the statement of the theorem. ∎

As already pointed out, an essential property of the boundary operator is that it represents the geometrical fact that a boundary should be a closed surface. This is established algebraically, first by extending the boundary to a homomorphism on chain groups over any coefficient groups. Then the geometric property just mentioned corresponds to the algebraic property $d^2 = 0$. That is, two successive applications of d give the zero homomorphism.

Definition 1-13. Let $\alpha = \sum a_i \sigma_i$ be a p chain on a topological space; the σ_i are singular simplexes, and the a_i are elements of the coefficient group. Then the *boundary* $d\alpha$ *of* α is defined to be $\sum a_i d\sigma_i$.

This definition automatically makes d into a homomorphism

$$d\colon C_p(E; \mathcal{G}) \to C_{p-1}(E; \mathcal{G})$$

where E is the space in question and $\mathcal{G}$ the coefficient group. Note that there is really a d for each dimension p. If this is to be emphasized, the homomorphism just introduced will be denoted by d_p, but it is usually clear from the context what dimension is involved without putting in the subscript.

Theorem 1-4. *If d is as in Definition* 1-13, $d^2 = 0$. *Or, in the notation that displays the dimensions,* $d_{p-1} d_p = 0$.

Proof. Since $C_p(E; \mathcal{G})$ is generated over $\mathcal{G}$ by the singular p simplexes, it is sufficient to prove that $d^2\sigma = 0$ for any singular p simplex σ. As noted in the remark following Definition 1-10, $\sigma = \sigma_1(x_0 x_1 \cdots x_p)$, and d commutes with σ_1 (Theorem 1-2), whence

$$d^2\sigma = \sigma_1(d^2(x_0 x_1 \cdots x_p))$$

Therefore, it is sufficient to show that $d^2(x_0 x_1 \cdots x_p) = 0$, and this will now be done by a simple computation:

$$\begin{aligned} d^2(x_0 x_1 \cdots x_p) &= d(\sum(-1)^i(x_0 x_1 \cdots \hat{x}_i \cdots x_p)) \\ &= \sum(-1)^i\, d(x_0 x_1 \cdots \hat{x}_i \cdots x_p) \\ &= \sum_{j<i}(-1)^{i+j}(x_0 x_1 \cdots \hat{x}_j \cdots \hat{x}_i \cdots x_p) \\ &\quad + \sum_{j>i}(-1)^{i+j-1}(x_0 x_1 \cdots \hat{x}_i \cdots \hat{x}_j \cdots x_p) \end{aligned}$$

The last step was made by using Theorem 1-3. An examination of the two sums appearing in that last step shows that each term in the one sum appears in the other with the opposite sign, so the whole expression reduces to zero, as was to be shown.

1-5. CYCLES AND HOMOLOGY

With the ideas so far introduced, the notions of closed surfaces and boundaries in a space can be put on a purely algebraic footing; that is, closed surfaces (called *cycles*) can be replaced by chains with boundary zero, and the geometric notion of boundary can be replaced by the algebraic notion of the boundary operator d.

Returning to a remark made at the beginning of this chapter, a sphere is distinguished from a torus by the fact that on the latter a closed curve can be drawn without forming a boundary, while this is not true for the former. In fact, two curves (α, β in Fig. 2) can be drawn so that their union does not form

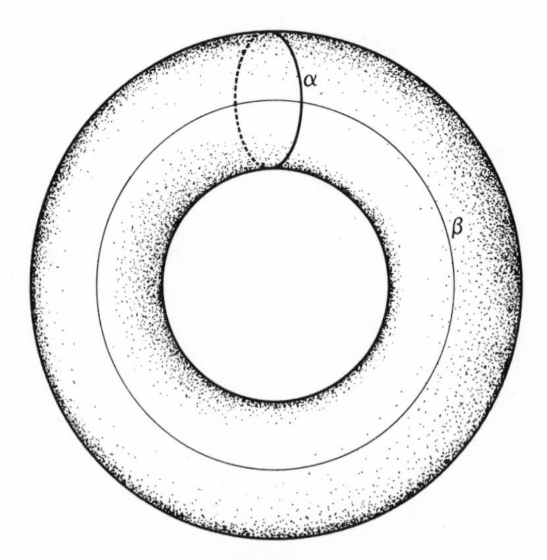

FIGURE 2

a boundary; that is, when the surface is cut along them it remains in one piece. On more complicated surfaces (cf. Fig. 3) more than two cuts can be made without breaking the surface into separate pieces, and the maximum number of such cuts is a topological property of the surface. Algebraically, this means considering the maximum number of one cycles (one chains with zero boundary) on the surface such that no linear combination is a boundary, and this number is the same as that of a set of generators of the kernel of the boundary operator d_1, linearly independent modulo the image of d_2. Thus, to give an algebraic account of the question of closed surfaces and boundaries in a space, the appropriate object of study is the quotient group of the kernel of d_p modulo the image of d_{p+1}, for each p. The necessary definitions will now be set up. Let E be a topological space and $\mathscr{G}$ the coefficient group.

Definition 1-14. The kernel of

$$d = d_p : C_p(E; \mathscr{G}) \to C_{p-1}(E; \mathscr{G})$$

is denoted by $Z_p(E; \mathscr{G})$ and is called the *p-dimensional cycle group of E over* $\mathscr{G}$. Its elements are *p-dimensional cycles* or *p cycles over* $\mathscr{G}$. When $p = 0$, $C_0(E; \mathscr{G}) = Z_0(E; \mathscr{G})$.

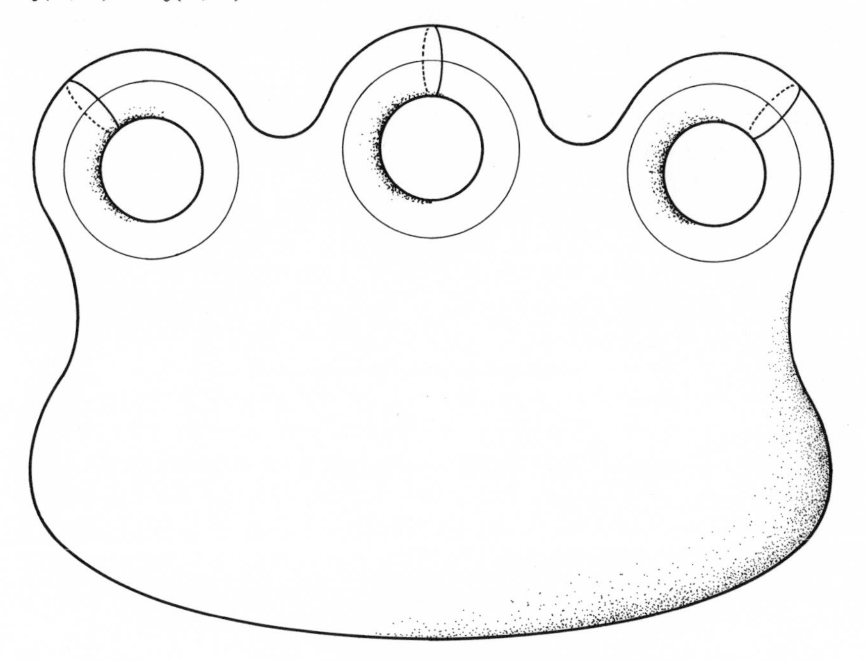

FIGURE 3

Definition 1-15. The image of

$$d = d_{p+1} \colon C_{p+1}(E; \mathscr{G}) \to C_p(E; \mathscr{G})$$

is denoted by $B_p(E; \mathscr{G})$ and is called the *p-dimensional boundary group* of E over $\mathscr{G}$. Its elements are *p-dimensional boundaries* or *p boundaries over* $\mathscr{G}$.

Definition 1-16. The quotient group $Z_p(E; \mathscr{G})/B_p(E; \mathscr{G})$ is denoted by $H_p(E; \mathscr{G})$ and is called the *p-dimensional homology group of E over* $\mathscr{G}$. Its elements are called *p-dimensional homology classes.* Two cycles α and β representing the same homology class are said to be *homologous*; this relation is written as $\alpha \sim \beta$.

In all the preceding definitions $\mathscr{G}$ can simply be dropped from the notation and terminology if this does not lead to confusion. In particular, if we are working consistently over Z, the group of integers, as coefficient group, this group is not normally mentioned.

A somewhat more general notion is needed in the development of the theory, namely, the notion of relative homology. Here chains in a certain subspace F of E are to be identified with zero. Thus, to be a cycle of E modulo F a chain must have a boundary that is a chain on F, rather than a boundary equal to zero, as in Definition 1-14. There are two equivalent approaches to this notion. Both are described, and the equivalence proof is given as an exercise.

Let E be a topological space, F a subspace, and $\mathscr{G}$ the coefficient group. Note that, for each p, $C_p(F; \mathscr{G})$ is a subgroup of $C_p(E; \mathscr{G})$.

Definition 1-17. The inverse image of $C_{p-1}(F; \mathscr{G})$ under

$$d = d_p \colon C_p(E; \mathscr{G}) \to C_{p-1}(E; \mathscr{G})$$

is denoted by $Z_p(E, F; \mathscr{G})$ and is called the *p-dimensional relative cycle group of E modulo F over* $\mathscr{G}$. Its elements are *relative p cycles.*

Definition 1-18. The subgroup of $C_p(E; \mathscr{G})$ generated by $C_p(F; \mathscr{G})$ and $B_p(E; \mathscr{G})$ is denoted by $B_p(E, F; \mathscr{G})$ and is called the *p-dimensional relative boundary group of E modulo F over* $\mathscr{G}$. Its elements are called *relative p boundaries.*

Since $d^2 = 0$, it is easy to see that $B_p(E, F; \mathscr{G})$ is a subgroup of $Z_p(E, F; \mathscr{G})$.

Definition 1-19. The quotient $Z_p(E, F; \mathscr{G})/B_p(E, F; \mathscr{G})$ is denoted by $H_p(E, F; \mathscr{G})$ and is called the *p-dimensional relative homology group of E* *modulo F over* $\mathscr{G}$. Its elements are *p-dimensional relative homology classes* and two relative cycles α, β representing the same relative homology class are said to be *homologous modulo F*. This relation is written $\alpha \sim \beta$ modulo F.

As usual, $\mathscr{G}$ is only included in the notation and terminology when it is necessary to avoid misunderstanding. Note that if F is the empty set, the above three definitions reduce to Definitions 1-14, 1-15, and 1-16; so homology (unqualified) is a special case of relative homology.

A second definition of the notion of relative homology will now be given which is more in line with some of the algebraic topics that arise in Chapter 3. The homomorphism

$$d\colon C_p(E; \mathscr{G}) \to C_{p-1}(E; \mathscr{G})$$

clearly has the property that it maps the subgroup $C_p(F; \mathscr{G})$ into $C_{p-1}(F; \mathscr{G})$. This can be seen by considering the action of d on a singular simplex on F. Hence d induces a homomorphism

$$d'\colon \frac{C_p(E; \mathscr{G})}{C_p(F; \mathscr{G})} \to \frac{C_{p-1}(E; \mathscr{G})}{C_{p-1}(F; \mathscr{G})}$$

for each p; that is, $d'\bar{\alpha}$, for an element $\bar{\alpha}$ of the first quotient group, represented by α, for example, in $C_p(E; \mathscr{G})$, is defined as the coset of $d\alpha$ in the second quotient group. It is clear that $d'^2 = 0$. If it is necessary to draw attention to the dimension of the elements on which d' is operating, the homomorphism just introduced is denoted by d'_p. The relation $d'^2 = 0$ then becomes $d'_{p-1}d'_p = 0$, which means that the image of d'_p is contained in the kernel of d'_{p-1}.

Definition 1-20. In the notation just introduced the kernel of d'_p is denoted by $Z'_p(E, F; \mathscr{G})$.

Definition 1-21. The image of d'_{p+1} is denoted by $B'_p(E, F; \mathscr{G})$.

Definition 1-22. The quotient group $Z'_p(E, F; \mathscr{G})/B'_p(E, F; \mathscr{G})$ is denoted by $H_p(E, F; \mathscr{G})$.

Exercise. 1-10. Prove that the group $H_p(E, F; \mathscr{G})$ defined by Definition 1-22 is isomorphic to that defined by Definition 1-19.

Definition 1-22 is algebraically more useful, lending itself easily to abstraction and generalization. From the geometric point of view, however, it is often convenient to speak of relative cycles and boundaries, implying the use of Definition 1-19.

1-6. INDUCED HOMOMORPHISMS

The induced homomorphisms already introduced for chain groups now will be carried over to homology groups, and a theorem similar to Theorem 1-1 will be proved. In particular, the topological invariance of the homology groups will be proved. The discussion will be carried out for relative homology, so a preliminary remark on mappings must be made.

Definition 1-23. Let E be a space and F a subspace. Then (E, F) is called a *pair of spaces*. Let (E, F) and (E', F') be two pairs of spaces, and let f be a continuous map of E into E' such that $f(F)$ is contained in F'. Then f is called a *map of the pair* (E, F) *into the pair* (E', F').

This reduces to the usual notion of a continuous map of E into E' when F and F' are both empty.

Another notion that will be used considerably from now on is that of a commutative diagram.

Suppose that a number of sets and maps are arranged in a diagram, a map from a set A to a set B being indicated by an arrow going from A to B. The diagram is said to be commutative if the composition of maps corresponding to any path in the diagram (i.e., sequence of arrows) depends only on the initial set and the final set and not on the particular path joining these sets. For example, to say that the following diagrams are commutative

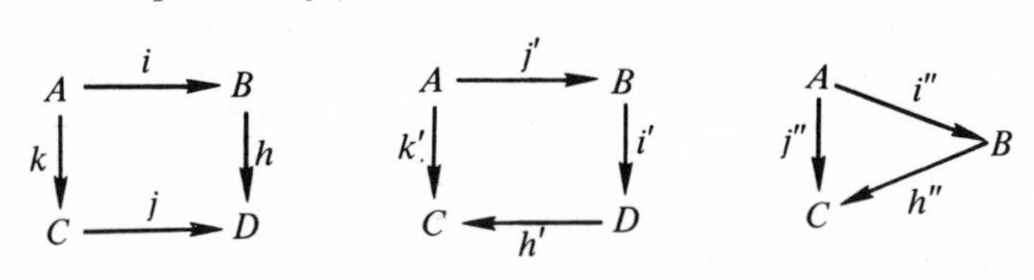

means, respectively, that $hi = jk$, $h'i'j' = k'$, and $h''i'' = j''$. Most of the diagrams used here are made up of squares and triangles like the above examples, and it is not difficult to see that such a diagram is commutative if and only if each square and triangle forms itself a commutative diagram.

Consider the following situation. If (E, F) and (E', F') are two pairs of topological spaces and f is a continuous map from the first into the second, then Theorem 1-2 implies that the following diagram is commutative. Note that the coefficient group is omitted in order to simplify the notation.

$$\begin{array}{ccc} C_p(E) & \xrightarrow{d} & C_{p-1}(E) \\ {\scriptstyle f_1}\downarrow & & {\scriptstyle f_1}\downarrow \\ C_p(E') & \xrightarrow{d} & C_{p-1}(E') \end{array}$$

At the same time, for each p, f_1 maps $C_p(F)$ into $C_p(F')$ and so induces a homomorphism

$$f'_1: \frac{C_p(E)}{C_p(F)} \to \frac{C_p(E')}{C_p(F')}$$

The commutativity of the above diagram then leads at once to the commutativity of the following:

$$\begin{array}{ccc} C_p(E)/C_p(F) & \xrightarrow{d'} & C_{p-1}(E)/C_{p-1}(F) \\ \downarrow{\scriptstyle f'_1} & & \downarrow{\scriptstyle f'_1} \\ C_p(E')/C_p(F') & \xrightarrow{d'} & C_{p-1}(E')/C_{p-1}(F') \end{array}$$

Here d' is the homomorphism used in Definitions 1-20 and 1-21.

Using the commutativity of this diagram, it is a straightforward matter to check that f'_1 maps $Z'_p(E, F)$ into $Z'_p(E', F')$ and also $B'_p(E, F)$ into $B'_p(E', F')$. Therefore f'_1 induces a homomorphism of the quotient groups

$$f_*: H_p(E, F) \to H_p(E', F')$$

Definition 1-24. The homomorphism f_* is called the *induced homomorphism on relative homology groups* corresponding to the continuous map f.

Note that if $F = F' = \emptyset$, f_* reduces to a homomorphism of $H_p(E)$ into $H_p(E')$.

It is sometimes convenient to think of f_* in a different way. By an argument similar to the above with f_1 instead of with f'_1, it can be seen that f_1 maps $Z_p(E, F)$ into $Z_p(E', F')$ and also $B_p(E, F)$ into $B_p(E', F')$. It therefore induces a homomorphism of $H_p(E, F)$ into $H_p(E', F')$ (using Definition 1-19), and it is an easy exercise to check that this homomorphism coincides with f_*. With this in mind, the operation of f_* can be described as follows. If $\bar{\alpha}$ is an element of $H_p(E, F)$ (i.e., a coset of $B_p(E, F)$ in $Z_p(E, F)$), it can be represented by a relative cycle α. Then $f_*(\bar{\alpha})$ is the relative homology class, modulo F', of $f_1(\alpha)$.

Theorem 1-5. (1) *If f is the identity map then f_* is the identity homomorphism.*

(2) *If $f: (E, F) \to (E', F')$, $g: (E', F') \to (E'', F'')$ are two maps of pairs, then $(gf)_* = g_* f_*$.*

Proof. These two results follow at once from the corresponding parts of Theorem 1-1, and the necessary verification is left as an exercise. ∎

The last theorem leads at once to the topological invariance of the homology groups, namely, the result that homeomorphic pairs of spaces have isomorphic homology groups. This is stated more formally as follows.

Theorem 1-6. *Let $f\colon (E, F) \to (E', F')$ be a homeomorphism, that is to say $f\colon E \to E'$ is a homeomorphism and $f(F) = F'$. Then $f_*\colon H_p(E, F) \to H_p(E', F')$ is an isomorphism onto for each p.*

Proof. Let g be the inverse of f. Thus gf and fg are both identity maps, so Part (1) of Theorem 1-5 implies that $(gf)_*$ and $(fg)_*$ are both identity homomorphisms. On the other hand, Part (2) of the same theorem implies that g_*f_* and f_*g_* are both identities, so g_* is the inverse of f_*, which is therefore an isomorphism, as was to be shown. ∎

1-7. THE MAIN THEOREMS

So far the homology groups have presented a somewhat unmanageable appearance since they are quotient groups of Abelian groups with infinitely many generators. Four basic properties of these groups will now be established, however, which, in many cases, yield a method of computation of the groups; in fact, for a certain class of spaces they provide a systematic algorithm. These properties are the Dimension, Exactness, Homotopy, and Excision theorems. In the subsequent sections these theorems are stated and proved.

1-8. THE DIMENSION THEOREM

The dimension theorem is concerned with the homology groups of a single point. It is so-called because, in the abstract axiomatic setting for homology theory, the property stated in this theorem singles out the dimension zero in an invariant manner.

Theorem 1-7. *If E is a space consisting of one point, then*

$$H_p(E; \mathcal{G}) = 0 \qquad (p \neq 0)$$

$$H_0(E; \mathcal{G}) \cong \mathcal{G}$$

where $\mathcal{G}$ is the coefficient group.

Proof. Since E consists of only one point, there is, for each p, exactly one singular p simplex $\sigma_p\colon \Delta_p \to E$ (i.e., mapping all of Δ_p on the one point of E). By Definition 1-12, it is clear that $d\sigma_p = \sum \pm\sigma_{p-1}$, where on the right-hand side there are $p + 1$ terms, all equal to σ_{p-1}, and with alternately plus and minus signs. It follows that

$$d\sigma_p = \sigma_{p-1} \qquad \text{if } p \text{ is even and} \neq 0$$
$$d\sigma_p = 0 \qquad \text{if } p \text{ is odd}$$
$$d\sigma_0 = 0$$

For each p, $C_p(E; \mathscr{G})$ is the group of multiples of σ_p by elements of $\mathscr{G}$ so it is isomorphic to $\mathscr{G}$.

If p is even and not zero, $a\sigma_p$ is a cycle if and only if $d(a\sigma_p) = a\sigma_{p-1} = 0$; that is, $a = 0$. Hence $Z_p(E; \mathscr{G}) = 0$, and $H_p(E; \mathscr{G}) = 0$.

If p is odd, $a\sigma_p$ is a cycle for all a since $d\sigma_p = 0$. Hence $Z_p(E; \mathscr{G}) = C_p(E; \mathscr{G})$. But also $d\sigma_{p+1} = \sigma_p$, so every p cycle is a boundary. Again $H_p(E; \mathscr{G}) = 0$.

Finally, when $p = 0$, $Z_0(E; \mathscr{G}) = C_0(E; \mathscr{G})$ and $B_0(E; \mathscr{G}) = 0$. Hence $H_0(E; \mathscr{G}) \cong \mathscr{G}$. ∎

1-9. THE EXACTNESS THEOREM

Let X, Y, and Z be three spaces such that $X \supset Y \supset Z$. Such a set of spaces is usually called a *triple*. The exactness theorem describes relations between the three sets of relative homology groups associated with the triple, namely, the $H_p(X, Y)$, $H_p(X, Z)$, and $H_p(Y, Z)$ for various values of p. Here some coefficient group (the same for all the homology groups) is assumed permanently fixed.

Reasoning geometrically, a relative p cycle of Y modulo Z can be thought of also as a relative p cycle of X modulo Z, and a relative p cycle of X modulo Z can be thought of as a relative p cycle of X modulo Y. Finally, the boundary of a relative p cycle of X modulo Y is a $(p - 1)$ cycle of Y and so a relative $(p - 1)$ cycle of Y modulo Z. Thus, there are three maps:

$$H_p(Y, Z) \to H_p(X, Z) \to H_p(X, Y) \to H_{p-1}(Y, Z)$$

Of course this sequence can be continued by starting with $H_{p-1}(Y, Z)$ instead of $H_p(Y, Z)$. Closer examination of the geometry of the situation suggests that in this sequence the image of each map coincides with the kernel of the next. Formulated in this way, the result is purely algebraic, and the proof about to be given is purely algebraic in form. The resulting theorem, with the appropriate changes in notation, will appear in various situations later.

To proceed now with the algebraic discussion, a special name will be given to the kind of sequence that has just been introduced.

Definition 1-25. Let $\xrightarrow{\phi_{i-1}} G_i \xrightarrow{\phi_i} G_{i+1} \xrightarrow{\phi_{i+1}}$ be a sequence of groups and homomorphisms with the property that the kernel of ϕ_i is equal to the image of ϕ_{i-1} for all i. This is called an *exact sequence*. The sequence can be assumed to extend infinitely in either direction, or it may terminate. Whether the indices $(i, i+1, \ldots)$ ascend, as here, or descend is clearly immaterial.

Now let

$$i: (Y, Z) \to (X, Z)$$
$$j: (X, Z) \to (X, Y)$$

be inclusion maps, that is, maps defined by $i(x) = x$ and $j(x) = x$. These maps are continuous; hence they induce homomorphisms on the chain groups and in turn induce homomorphisms

$$i_1: \frac{C_p(Y)}{C_p(Z)} \to \frac{C_p(X)}{C_p(Z)}$$

$$j_1: \frac{C_p(X)}{C_p(Z)} \to \frac{C_p(X)}{C_p(Y)}$$

For convenience write $C_p(X, Y)$ for $C_p(X)/C_p(Y)$, and use similar notations for the other pairs. Then the following diagram of groups and homomorphisms can be constructed.

$$\begin{array}{ccccccccc}
 & & \downarrow & & \downarrow & & \downarrow & & \\
0 & \longrightarrow & C_p(Y, Z) & \xrightarrow{i_1{}^p} & C_p(X, Z) & \xrightarrow{j_1{}^p} & C_p(X, Y) & \longrightarrow & 0 \\
 & & \downarrow d'_p & & \downarrow d_p & & \downarrow d''_p & & \\
0 & \longrightarrow & C_{p-1}(Y, Z) & \xrightarrow{i_1^{p-1}} & C_{p-1}(X, Z) & \xrightarrow{j_1^{p-1}} & C_{p-1}(X, Y) & \longrightarrow & 0 \\
 & & \downarrow & & \downarrow & & \downarrow & &
\end{array} \tag{1-1}$$

Here the d's are all boundary operators, marked so as to name their positions in the diagram. The map $i_1{}^p$ is the inclusion map, while $j_1{}^p$ can be identified with the natural map onto the quotient group. Superscripts have been inserted to mark positions in the diagram. Zero is mapped into the first term of each line and the last term is mapped onto zero, so that each horizontal line in the diagram is exact. All the squares in the diagram are commutative.

In the terminology of Definition 1-24, the construction of i_* and j_*, induced by i and j, can be described as follows. If $\bar{\alpha} \in H_p(Y, Z)$ is represented by the relative cycle α in $C_p(Y, Z)$, then α is in the kernel of d'_p, so by the commutativity of the diagram $i_1{}^p(\alpha)$ is in the kernel of d_p. The coset of $i_1{}^p(\alpha)$ modulo the image of d_{p+1} is then $i_*(\bar{\alpha})$. Similarly, if $\bar{\beta} \in H_p(X, Z)$ is represented by β in the kernel of d_p, then $j_1{}^p(\beta)$ is in the kernel of d_p'' and its coset modulo the image of d_{p+1}'' is $j_*(\bar{\beta})$.

Another homomorphism will now be described, corresponding to the operation of taking the boundary of a relative cycle on X modulo Y. Take $\bar{\alpha} \in H_p(X, Y)$ and let it be represented by α in the kernel of d_p''. $j_1{}^p$ is onto, so $\alpha = j_1{}^p(\beta)$ for some $\beta \in C_p(X, Z)$. The relation $d_p''\alpha = 0$ implies that $d_p'' j_1{}^p(\beta) = 0$, and by the commutativity of the diagram this is the same as the relation $j_1^{p-1} d_p(\beta) = 0$. Thus $d_p(\beta)$ is in the kernel of j_1^{p-1}, and so by the exactness of the horizontal lines, $d_p(\beta) = i_1^{p-1}(\gamma)$ for some γ in $C_{p-1}(Y, Z)$. Since $d_{p-1}d_p = 0$, it follows that $d_{p-1} i_1^{p-1}(\gamma) = d_{p-1}d_p(\beta) = 0$ and so, by commutativity, $i_1^{p-2} d_{p-1}'(\gamma) = d_{p-1} i_1{}^p(\gamma) = 0$. Thus $d_{p-1}'(\gamma)$ is in the kernel of i_1^{p-2} but the exactness of the horizontal lines in the diagram implies that the kernel of i_1^{p-2} is zero, so $d_{p-1}'(\gamma) = 0$. Thus γ is in the kernel of d_{p-1}'. Its coset modulo the image of d_p' is denoted by $\partial\bar{\alpha}$.

It must be checked, of course, that $\partial\bar{\alpha}$ depends only on $\bar{\alpha}$ and not on the choices of α and β made in the above construction. Suppose that α' is a second representative of the class $\bar{\alpha}$, and repeat the above construction, starting with α'. Then $\alpha' = j_1{}^p(\beta')$ for some β' in $C_p(X)$, and $d_p(\beta') = i_1^{p-1}(\gamma')$ for some γ' in $C_{p-1}(Y)$. The point now is to see that γ and γ' belong to the same coset modulo the image of d_{p-1}'. α and α' represent the same element $\bar{\alpha}$, so $\alpha - \alpha'$ is in the image of d_{p+1}''. Write $\alpha - \alpha' = d_{p+1}''\gamma$ with γ in $C_{p+1}(X, Y)$. The exactness of the horizontal lines implies that $\gamma = j_1^{p+1}\mu$ for some μ in $C_{p+1}(X)$, thus

$$\alpha - \alpha' = d_{p+1}'' j_1^{p+1} = j_1{}^p d_{p+1}\mu$$

the last step following from the commutativity of the diagram. That is,

$$j_1{}^p(\beta) - j_1{}^p(\beta') = j_1{}^p d_{p+1}\mu$$

and so $\beta - \beta'' - d_{p+1}\mu$ is in the kernel of $j_1{}^p$, which is the same as the image of $i_1{}^p$. Write

$$\beta - \beta' - d_{p+1}\mu = i_1{}^p v$$

Applying d_p and remembering that $d_p d_{p+1} = 0$ it follows that

$$d_p(\beta) - d_p(\beta') = d_p i_1{}^p v = i_1^{p-1} d_p' v$$

Since $d_p(\beta) = i_1^{p-1}(\gamma)$ and $d_p(\beta') = i_1^{p-1}(\gamma')$ the last equation implies that

$$i_1^{p-1}(\gamma) - i_1^{p-1}(\gamma') = i_1^{p-1} d_p' v$$

But i_1^{p-1} has zero kernel so this equation implies that $\gamma - \gamma'$ is in the image of d'_p, which is the required result.

Thus $\partial\bar{\alpha}$, the coset of γ, is a well-defined element of $H_{p-1}(Y, Z)$ depending only on $\bar{\alpha}$. It is easy also to check that ∂ is a homomorphism.

Definition 1-26. The homomorphism $\partial\colon H_p(X, Y) \to H_{p-1}(Y, Z)$ constructed as above is called the *boundary homomorphism* or *homology boundary homomorphism* (to emphasize the fact that it operates on homology groups).

Note that although the operator d on chains is also a homomorphism it is always called the boundary operator to avoid confusion.

An important property of ∂ is that it commutes with homomorphisms induced by continuous maps. This is really just a reflection of the fact that the boundary operator commutes with such homomorphisms (cf. Theorem 1-2). The formal statement of the result is as follows.

Theorem 1-8. *Let (X, Y, Z) and (E, F, G) be two triples and let $f\colon X \to E$ be a continuous map such that $f(Y) \subset F$ and $f(Z) \subset G$. Let ∂ and $\bar{\partial}$ be the boundary homomorphisms for the two triples. Then the following diagram is commutative:*

$$\begin{array}{ccc} H_p(X, Y) & \xrightarrow{\partial} & H_{p-1}(Y, Z) \\ {\scriptstyle f_*}\downarrow & & \downarrow{\scriptstyle f_*} \\ H_p(E, F) & \xrightarrow{\bar{\partial}} & H_{p-1}(F, G) \end{array}$$

Here the f_* on the left is induced by the map f of (X, Y) into (E, F), that on the right is induced by the restriction of f to Y, regarded now as a map of the pair (Y, Z) into (F, G).

Proof. Construct the diagram (1-1) for the triple (X, Y, Z) and a similar one for (E, F, G), marking all the maps of the second with horizontal bars. The following two commutative diagrams result:

$$\begin{array}{ccccccccc} 0 & \longrightarrow & C_p(Y, Z) & \xrightarrow{i_1{}^p} & C_p(X, Z) & \xrightarrow{j_1{}^p} & C_p(X, Y) & \longrightarrow & 0 \\ & & \downarrow{\scriptstyle d'_p} & & \downarrow{\scriptstyle d_p} & & \downarrow{\scriptstyle d''_p} & & \\ 0 & \longrightarrow & C_{p-1}(Y, Z) & \xrightarrow{i_1^{p-1}} & C_{p-1}(X, Z) & \xrightarrow{j_1^{p-1}} & C_{p-1}(X, Y) & \longrightarrow & 0 \end{array} \tag{1-2}$$

$$\begin{array}{ccccccccc} 0 & \longrightarrow & C_p(F, G) & \xrightarrow{\bar{i}_1{}^p} & C_p(E, G) & \xrightarrow{\bar{j}_1{}^p} & C_p(E, F) & \longrightarrow & 0 \\ & & \downarrow{\scriptstyle \bar{d}'_p} & & \downarrow{\scriptstyle \bar{d}_p} & & \downarrow{\scriptstyle \bar{d}''_p} & & \\ 0 & \longrightarrow & C_{p-1}(F, G) & \xrightarrow{\bar{i}_1^{p-1}} & C_{p-1}(E, G) & \xrightarrow{\bar{j}_1^{p-1}} & C_{p-1}(E, F) & \longrightarrow & 0 \end{array} \tag{1-3}$$

Of course the two diagrams are connected by the homomorphisms induced by f with the commutativity relations given by Theorems 1-1 and 1-2. Take $\bar{\alpha} \in H_p(X, Y)$ represented by $\alpha \in C_p(X, Y)$ as in the description of the construction of ∂. Then $\partial\bar{\alpha}$ is represented by an element $\gamma \in C_{p-1}(Y, Z)$ satisfying the relations

$$\alpha = j_1{}^p(\beta),\ d_p(\beta) = i_1^{p-1}(\gamma),\ d'_{p-1}(\gamma) = 0$$

for some $\beta \in C_p(X, Z)$. Now f_1 will denote the homomorphism induced by f mapping each entry of (1-2) into the corresponding entry of (1-3) (it seems unnecessary to distinguish the various occurrences of f_1 from each other). The relations connecting α, β, γ become, on applying f_1,

$$\begin{aligned} f_1(\alpha) &= f_1 j_1{}^p(\beta) = \bar{j}_1{}^p f_1(\beta) \\ \bar{d}_p f_1(\beta) &= f_1 d_p(\beta) = f_1 i_1^{p-1}(\gamma) = \bar{i}_1^{p-1} f_1(\gamma) \\ \bar{d}'_{p-1} f_1(\gamma) &= f_1 d'_{p-1}(\gamma) = 0 \end{aligned}$$

using the commutativity relations satisfied by f_1. These relations are equivalent to the statement that $\bar{\partial} f_*(\bar{\alpha})$ is represented by $f_1(\gamma)$. That is,

$$\bar{\partial} f_*(\bar{\alpha}) = f_* \partial \bar{\alpha}$$

as was to be shown. ∎

Definition 1-27. The sequence

$$\xrightarrow{\partial} H_p(Y, Z) \xrightarrow{i_*} H_p(X, Z) \xrightarrow{j_*} H_p(X, Y) \xrightarrow{\partial} H_{p-1}(Y, Z) \xrightarrow{i_*}$$

is called the *homology sequence of the triple* (X, Y, Z), or the *singular homology sequence* (if attention is to be drawn to the particular homology theory under discussion). If Z is empty then $H_p(X, Z)$ and $H_p(Y, Z)$ are replaced by $H_p(X)$ and $H_p(Y)$ and the resulting sequence is called the *homology sequence of the pair* (X, Y). In either case the sequence extends indefinitely to the left and terminates with $H_0(X, Y)$ on the right.

Theorem 1-9. *The homology sequence of a triple is exact. So, in particular, the homology sequence of a pair is exact.*

Proof. The proof is in six parts corresponding to the three places where exactness has to be checked. The following commutative diagram has appeared before but is repeated here for convenience:

$$\begin{array}{ccccccccc} 0 & \longrightarrow & C_p(Y,Z) & \xrightarrow{i_1{}^p} & C_p(X,Z) & \xrightarrow{j_1{}^p} & C_p(X,Y) & \longrightarrow & 0 \\ & & \downarrow d'_p & & \downarrow d_p & & \downarrow d''_p & & \\ 0 & \longrightarrow & C_{p-1}(Y,Z) & \xrightarrow{i_1^{p-1}} & C_{p-1}(X,Z) & \xrightarrow{j_1^{p-1}} & C_{p-1}(X,Y) & \longrightarrow & 0 \end{array}$$

Part (1). Since $j_1{}^p i_1{}^p = 0$ it follows that $j_* i_* = 0$. Thus the image of i_* is contained in the kernel of j_*.

Part (2). Take $\bar{\alpha}$ in the kernel of j_*. $\bar{\alpha}$ is represented by an element α in $C_p(X, Z)$ which satisfies $d_p \alpha = 0$. To say that $j_* \bar{\alpha} = 0$ means that $j_1{}^p \alpha = d''_{p+1}\beta$, with $\beta \in C_{p+1}(X, Y)$. But j_1^{p+1} is onto and so $\beta = j_1^{p+1}\gamma$ with $\gamma \in C_{p+1}(X, Z)$, whence $j_1{}^p \alpha = d''_{p+1} j_1^{p+1} \gamma = j_1{}^p d_{p+1} \gamma$ (by commutativity). Thus $j_1{}^p(\alpha - d_{p+1}\gamma) = 0$ so $\alpha = d_{p+1}\gamma$ is in the image of $i_1{}^p$. But $\alpha - d_{p+1}\gamma$ also represents $\bar{\alpha}$ and so $\bar{\alpha}$ is in the image of i_*.

Parts (1) and (2) show that the image of i_* is equal to the kernel of j_*.

Part (3). Let $\bar{\alpha}$ be in the image of j_*. $\bar{\alpha}$ is represented by $j_1{}^p \beta$ with $\beta \in C_p(X, Z)$. Following through the construction of ∂, it is apparent that $\partial \bar{\alpha} = 0$. So the image of j_* is contained in the kernel of ∂.

Part (4). Take $\bar{\alpha} \in H_p(X, Y)$ and suppose that $\partial \bar{\alpha} = 0$. Referring back to the construction of ∂ this means that if $\bar{\alpha}$ is represented by $\alpha \in C_p(X, Y)$ then $\alpha = j_1{}^p(\beta)$ and $d_p \beta = i_1^{p-1}(\gamma)$ and here γ is homologous to 0. Thus $\gamma = d'_p \theta$ with $\theta \in C_p(Y, Z)$. Hence $d_p \beta = i_1^{p-1} d'_p \theta = d_p i_1{}^p \theta$ (by commutativity). Hence $d_p(\beta - i_1{}^p \theta) = 0$. Thus $\beta - i_1{}^p \theta$ represents an element $\bar{\beta}$ of $H_p(Y, Z)$ and $j_1{}^p(\beta - i_1{}^p \theta) = j_1{}^p \beta = \alpha$, so that $\bar{\alpha} = j_* \bar{\beta}$. Hence the kernel of ∂ is contained in the image of j_*.

Parts (3) and (4) imply that the image of j_* is equal to the kernel of ∂.

Part (5). Take an element $\bar{\gamma}$ in the image of ∂, say $\bar{\gamma} = \bar{\alpha}$, with $\bar{\alpha}$ and $\bar{\gamma}$ represented by α and γ. Here $\alpha \in C_p(X, Y)$, $\alpha = j_1{}^p(\beta)$, and $d_p \beta = i_1^{p-1}\gamma$. Then $i_* \bar{\gamma}$ is represented by $d_p \beta$ and so $i_* \bar{\gamma} = 0$. Thus the image of ∂ is contained in the kernel of i_*.

Part (6). Suppose $\bar{\gamma}$ is in the kernel of i_*. Then $\bar{\gamma}$ is represented by an element $\gamma \in C_{p-1}(Y, Z)$ such that $i_1^{p-1}\gamma = d_p \beta$ for some $\beta \in C_p(X, Z)$. Write $\alpha = j_1{}^p(\beta)$. It is easy to check that $d''_p \alpha = 0$ and that α represents an element $\bar{\alpha}$ such that $\bar{\alpha} = \partial \bar{\gamma}$. Hence the kernel of i_* is contained in the image of ∂.

Parts (5) and (6) show that the image of ∂ is equal to the kernel of i_*. ∎

Note. Some special attention should be given to the lower end of the homology sequence. The last terms will be

$$\rightarrow H_0(X, Z) \overset{j_*}{\rightarrow} H_0(X, Y)$$

An element $\bar{\alpha} \in H_0(X, Y)$ is represented by an element $\alpha \in C_0(X, Y)$. Since $j_1{}^0$ is onto, $\alpha = j_1{}^0\beta$ for some $\beta \in C_0(X, Z)$ but since there are no elements of dimension -1, $d_p\beta$ must be 0. Thus β represents an element $\bar{\beta} \in H_0(X, Z)$ and $\bar{\alpha} = j_*\bar{\beta}$. Thus j_* is onto at dimension 0. The homology sequence can thus be completed by adding a 0 at the lower end:

$$\rightarrow H_0(X, Z) \overset{j_*}{\rightarrow} H_0(X, Y) \rightarrow 0$$

and it will be exact up to the last term.

Exercise. 1-11. Prove that for any space X, $H_r(X, X) = 0$ for all r.

1-10. THE HOMOTOPY THEOREM

Intuitively, if a one cycle on a space is thought of as a closed curve, then a continuous deformation of the curve from one position to another will trace out a surface whose boundary is formed by the initial and final positions of the curve. In this way, homologous cycles, should be represented by the initial and final positions of the curve.

The object of the homotopy theorem is to give precise expression to the idea just sketched; that is, if f and g are continuous maps of a space E into a space D', and if they are homotopic (i.e., if one can be continuously deformed into the other) then for any cycle α on E the cycles $f_1(\alpha)$ and $g_1(\alpha)$ in E' can be thought of as obtained from each other by continuous deformation. The homotopy theorem will show that these cycles are homologous. This comes to the same thing as showing that the induced homomorphisms f_* and g_* are the same, and this is the formulation which will be used in the theorem. In fact, the result will be treated more generally, in terms of relative homology.

Definition 1-28. Let f and g be continuous maps of a pair (E, F) into a pair (E', F'). f and g will be called *homotopic* if there exists a continuous map

$$h: (E \times I, F \times I) \rightarrow (E', F')$$

where I is the unit interval $[0, 1]$, such that

$$h(x, 0) = f(x)$$
$$h(x, 1) = g(x)$$

for all x in E. This relation will be denoted by $f \simeq g$.

Geometrically, this means that, as t varies on I from 0 to 1, h deforms f continuously into g in such a way that at each stage of the deformation the subspace F is carried into F'.

In particular, if $F = \emptyset$, one simply speaks of the maps f and g of E into E' as being homotopic.

The homotopy theorem can now be stated as follows.

Theorem 1-10. *Let f and g be homotopic maps of the pair (E, F) into the pair (E', F'). Then $f_* = g_*$, where these are the induced homomorphisms on the relative homology groups, for any coefficient group. If F and F' are empty this reduces to a statement about homology groups of E and E'.*

The proof will be carried out by first making a reformulation in such a way that only a special case needs to be considered. Let $h: (E \times I, F \times I) \to (E', F')$ be the map that defines the homotopy of f and g (cf. Definition 1-28), and let f' and g' be maps of (E, F) into $(E \times I, F \times I)$ defined by the formulas

$$f'(x) = (x, 0)$$
$$g'(x) = (x, 1)$$

for all x in E. Then it is clear that $f = hf'$ and $g = hg'$. Thus, in order to prove that $f_* = g_*$ it will be sufficient to show that $f'_* = g'_*$. This amounts to proving a special case of Theorem 1-10 with (E', F') replaced by $(E \times I, F \times I)$ and f, g replaced by f', g'. Changing notation now so that the primes can be dropped, the following is the result to be proved.

Theorem 1-11. *Let f and g be maps of (E, F) into $(E \times I, F \times I)$ defined by $f(x) = (x, 0)$, $g(x) = (x, 1)$ for all x in E. Then $f_* = g_*$, where these are the induced homomorphisms on the relative homology groups, for any coefficient group.*

The idea of the proof is to take a relative cycle α in E modulo F and to construct a chain in $E \times I$ whose boundary will be $g_1(\alpha) - f_1(\alpha)$ along with a chain on $F \times I$. The construction is carried out by forming a prism over each singular simplex in E, identified with $E \times \{0\}$. If a singular simplex is regarded as a simplex S embedded in $E \times \{0\}$, this prism would be represented by an embedding of $S \times I$ in $E \times I$. The boundary of this prism would then consist of S, a corresponding simplex S' in $E \times \{1\}$, and the vertical sides. When a number of simplexes are put together to form a cycle these vertical sides cancel out, so the various prisms fit together to form a chain of the required type.

The formal definitions will now be given. Let $[x_0 x_1 \cdots x_p]$ be the standard Euclidean simplex Δ_p in $(p + 1)$ space. $\Delta_p \times I$ can be embedded in a natural way in $(p + 2)$ space so that the first $p + 1$ coordinates of (x, t) are those of x, while the last coordinate is t. Δ_p can be identified with $\Delta_p \times \{0\}$. If $(x_i, 0) = x_i$ is a vertex of Δ_p the vertex $(x_i, 1)$ of $\Delta_p \times I$ will be denoted by y_i.

Definition 1-29. Let $(x_0 x_1 \cdots x_p)$ be the identity map of Δ_p on itself, a singular simplex on Δ_p. Then *the prism over* $(x_0 x_1 \cdots x_p)$ is the chain

$$P(x_0 x_1 \cdots x_p) = \sum (-1)^i (x_0 x_1 \cdots x_i y_i \cdots y_p)$$

It is clear that $P(x_0 x_1 \cdots x_p)$ is a $(p + 1)$ chain on $\Delta_p \times I$. Examination of lower dimensional cases shows that, intuitively at least, this chain corresponds to the subdivision of the prism (using the word in a geometric sense now) $\Delta_p \times I$ into simplexes. As usual, the actual signs in the defining formula are selected so that the boundary relations will work out suitably.

The operator P will now be extended to singular simplexes on any space by using induced homomorphisms.

Definition 1-30. Let σ be a singular p simplex on a space E, that is, a continuous map of Δ_p into E. Define the continuous map $\sigma' : \Delta_p \times I \to E \times I$ by the formula

$$\sigma'(x, t) = (\sigma(x), t)$$

Then $P\sigma$, *the prism over* σ, is the $(p + 1)$ chain on $E \times I$ defined by the formula

$$P\sigma = \sigma_1'(P(x_0 x_1 \cdots x_p))$$

If $\alpha \in C_p(E; \mathscr{G})$, $\alpha = \sum a_i \sigma_i$, then define $P\alpha$ to be

$$P\alpha = \sum a_i P\sigma_i$$

This definition clearly makes P a homomorphism of $C_p(E; \mathscr{G})$ into $C_{p+1}(E \times I; \mathscr{G})$.

The interaction of P with induced homomorphisms will now be examined. Let $f: E \to E'$ be a continuous map and define $f' : E \times I \to E' \times I$ by $f'(x, t) = (f(x), t)$.

Theorem 1-12. *For any coefficient group the following diagram is commutative:*

$$\begin{array}{ccc} C_p(E) & \xrightarrow{P} & C_{p+1}(E \times I) \\ \Big\downarrow{\scriptstyle f_1} & & \Big\downarrow{\scriptstyle f_1'} \\ C_p(E') & \xrightarrow{P} & C_{p+1}(E' \times I) \end{array}$$

Proof. This expresses exactly what would be expected from geometric intuition: given a simplex in $E = E \times \{0\}$ then the result of constructing a prism over it and then mapping into $E' \times I$ by f' is the same as that of first carrying the simplex over to E', by means of f, and then constructing the prism over the result. The proof is a straightforward verification. It is obviously sufficient to check the commutativity of the diagram by applying the maps to a generator of $C_p(E)$, namely, to a singular simplex on E. Let σ be a singular p simplex on E.

$$\begin{aligned} f'_1 P\sigma &= f'_1 \sigma'_1 P(x_0 x_1 \cdots x_p) \qquad \text{(Definition 1-30)} \\ &= (f'\sigma')_1 P(x_0 x_1 \cdots x_p) \\ &= (f\sigma)'_1 P(x_0 x_1 \cdots x_p) \end{aligned}$$

where $(f\sigma)'$ is related to $f\sigma$ as σ' is related to σ. The last expression is therefore $P(f\sigma)$ (Definition 1-30), so

$$f'_1 P\sigma = P(f\sigma) = Pf_1\sigma$$

as was to be shown. ∎

The most important feature of the operator P is its behavior relative to the boundary operator d. First, however, a special case of the operation of P must be examined in which it acts on a singular simplex $\sigma = (u_0 u_1 \cdots u_p)$ defined by a linear map. Here $P\sigma$ must be defined by using Definition 1-30. It is natural to ask whether $P\sigma$ could also be given by a formula similar to that which defines $P(x_0 x_1 \cdots x_p)$. Let $S = [u_0 u_1 \cdots u_p]$ be contained in a Euclidean N space E_N and embed $S \times I$ in E_{N+1}, so that the first N coordinates of (x, t) are those of x and the last coordinate is t. As usual, S is to be identified with $S \times 0$. For each vertex $(u_i, t) = u_i$ of S let v_i be the vertex $(u_i, 1)$ of $S \times I$.

Theorem 1-13. *With the notation just given*

$$P(u_0 u_1 \cdots u_p) = \sum (-1)^i (u_0 u_1 \cdots u_i v_i \cdots v_p)$$

Proof. Writing $\sigma = (u_0 u_1 \cdots u_p)$, Definition 1-30 states that

$$\begin{aligned} P\sigma &= \sigma'_1 P(x_0 x_1 \cdots x_p) \\ &= \sum (-1)^i \sigma'_1 (x_0 x_1 \cdots x_i y_i \cdots x_p) \end{aligned}$$

The essential point of the proof is to check that

$$\sigma'_1(x_0 x_1 \cdots x_i y_i \cdots y_p) = \sigma'(x_0 x_1 \cdots x_i y_i \cdots y_p)$$

coincides with the linear map $(u_0 u_1 \cdots u_i v_i \cdots v_p)$.

Note first that, since $(x_0 x_1 \cdots x_i y_i \cdots y_p)$ is a linear map of Δ_{p+1} onto $[x_0 x_1 \cdots x_i y_i \cdots y_p]$ the Euclidean coordinates of the image of a point z are homogeneous linear functions of the Euclidean coordinates of z. Obviously, this property also is satisfied by σ', and so by the composition $\sigma'(x_0 x_1 \cdots x_i y_i \cdots y_p)$. In addition, $\sigma'(x_0 x_1 \cdots x_i y_i \cdots y_p)$ clearly maps the vertices of Δ_{p+1} in standard order, on the vertexes of $[u_0 u_1 \cdots u_i v_i \cdots v_p]$ in the stated order. It follows at once from Exercise 1-6 that

$$\sigma'(x_0 x_1 \cdots x_i y_i \cdots y_p)$$

is the linear map $(u_0 u_1 \cdots u_i v_i \cdots v_p)$, as required. ∎

With this preparation the formula for dP can be worked out. This will be done in two stages, first for the operation on $(x_0 x_1 \cdots x_p)$, then for the operation on any singular simplex.

Theorem 1-14. *Using the notation of Definition* 1-29,

$$dP(x_0 x_1 \cdots x_p) = (y_0 y_1 \cdots y_p) - (x_0 x_1 \cdots x_p) - Pd(x_0 x_1 \cdots x_p)$$

Proof.
$$\begin{aligned} dP(x_0 x_1 \cdots x_p) &= \sum (-1)^i d(x_0 x_1 \cdots x_i y_i \cdots x_p) \\ &= \sum_{j \le i} (-1)^{i+j}(x_0 x_1 \cdots \hat{x}_j \cdots x_i y_i \cdots y_p) \\ &\quad + \sum_{j \ge i} (-1)^{i+j+1}(x_0 x_1 \cdots x_i y_i \cdots \hat{y}_j \cdots y_p) \end{aligned}$$

The term $(y_0 y_1 \cdots y_p)$ can be picked from the first sum and the term $-(x_0 x_1 \cdots x_p)$ from the second. Then, if the remaining terms in both sums for which $i = j$ are examined, it is found that they cancel in pairs. The remaining terms are

$$\sum_{j<i} (-1)^{i+j}(x_0 x_1 \cdots \hat{x}_j \cdots x_i y_i \cdots y_p) + \sum_{j>i} (-1)^{i+j+1}(x_0 x_1 \cdots x_i y_i \cdots \hat{y}_j \cdots y_p) \tag{1-4}$$

If the theorem is to come out as stated, this should be equal to

$$-Pd(x_0 x_1 \cdots x_p)$$

This expression will now be computed, using Theorem 1-13.

$$\begin{aligned} Pd(x_0 x_1 \cdots x_p) &= \sum (-1)^j P(x_0 x_1 \cdots \hat{x}_j \cdots x_p) \\ &= \sum_{i<j} (-1)^{i+j}(x_0 x_1 \cdots x_i y_i \cdots \hat{y}_j \cdots y_p) \\ &\quad + \sum_{i>j} (-1)^{i+j-1}(x_0 x_1 \cdots \hat{x}_j \cdots x_i y_i \cdots y_p) \end{aligned}$$

It is at once clear that this is equal to minus the expression (1-4). The required result follows. ∎

With the notation of Theorem 1-11, Theorem 1-14 can be extended to any singular simplex.

Theorem 1-15. *If σ is a singular p simplex on E, then*

$$dP\sigma = f_1(\sigma) - f_1(\sigma) - Pd\sigma$$

Proof. σ is a continuous map of Δ_p into E, and σ' is the corresponding map of $\Delta_p \times I$ into $E \times I$ such that $\sigma'(x, t) = (\sigma(x), t)$. Apply the induced homomorphism σ_1' to both sides of the result of Theorem 1-14. Then

$$\sigma_1' \, dP(x_0 x_1 \cdots x_p) = \sigma_1'(y_0 y_1 \cdots y_p) - \sigma_1'(x_0 x_1 \cdots x_p) - \sigma_1' Pd(x_0 x_1 \cdots x_p) \tag{1-5}$$

On the left σ_1' commutes with d (Theorem 1-2) and $\sigma_1' P = P\sigma_1$ (Theorem 1-12). Also, $\sigma_1(x_0 x_1 \cdots x_p) = \sigma$, so the left side is equal to $dP\sigma$. Similarly, the last term on the right is $Pd\sigma$. It is clear that $\sigma_1'(y_0 y_1 \cdots y_p) = g_1(\sigma)$ and $\sigma_1'(x_0 x_1 \cdots x_p) = f_1(\sigma)$. The result stated in the theorem is obtained by making all these substitutions in (1-5). ∎

Since d, P, f_1, g_1 are all homomorphisms on chain groups over any coefficient group, Theorem 1-15 leads at once to the following result.

Theorem 1-16. *The homomorphisms d, P, f_1, g_1 are connected by the relation*

$$g_1 - f_1 = dP + Pd$$

It is now possible to prove the homotopy theorem in its second formulation, namely, Theorem 1-11.

Proof of Theorem 1-11. Let $\bar{\alpha} \in H_p(E, F)$ and let α be a relative cycle of E modulo F representing $\bar{\alpha}$. Then, by Theorem 1-16,

$$g_1(\alpha) - f_1(\alpha) = dP\alpha + Pd\alpha$$

Since $d\alpha \in C_{p-1}(F)$, $Pd\alpha \in C_p(F \times I)$. Hence this equation means that $g_1(\alpha)$ and $f_1(\alpha)$ are homologous in $E \times I$ modulo $F \times I$. Since these are relative cycles representing $g_*(\bar{\alpha})$ and $f_*(\bar{\alpha})$, respectively, it follows that $g_*(\bar{\alpha}) = f_*(\bar{\alpha})$. This holds for any $\bar{\alpha}$ in $H_p(E, F)$, hence $f_* = g_*$, as was to be shown. ∎

The proof of Theorem 1-11 is thus completed, and as already indicated, this automatically proves Theorem 1-10.

Exercises. 1-12. Let E be a solid n sphere (that is the set of points in Euclidean n space satisfying the inequality $\sum_{i=1}^{r} x_i^2 \leqq 1$) and let P be a point. Let $f: E \to P$ map E on the point P and let $g: P \to E$ map P on, for example, the center of E. Show that fg and gf are both homotopic to the identity. Then show that, for any coefficient group, $H_r(E) = 0 (r > 0)$ and that $H_0(E)$ is isomorphic to the coefficient group.

1-13. Exercise 1-12 illustrates, in a simple way, the notion of deformation retraction. Let E be a space and F a subspace. Let g be the inclusion map of F into E and suppose that there is a map $f: E \to F$ such that fg and gf are both homotopic to the identity. Then F is called a deformation retract of E. For example, if P in Exercise 1-12 is already taken as the center of E, then P is a deformation retract of E. Show that, if F is a deformation retract of E, then the inclusion map induces an isomorphism of $H_r(F)$ onto $H_r(E)$ for each r and for any coefficient group.

1-14. There is yet a further generalization of Exercises 1-12 and 1-13. Let (E, F) and (X, Y) be two pairs of spaces, and suppose that there are continuous maps

$$f: (E, F) \to (X, Y)$$
$$g: (X, Y) \to (E, F)$$

such that fg and gf are both homotopic to the identity (as maps of pairs). Then show that f induces an isomorphism of $H_r(E, F)$ onto $H_r(X, Y)$ for all r and for any coefficient group. Pairs of spaces with maps f and g having this property are said to be *of the same homotopy type*, or *homotopically equivalent*. The maps f and g are said to be homotopy inverses of each other. Note that two such pairs of spaces are not necessarily homeomorphic (cf. Exercise 1-12), but the present exercise shows that they have the same homology groups. This already shows a weakness of homology groups.

1-11. THE EXCISION THEOREM

In the relative singular homology theory of a space modulo a subspace F, chains on F are identified with zero. This suggests that the removal of a set A from F should not affect the homology groups; that is, it might be expected that the homology groups of $E - A$ modulo $F - A$ would be the same as those of E modulo F. This is so provided that the condition $\bar{A} \subset \mathring{F}$ is imposed. The reason for imposing this condition will become clear if the proof of Theorem 1-23 is examined; in other words, if A and $E - F$ have limit points in common it may not be possible to drop simplexes on A in the manner required. The main theorem of this section, Theorem 1-23, is called the excision theorem, excision being the process of removing the set A.

The technique used in proving the excision theorem consists of cutting down the size, in a suitable sense, of the singular simplexes so that any element $\bar{\alpha}$ of $H_p(E, F)$ is represented by a relative cycle, all of whose singular simplexes are either on F or on the complement of A. The part on F can then be identified with zero, so that $\bar{\alpha}$ is represented by a relative cycle of $E - A$ modulo $F - A$. A similar argument shows that, if a relative cycle of $E - A$ modulo $F - A$ is homologous to zero in E modulo F, then it is already homologous to zero in $E - A$ modulo $F - A$. These two results together yield the excision theorem.

Cutting down the size of the singular simplexes is effected by the operation of barycentric subdivision, the description of which occupies a large part of this section.

Geometrically speaking, the barycentric subdivision of a one simplex is bisection into two subsimplexes; that of a two simplex (i.e., triangle) is subdivision into six smaller triangles obtained by cutting along the medians. Inductively, this process can be continued, assuming that the faces of a p simplex S are already subdivided. The subdivision of S is then obtained by constructing all the simplexes with one vertex at the mass center, or barycenter, of S and by having as opposite face one of the subdivisions of the boundary of S. The definitions will now be set up to give a formulation of this idea as an algebraic operation on singular simplexes.

Definition 1-31. Let $\sigma = (y_0 y_1 \cdots y_p)$ be a singular simplex defined by a linear map of Δ_p into some subset of Euclidean space and let z be a point of that space. Then $z\sigma$ will denote the singular $(p + 1)$ simplex $(zy_0 y_1 \cdots y_p)$ and will be called the *join of z to σ*. If $\alpha = \sum a_i \sigma_i$ is a linear combination of singular simplexes defined by linear maps, then $z\alpha$ will denote $\sum a_i(z\sigma_i)$ and will be called the *join of z to α*.

Theorem 1-17. *Using the notation just introduced*

$$d(z\alpha) = \alpha - z(d\alpha)$$

Proof. It is clearly sufficient to prove this formula in the case in which $\alpha = \sigma = (y_0 y_1 \cdots y_p)$. In this case

$$\begin{aligned} d(z\sigma) &= d(zy_0 y_1 \cdots y_p) \\ &= (y_0 y_1 \cdots y_p) - \sum (-1)^i (zy_0 y_1 \cdots \hat{y}_i \cdots y_p) \qquad \text{(by Theorem 1-3)} \\ &= \sigma - z \sum (-1)^i (y_0 y_1 \cdots \hat{y}_i \cdots y_p) \\ &= \sigma - z(d\sigma) \end{aligned}$$

which is the required result. ∎

Definition 1-32. Let S be a Euclidean p simplex. The *barycenter* of S is the point whose barycentric coordinates in S are

$$\left(\frac{1}{p+1}, \frac{1}{p+1}, \ldots, \frac{1}{p+1}\right)$$

Note that if p is 1, 2, or 3, the barycenter of S coincides with the midpoint of a segment or centroid of a triangle or tetrahedron, respectively.

The barycentric subdivision operator will now be defined inductively by dimension, using the geometric notion of subdivision indicated above as motivation. The coefficient group throughout is arbitrary.

Definition 1-33. (1) If α is a chain of dimension zero on any space, $B\alpha = \alpha$.

(2) Assume that B has already been defined for any $(p-1)$ chain α on any space.

(3) If $(x_0 x_1 \cdots x_p)$ is the identity map of Δ_p on itself, regarded as a singular simplex on Δ_p, and if b is the barycenter of Δ_p,

$$B(x_0 x_1 \cdots x_p) = bBd(x_0 x_1 \cdots x_p)$$

using the notation of Definition 1-31 on the right. Note that the operation of B on the right has already been defined by (2).

(4) If σ is a singular p simplex on a space E, that is, a continuous map of Δ_p into E, and if σ_1 is the corresponding induced homomorphism on chain groups,

$$B\sigma = \sigma_1 B(x_0 x_1 \cdots x_p)$$

(5) If $\alpha = \sum a_i \sigma_i$ is a p chain on E then

$$B\alpha = \sum a_i B\sigma_i$$

The operation B so defined is clearly a homomorphism of $C_p(E)$ into itself for each p and for any coefficient group.

In terms of the geometry of the situation it is reasonable to expect the boundary of a subdivided simplex to be the same as the original boundary subdivided. In other words, the extra faces that cut the simplex into the pieces of the subdivision all cancel one another out. Algebraically, this means that the operator B should commute with the boundary operator d. This will now be proved.

At the same time, B can be expected to commute with the homomorphisms induced by continuous maps. Geometrically, this would mean that, if a simplex S is embedded in E and if E is continuously mapped into F, then subdividing S and mapping into F gives the same result as carrying S over into F by the continuous map and then subdividing. This will also be proved now.

Theorem 1-18. (1) *If $f: E \to F$ is a continuous map and f_1 is the induced homomorphism on chain groups then $f_1 B = B f_1$.*
(2) *The operator B satisfies $Bd = dB$.*

Proof. For the first part it is sufficient to show that $f_1 B$ and $B f_1$ give the same result when they operate on a singular simplex. So let σ be a singular p simplex on E. Then

$$\begin{aligned} f_1 B\sigma &= f_1 \sigma_1 B(x_0 x_1 \cdots x_p) && \text{(Definition 1-33)} \\ &= (f\sigma)_1 B(x_0 x_1 \cdots x_p) && \text{(Theorem 1-1)} \\ &= B(f\sigma) && \text{(Definition 1-33, using the singular simplex } f\sigma \text{ on } F) \\ &= B f_1 \sigma && \text{(Definition 1-9)} \end{aligned}$$

This completes the proof of the first part of the theorem.

The second part is more complicated. Suppose it is already known that $dB\alpha = Bd\alpha$ for any chain α of dimension less than p. This is obviously so for chains of dimension 0. Now let σ be a singular p simplex on E.

$$\begin{aligned} dB\sigma &= d\sigma_1 B(x_0 x_1 \cdots x_p) && \text{(Definition 1-33)} \\ &= \sigma_1 dB(x_0 x_1 \cdots x_p) && \text{(Theorem 1-2)} \\ &= \sigma_1 d(bBd(x_0 x_1 \cdots x_p)) && \text{(Definition 1-33)} \\ &= \sigma_1 [Bd(x_0 x_1 \cdots x_p) - bdBd(x_0 x_1 \cdots x_p)] && \text{(Theorem 1-17)} \end{aligned}$$

By the induction hypothesis $dBd(x_0 x_1 \cdots x_p) = Bdd(x_0 x_1 \cdots x_p)$, since $d(x_0 x_1 \cdots x_p)$ is a chain of dimension $p - 1$. And $d^2 = 0$, so that

$$dBd(x_0 x_1 \cdots x_p) = 0$$

Hence

$$\begin{aligned} dB\sigma &= \sigma_1 Bd(x_0 x_1 \cdots x_p) \\ &= B\sigma_1 d(x_0 x_1 \cdots x_p) && \text{[Part (1) of this theorem]} \\ &= Bd\sigma && \text{(Definition 1-12)} \end{aligned}$$

This shows that Bd and dB have the same operation on singular simplexes, completing the proof of the second part of the theorem. ∎

Now suppose that α is a cycle on E. Geometrically, α is a closed surface and, in some sense, $B\alpha$ should be the same closed surface but differently subdivided into simplexes. Algebraically, then, α and $B\alpha$ should be homologous. To prove this a chain $H\alpha$ must be constructed such that $B\alpha - \alpha = dH\alpha$. Thinking geometrically again for a moment, think of the surface $B\alpha$ as being displaced slightly from the surface α. The situation then becomes reminiscent of the homotopy theorem. Thus prisms should be constructed over the simplexes forming α so that the boundary of a prism $H\sigma$ over σ will consist of σ, $B\sigma$, and some chain over the boundary of σ; if the construction is done inductively the latter chain should be $Hd\sigma$. In algebraic terms, for each p, a homomorphism

$$H_p: C_p(E) \to C_{p+1}(E)$$

should be constructed such that, for a singular simplex on E,

$$dH_p\sigma = B\sigma - \sigma - H_{p-1}d\sigma \tag{1-6}$$

a formula of the same type as that appearing in Theorem 1-16.

In *Introduction* an explicit form is given for H. It is more suitable here to use a different procedure, which will prepare the way for a more general technique to be introduced later.

The construction of H will be carried out inductively. Suppose that H_q has been constructed for q chains on any space with $q < p$, satisfying the conditions

$$dH_q\alpha = B\alpha - \alpha - H_{q-1}d\alpha \tag{1-7}$$

$$f_1H_q = H_qf_1 \tag{1-8}$$

where f_1 is an induced homomorphism of chain groups corresponding to a continuous map of one space into another. The construction can be begun by taking H_0 to be the zero homomorphism. Now consider the singular simplex $(x_0 x_1 \cdots x_p)$ on Δ_p, the standard Euclidean p simplex. $H_p(x_0 x_1 \cdots x_p)$ is constructed so that (1-6) is satisfied. The right-hand side of (1-6) will be a cycle in this case. For

$$\begin{aligned} d[B(x_0 x_1 \cdots x_p) &- (x_0 x_1 \cdots x_p) - H_{p-1}d(x_0 x_1 \cdots x_p)] \\ &= dB(x_0 x_1 \cdots x_p) - d(x_0 x_1 \cdots x_p) - dH_{p-1}d(x_0 x_1 \cdots x_p) \\ &= Bd(x_0 x_1 \cdots x_p) - d(x_0 x_1 \cdots x_p) \\ &\quad - [Bd(x_0 x_1 \cdots x_p) - d(x_0 x_1 \cdots x_p) - H_{p-2}d^2(x_0 x_1 \cdots x_p)] \\ &= 0 \end{aligned}$$

Note that in the last step Theorem 1-18 is used as well as the induction hypothesis on H_{p-1}, operating on the $(p-1)$ chain $d(x_0 x_1 \cdots x_p)$. But if $B(x_0 x_1 \cdots x_p) - (x_0 x_1 \cdots x_p) - H_{p-1}d(x_0 x_1 \cdots x_p)$ is a cycle on Δ_p then it is a boundary (Exercise 1-11), so it can be written as $dH_p(x_0 x_1 \cdots x_p)$ for some chain $H_p(x_0 x_1 \cdots x_p)$. Choose such a chain. The choice is not unique, but once made it gives a definition of $H_p(x_0 x_1 \cdots x_p)$ which automatically satisfies (1-6). Now, if σ is a singular p simplex on any space define

$$H_p \sigma = \sigma_1 H_p(x_0 x_1 \cdots x_p)$$

It must be checked that conditions (1-7) and (1-8) are satisfied by this definition. So let σ be a singular p simplex on a space E.

$$\begin{aligned}
dH_p \sigma &= d\sigma_1 H_p(x_0 x_1 \cdots x_p) \\
&= \sigma_1 dH_p(x_0 x_1 \cdots x_p) \qquad \text{(Theorem 1-2)} \\
&= \sigma_1[(B(x_0 x_1 \cdots x_p) - (x_0 x_1 \cdots x_p) - H_{p-1}d(x_0 x_1 \cdots x_p)] \\
&\qquad \text{(above definition for } H_p(x_0 x_1 \cdots x_p)) \\
&= B\sigma - \sigma - \sigma_1 H_{p-1} d(x_0 x_1 \cdots x_p) \\
&= B\sigma - \sigma - H_{p-1}\sigma_1 d(x_0 x_1 \cdots x_p) \\
&\qquad \text{(Condition (1-8) for dimension } p-1) \\
&= B\sigma - \sigma - H_{p-1} d\sigma
\end{aligned}$$

This verifies Eq. (1-7) on a singular p simplex and, consequently, on any p chain.

Next let $f\colon E \to F$ be a continuous map and σ a singular p simplex on E. Then

$$\begin{aligned}
f_1 H_p \sigma &= f_1 \sigma_1 H_p(x_0 x_1 \cdots x_p) \qquad \text{(Definition of } H_p \sigma) \\
&= (f\sigma)_1 H_p(x_0 x_1 \cdots x_p) \qquad \text{(Theorem 1-1)} \\
&= H_p(f\sigma) \qquad \text{(Definition of } H_p \text{ applied for } f\sigma) \\
&= H_p f_1 \sigma
\end{aligned}$$

This verifies condition (1-8). Both (1-7) and (1-8) thus are satisfied for dimension p, on singular p simplexes. H_p is now defined on a p chain $\alpha = \sum a_i \sigma_i$ by setting

$$H_p \alpha = \sum a_i H_p \sigma_i$$

The result so obtained can be summed up as follows:

Theorem 1-19. *For each p there is a homomorphism*

$$H_p\colon C_p(E) \to C_{p+1}(E)$$

for any topological space and any coefficient group, satisfying the conditions

$$B\alpha - \alpha = dH\alpha + Hd\alpha$$
$$f_1 H_p = H_p f_1$$

where f_1 is induced by a continuous map $f\colon E \to F$.

If the dimension can be seen from the context, H_p will simply be written as H.

The main consequence of the properties of H is the following theorem.

Theorem 1-20. *Let (E, F) be a pair of spaces and let α be a relative p cycle of E modulo F. Then α is homologous to $B\alpha$ modulo F.*

Proof. Since $dB = Bd$ it is clear that $B\alpha$ is also a relative p cycle of E modulo F. In the formula $B\alpha - \alpha = dH\alpha + Hd\alpha$ the last term is a p chain on F, which means that α and $B\alpha$ are homologous modulo F as required. ▌

The following lemma is now proved by an easy induction (cf. *Introduction*, Theorem 42). The proof should be carried out here as an exercise.

Lemma 1-21. *Let δ be the diameter of Δ_p. Then $B^r(x_0 x_1 \cdots x_p)$ is a chain each of whose singular simplexes is on a set of diameter not greater than $(p/p + 1)\delta^r$.*

Note that this means that the simplexes forming $B^r(x_0 x_1 \cdots x_p)$ can be made arbitrarily small by making r sufficiently large. Since a singular simplex σ is a continuous map, a well-known property of continuous maps on compact metric spaces (cf. [2], p. 154) gives the following result.

Lemma 1-22. *Let σ be a singular simplex on E and let $U_1, U_2, \ldots, U_k$ be a collection of open sets in E whose union is E. Then there is an integer r such that each singular simplex of the chain $B^r\sigma$ is on one of the U_i.*

Note that the result can be applied simultaneously to the simplexes of a chain α, giving the same statement about $B^r\alpha$. This, then, is the sense in which the operator B can be used to make the simplexes of a chain "small." The excision theorem can now be proved.

Theorem 1-23. *Let (E, F) be a pair of spaces and let A be a set such that $\bar{A} \subset \mathring{F}$. Then the inclusion map*

$$i: (E - A, F - A) \to (E, F)$$

induces isomorphisms onto of the homology groups, for any coefficient group.

Proof. Take an element $\bar{\alpha} \in H_p(E, F)$, represented by a relative cycle α of E modulo F. By Theorem 1-20 and Lemma 1-22 α can be replaced by $B^r\alpha$ for any r, and r can be chosen so that each singular simplex of $B^r\alpha$ is on either $\mathring{F}$ or $E - \bar{A} \subset E - A$. It follows that $\bar{\alpha}$ can be represented by a relative cycle of $E - A$ modulo $F - A$, so i_*, induced by i, is onto.

Next, take $\bar{\alpha}$ in the kernel of i_*. If $\bar{\alpha}$ is represented by α it means that $\alpha = d\beta + \gamma$ with β a chain on E and γ a chain on F. Again, by repeated application of B, if necessary, it can be assumed that the singular simplexes of β are all either on $\mathring{F}$ or on $E - \bar{A}$. Write $\beta = \beta_1 + \beta_2$ where β_1 is a chain on $\mathring{F}$ and β_2 is on $E - \bar{A}$. Then $\alpha - d\beta_2 = d\beta_1 + \gamma$. Here, however, the left side is on $E - \bar{A}$ and the right side is on $\mathring{F}$, so both are on $F - A$. Hence $\alpha - d\beta_2$ is a chain on $F - A$, so $\bar{\alpha} = 0$. Thus, the kernel of i_* is zero and the proof is complete. ∎

2

Singular and Simplicial Homology

The main theorems of the last chapter are the basic properties of the singular homology groups. With these properties as a foundation, this chapter develops a systematic method for calculating the homology groups of certain types of space. At the same time, this procedure indicates the possibility of an axiomatic approach to the subject.

2-1. THE AXIOMATIC APPROACH

The basic properties of singular homology theory, as developed previously, can be summarized as follows:

(1) The theory assigns to each pair of spaces (E, F) and each integer $p \geqq 0$ an Abelian group $H_p(E, F)$ and to each continuous map $f: (E, F) \to (E', F')$ a homomorphism

$$f_*: H_p(E, F) \to H_p(E', F')$$

for each p.

(2) For a composition of continuous maps fg, the formula

$$(fg)_* = f_* g_*$$

holds, and if f is the identity map than f_* is the identity homomorphism (Theorem 1-5).

(3) If E consists of one point and F is empty, $H_p(E, F) = 0$ for all $p > 0$, while $H_0(E, F)$ is a fixed, preassigned group, the coefficient group (Theorem 1-7).

(4) If f and g are homotopic maps of the pair (E, F) into the pair (E', F'), then the induced homomorphisms f_* and g_* are equal (Theorem 1-10).

(5) If $E \supset F \supset G$, there is a homomorphism $\partial: H_p(E, F) \to H_{p-1}(F, G)$ which commutes with the homomorphisms associated with continuous maps (Theorem 1-8) and the sequence

$$\to H_p(F, G) \xrightarrow{i_*} H_p(E, G) \xrightarrow{j_*} H_p(E, G) \xrightarrow{\partial} H_{p-1}(F, G) \to$$

where i and j are the appropriate inclusion maps, is exact (Theorem 1-9).

(6) If (E, F) is any pair of spaces and A is a set in E such that $\bar{A} \subset \mathring{F}$ then the homomorphism

$$i_*: H_p(E - A, F - A) \to H_p(E, F)$$

induced by inclusion is an isomorphism onto for all p (Theorem 1-22).

Now it has been discovered that there are methods other than that of singular homology theory by which groups and homomorphisms can be assigned to spaces and maps in such a way that properties like the above are satisfied. Some such systems will be seen later. As the subject of algebraic topology developed, however, it was observed that, for certain classes of space, the above properties are characteristic for homology groups, regardless of the particular mode of construction. In other words, a family of groups and homomorphisms is defined up to isomorphisms for spaces of this class, simply by virtue of satisfying these properties. Another way of expressing this is to say that these properties can be taken as axioms for homology theory, and at least on a certain class of spaces, these axioms define uniquely the homology theory for a given coefficient group. This point of view is followed systematically in [1]. In fact the axiom system can be weakened a bit. For example, it is enough to state (5) with $G = \emptyset$ and (6) for open sets A only.

The class of spaces just referred to is the class of triangulable spaces or simplicial complexes. The proof that the above axioms define a unique homology theory for such spaces consists essentially in showing that they lead to an algorithm for computing the homology groups in terms of the geometric structure of the space. This algorithm will now be obtained. In this way, a stepping stone is furnished which leads to the purely axiomatic approach to homology theory, although no attempt is made here to take this as the basic framework of the present treatment.

Although the algorithm is derived entirely from the properties (1) to (6), it is sometimes convenient, and clarifies the geometric meaning of what is going on, to introduce some results in the exercises which depend on the construction of the singular homology groups.

Before proceeding with the homology theory of simplicial complexes there is one simple consequence of properties (1) to (6) which should be noted; that is, the groups $H_p(E, F)$ described in these properties are automatically topological invariants. In fact, the topological invariance proof (Theorem 1-6) uses only properties (1) and (2). In particular, the coefficient group $\mathcal{G}$ is actually defined up to isomorphism by (3), since any two points are homeomorphic topological spaces, and in fact, by a unique homeomorphism.

2-2. SIMPLICIAL COMPLEXES

As indicated in the last section, it is possible to set up an algorithm for computing the homology groups of a certain class of spaces known as simplicial complexes. The necessary definitions and some preliminary results on such spaces are given in this section. A simplicial complex is a space built up from simplexes and is, historically, the type of space to which notions of homology were first applied. The concept may appear somewhat restricted. Indeed it is, but nevertheless it is extremely useful, since many spaces occurring in the applications of topology are of this type.

Definition 2-1. A *simplicial complex* K is a subspace of some Euclidean space, consisting of a finite number of Euclidean simplexes such that the intersection of any two is a face of each. The *dimension* of K is the largest of the dimensions of the simplexes making up K.

Examples

2-1. Figures 4 and 5 show two-dimensional complexes.
2-2. An r simplex is itself an r-dimensional complex.
2-3. The boundary of an r simplex is an $(r-1)$-dimensional complex.

FIGURE 4

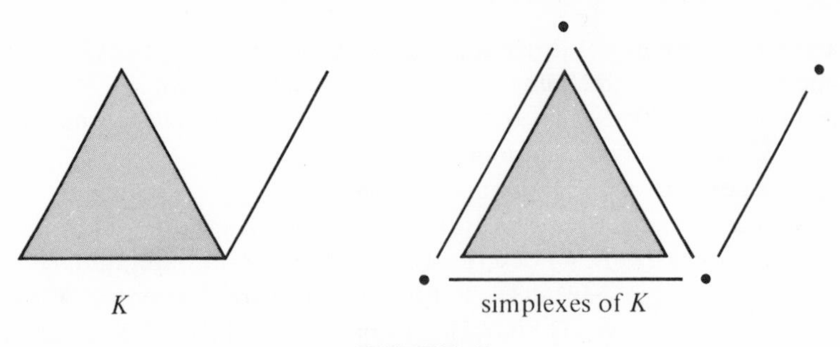

FIGURE 5

Definition 2-2. The *simplexes of a complex K* are all the simplexes making up K along with all their faces.

For example, in Fig. 5 the simplexes of K are as follows: one 2 simplex, four 1 simplexes, and four 0 simplexes.

Note that the list of simplexes of a complex K has the property that if a simplex belongs to the list so do all its faces. This suggests a notion that is not employed here, but which is frequently of use in considering the more algebraic features of the theory of simplicial complexes; that is, the complex can be defined abstractly as a set E of elements, called vertexes, along with certain specified subsets of E, called simplexes, that satisfy the condition that if S is a simplex so are all its subsets. As the discussion proceeds it will become clear that the homology groups and certain kinds of maps of simplicial complexes can be defined entirely in terms of this abstract notion. Here, however, the translation into such terms is regarded as an exercise for the reader, since the point of view taken here is being kept as geometrical as possible.

Definition 2-3. A *subcomplex L* of K is a simplicial complex each of whose simplexes is also a simplex of K.

Alternatively, a subcomplex is the point-set union of a subset of the set of simplexes of K.

Definition 2-4. A *simplicial pair* (K, L) is a simplicial complex K with a subcomplex L.

The following definition picks out from a complex a special sequence of subcomplexes which will be important in constructing inductive arguments.

Definition 2-5. The *r skeleton of a complex* is the union of all the simplexes of the complex of dimension less than or equal to r. If the complex is K, the usual notation for the r skeleton will be K_r.

Thus, the 0 skeleton of a complex K is the set of vertexes, the 1 skeleton is the union of the vertexes and edges, and so on. In particular, if K is of dimension n, then $K_n = K$.

It must be noted at this point that, when K is a simplicial complex, it is usually understood that not only a topological space is given, but also a particular representation as a union of simplexes. The same topological space, or a homeomorphic space, is capable of different representations as a complex. Generally speaking, it will be clear from the context whether attention is being paid to the space only, or to its representation as a union of simplexes. If some distinction is needed, different notations will be used for the space and for its representation. In particular, if different decompositions into simplexes are used for the same space, these will have to be distinguished by the notation.

Examples

2-4. If a diagonal is added to each face of a cube the surface of the cube appears as the union of triangles. This is easily seen to be a simplicial complex. On the other hand, the surface of a cube is homeomorphic to the surface of a tetrahedron, and the latter surface is also a simplicial complex. Thus, the two homeomorphic spaces have two entirely different representations as complexes.

2-5. Figure 6 shows two copies of the same space represented as a simplicial complex in two different ways. In the same spirit as the above remarks, note that a space may not be given as a complex at all, but nevertheless may be representable as one. For example, the surface of a sphere is homeomorphic to that of a tetrahedron, so it is homeomorphic to a complex.

Definition 2-6. If a topological space E is homeomorphic to a simplicial complex K, E is called a *triangulable space*. If, under this homeomorphism, a subspace F of E corresponds to a subcomplex L of K, then (E, F) will be called a *triangulable pair*. A homeomorphism $(E, F) \to (K, L)$ as described here is a *triangulation* of (E, F).

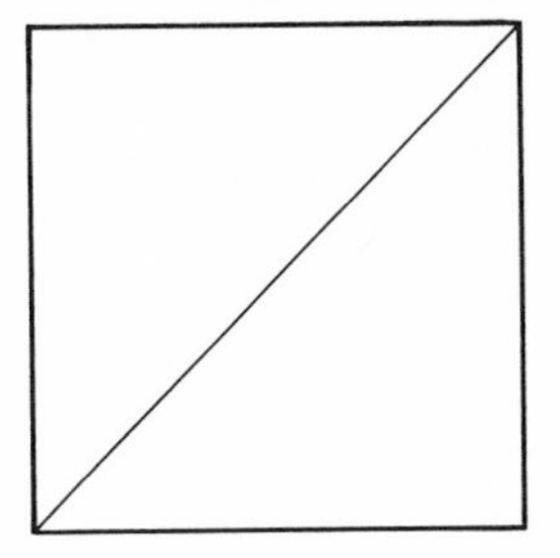

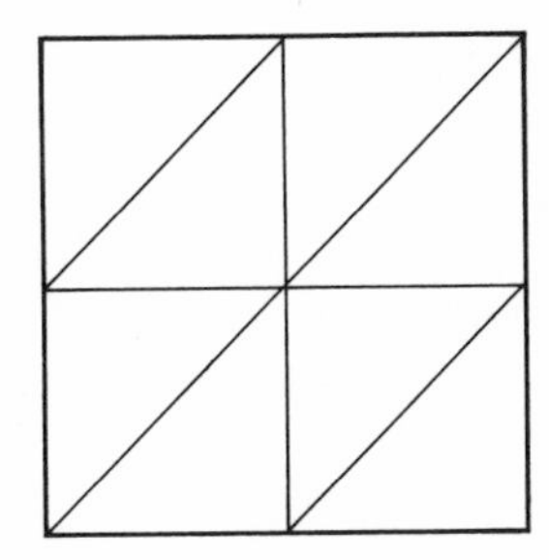

FIGURE 6

Exercises. 2-1. Prove that any simplicial complex K is homeomorphic to a subcomplex of the complex consisting of one simplex of sufficiently high dimension.

(*Hint:* Start by setting the vertexes of K into one-to-one correspondence with a set of linearly independent points of some Euclidean space.)

2-2. Let S be an r-dimensional Euclidean simplex. Express $S \times I$ as a simplicial complex in such a way that, for each face S' of S of any dimension, $S' \times I$ is a subcomplex.

(*Hint:* The $(r+1)$ simplexes of $S \times I$ should correspond to the terms in the formula for the prism operator in Definition 1-29.)

2-3. Let K be a simplicial complex. Show that $K \times I$ can be expressed as a simplicial complex in such a way that, if L is a subcomplex of K then $L \times I$ is a subcomplex of $K \times I$.

2-4. Let a barycentric subdivision operator B on Euclidean simplexes be defined as follows:

(a) If S is of dimension 0, $BS = S$.

(b) Assume that, for any simplex S' of dimension less than r, BS' is defined as a collection of Euclidean simplexes. Let b be the barycenter of an r simplex S and define BS to be the collection of all simplexes $[by_0 y_1 \cdots y_{r-1}]$ where $[y_0 y_1 \cdots y_{r-1}]$ is in the collection BS', S' running through all the faces of S.

Prove that BS is a simplicial complex with the same underlying space as S.

2-5. Let $[y_0 y_1 \cdots y_p]$ be a simplex of the complex BS, in the notation of Exercise 2-6. Show that $y_0, y_1 \cdots y_p$ can be ordered so that they are the barycenters of faces $S_0, S_1, \cdots, S_p$, respectively, of S with the property that $S_0 \subset S_1 \cdots \subset S_p$.

Show that in this way a one-to-one correspondence is set up between simplexes of BS and ascending sequences of faces of S (ordered by inclusion). In particular, the simplexes of maximum dimension in BS correspond to those sequences in which all dimensions appear from 0 up to dim S.

2-6. If K is a complex, let BK denote the set of all simplexes in all BS where S is a simplex of K. Show that the union of simplexes in BK is a complex. Also, if L is a subcomplex of K, show that BL is the subcomplex of BK that is obtained by taking all simplexes of BK lying in the space L.

Note that, combining this result with Exercise 2-1, the barycentric subdivision of any complex can be obtained by taking the barycentric subdivision of a simplex and then restricting to a subcomplex.

2-7. The star of a vertex of a complex K is defined as the union of all simplexes of K having x as a vertex. Prove that, if L is a subcomplex, the star of x in L is the intersection of L with the star of x in K. Prove that the same property holds for the open star of x, that is to say the interior of the star of x.

2-3. *THE DIRECT SUM THEOREM*

The setting up of an algorithm for computing homology groups depends on breaking up the given complex into its simplexes and, in a sense, letting each simplex make its own contribution. More precisely, this is done by means of the direct sum theorem of Section 2-4. In the meantime, a preliminary direct sum theorem will be obtained that corresponds to the breaking up of a space into separated subsets.

Theorem 2-1. *Let (E, F) be a pair of topological spaces and suppose that $E = E_1 \cup F_2$, where E_1 and E_2 are open and closed in E. Let $F_1 = E_1 \cap F$, $F_2 = E_2 \cap F$. Then, for all p,*

$$H_p(E, F) \cong H_p(E_1, F_1) \oplus H_p(E_2, F_2)$$

The summands on the right are the images of homomorphisms induced by the inclusions $(E_1, F_1) \to (E, F)$ and $(E_2, F_2) \to (E, F)$; both of these homomorphisms are being isomorphisms.

Proof. Consider the following diagram:

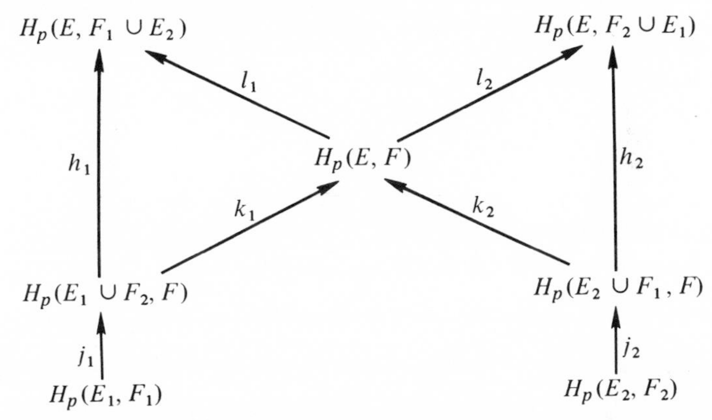

All the maps in the diagram are induced by inclusions, so it follows at once that the diagram is commutative. The hypotheses on E and F imply by the excision theorem (Theorem 1-23) that all the vertical maps are isomorphisms. For example, j_1 corresponds to the removal of the set F_2, whose closure in E is contained in the interior of F in E. Similar reasoning can be applied to the others. The exactness theorem (Theorem 1-9) implies that the two diagonal sequences in the diagram are exact.

Define i_1 and i_2 to be the compositions $k_1 j_1$ and $k_2 j_2$, respectively.

The commutativity of the diagram implies that $l_1 k_1 = h_1$, so since h_1 is an isomorphism onto, $h_1^{-1} l_1 k_1$ is the identity map of $H_p(E_1 \cup F_2, F)$ on itself. Thus k_1 has a left-sided inverse, hence $H_p(E, F)$ is isomorphic to the direct sum of the image of k_1 and the kernel of $h_1^{-1} l_1$. The kernel of $h_1^{-1} l_1$ is the same as that of l_1, since h_1^{-1} is an isomorphism; this is the same as the image of k_2, by the exactness of the diagonal of the diagram. Hence $H_p(E, F)$ is the direct sum of the images of k_1 and k_2, and since the j_i are isomorphisms, this is the same as the direct sum of the images of i_1 and i_2. To complete the proof it must be shown that i_1 and i_2 are isomorphisms into $H_p(E, F)$. Suppose, then, that $i_1(\alpha) = 0$; that is, $k_1 j_1(\alpha) = 0$. But k_1 has a left-sided inverse, so $j_1(\alpha) = 0$. j_1 is an isomorphism, and it follows that $\alpha = 0$. Thus it has been shown that i_1 has zero kernel. Similarly, i_2 has zero kernel, so the proof is complete. ∎

The above theorem can be extended to the decomposition of E into any number of sets that are both open and closed in E. Either induction or transfinite induction must be used.

As explained in Section 2-1, this argument is based on the properties of homology listed there. There is also however, a direct, more geometrical approach that can be used for singular homology, which uses the properties of singular simplexes. This is outlined in the following exercises.

Exercises. 2-8. Let E be a topological space expressed as $E_1 \cup E_2$. Suppose that it is not possible to join any point of E_1 to any point of E_2 by a continuous path. Prove that, for singular chain groups over any coefficient group,

$$C_p(E) \cong C_p(E_1) \oplus C_p(E_2)$$

for all p. Show that if, in the above decomposition, $\alpha \in C_p(E)$ is expressed as $\alpha_1 + \alpha_2$, $\alpha_1 \in C_p(E_1)$, $\alpha_2 \in C_p(E_2)$, then $\alpha \in Z_p(E)$ if and only if $\alpha_i \in Z_p(E_i)$, $i = 1, 2$, and $\alpha \in B_p(E)$ if and only if $\alpha_i \in B_p(E_i)$, $i = 1, 2$.

2-9. Prove that, with the conditions of Exercise 2-8,

$$H_p(E) \cong H_p(E_1) \oplus H_p(E_2)$$

and that the homomorphisms of $H_p(E_1)$ and $H_p(E_2)$ induced by the inclusion maps are isomorphisms.

2-10. Generalize the result of Exercise 2-9 to the homology groups of a pair (E, F), with $E = E_1 \cup E_2$ and $F_i = E_i \cap F$, $i = 1, 2$.

Note that this exercise implies that, for singular homology, $H_p(E)$ is the direct sum of the groups $H_p(E_i)$, where the E_i run through either the connected components or the arcwise connected components of E. On the other hand, the conclusion of Theorem 2-1, which applies to any homology theory satisfying conditions (1)-(6) of Section 2-1, is somewhat weaker because it is restricted to the situation in which each E_i is open and closed in E.

2-4. *THE DIRECT SUM THEOREM FOR COMPLEXES*

Theorem 2-1 will now be applied to the study of the homology groups of a simplicial pair. The following notation will be used: (K, L) is the simplicial pair to be studied; K_r denotes the r skeleton of K; and K_r' denotes $K_r \cup L$ for each r. The first step is to examine the homology of K_r' modulo K_{r-1}'. Throughout this book the boundary of a simplex S is denoted by $\dot{S}$.

The result to be proved here is that $H_p(K_r', K_{r-1}')$ is the direct sum of the groups $H_p(S, \dot{S})$, where S runs through the r simplexes of $K - L$. The method is to construct two neighborhoods of K_{r-1}', the closure of the smaller being contained in the interior of the larger so that the excision theorem can be used. When the smaller neighborhood has been cut out the remainder will be a disjoint union of simplexes to which Theorem 2-1 can be applied. With a few adjustments this will give the required result.

To carry out the details of the proof just sketched, take first an r simplex and define a function f on it as follows. If b is the barycenter of S then for any $x \neq b$ in S draw the segment bx and extend it to meet $\dot{S}$ in x'. Let $f(x)$ be the ratio in which x divides the segment bx'. In particular define $f(b) = 0$. It can be shown (this should be done as an exercise) that f is continuous on S. Note that f takes values between 0 and 1, and in fact, it is equal to 0 only at b and equal to 1 only on $\dot{S}$.

A similar function f will now be defined on the whole of K_r', in the notation introduced at the beginning of this section.

(1) Set $f(x) = 1$ for all $x \in K_{r-1}'$;

(2) If S is an r simplex of K not in L, define f on S in the manner described above.

It must be checked that the two parts of the definition of f do not contradict each other. Suppose, then, that x is in an r simplex S of $K - L$ and, at the same time, that x is in K_{r-1}'. Then x must be in some face of S. Therefore, both (1) and (2) give the same value for $f(x)$, namely 1. It follows that f is a continuous function on K_r' which is equal to 1 at points of K_{r-1}'. It is clear that the points of K_{r-1}' are the only points at which f takes the value 1.

Let V be the set of points of K_r' for which $f(x) > \frac{1}{2}$ and let W be the set for which $f(x) > \frac{3}{4}$. Since f is continuous on K_r', the sets V and W are open in K_r'. It is also easy to see that $\overline{W}$ is the set of points for which $f(x) \geqq \frac{3}{4}$ (the closure being taken in K_r'). Since $f(x) = 1$ only on K'_{r-1}, it follows that

$$K_{r-1}' \subset W \subset \overline{W} \subset \mathring{V} = V$$

The essential properties of V and W are that there is a deformation of K_r' which collapses V into K_{r-1}', and that the removal of W leaves a disjoint union of r simplexes. These, and some related properties, are established in the following lemmas.

It is convenient to take first the case in which K consists of just one r simplex S. The function f is defined on S as described before.

Lemma 2-2. *The set S' of points x in S for which $f(x) \leqq c$, where $0 < c \leqq 1$, is an r simplex.*

Proof. Let $y_0, y_1, \ldots, y_r$ be the vertexes of S, and if S is in n space let $(y_i^1, y_i^2, \ldots, y_i^n)$ be the coordinates of y_i. Let $(b^1, b^2, \ldots, b^n)$ be the barycenter of S. Thus

$$b^j = \frac{\sum_{i=0}^r y_i^j}{r+1}.$$

For each i take z_i on the segment $by\ i$, dividing this segment in the ratio c to 1. Thus $f(x_i) = c$, and the coordinates of z_i are

$$z_i^{\,j} = \frac{cy_i^{\,j} + b^j}{c+1}$$

It will now be shown that the set S' in S such that $f(x) \leqq c$ is exactly the simplex $[z_0 z_1 \cdots z_r]$. If x is in this simplex then its coordinates are given by

$$x^j = \sum_{i=0}^{r} \lambda_i z_i^{\,j}$$

$$\sum_{i=0}^{r} \lambda_i = 1$$

$$\lambda_i \geqq 0$$

Hence

$$x^j = \frac{cy^j + b^j}{c+1}$$

where $y^j = \sum_{i=0}^{r} \lambda_i y_i^{\,j}$. In other words, y is a point of S and x divides the segment by in the ratio c to 1. If by is now extended to meet the boundary of S in y', the ratio in which x divides the segment by' is less than or equal to c to 1. Hence $f(x) \leqq c$, so x is in S'; that is, $[z_0 z_1 \cdots z_r] \subset S'$.

In addition, x is in the boundary of $[z_0 z_1 \cdots z_r]$ if and only if one of the λ_i is 0, that is, if and only if y is in the boundary of S. When this happens $f(x) = c$.

Suppose that $x' \notin [z_0 z_1 \cdots z_r]$. Then the segment bx' meets the boundary of $[z_0 z_1 \cdots z_r]$ in a point $x \neq x'$, and the boundary of S in a point y. Clearly, $f(x') > f(x)$, and by the remark made in the last paragraph, $f(x) = c$. Thus $x \notin S'$. It follows that $[y_0 z_1 \cdots z_r] = S'$, as was to be shown. ∎

Continuing with the notation of the last lemma, construct the operation of expanding S' so that it fills the whole of S, while $S - \mathring{S}'$ is collapsed onto the boundary of S. This is done by constructing a homotopy of the identity map of S on itself. To do this note the following geometric point concerning the relation between S and S': if $S_1 = [y_0 y_1 \cdots \hat{y}_i \cdots y_r]$ is one of the faces of S, then $[by_0 \cdots \hat{y}_i \cdots y_r]$, which will be written as bS_1 for brevity, is the union of segments joining b to the points of S_1, and S is the union of simplexes obtained in this way from all its faces. The set of points of bS_1 for which $f(x) = c$ is easily seen to be a hyperplane section of bS_1, so the two subsets defined by $f(x) \leqq c$ and $f(x) \geqq c$ in bS_1 are both convex. In particular, the part of $S - \mathring{S}'$ in bS_1 is convex.

Consider the linear map g of S' on S that carries each vertex z_j onto the corresponding vertex y_j. In particular, g maps $[z_0 z_1 \cdots \hat{z}_i \cdots z_r]$ linearly onto $[y_0 y_1 \cdots \hat{y}_i \cdots y_r]$, and both of these simplexes are in the part of $S - \mathring{S}'$ in bS_1. The convexity property just noted implies that if

$$x \in [z_0 z_1 \cdots \hat{z}_i \cdots z_r]$$

then the segment $xg(x)$ is contained in $S - \mathring{S}'$. This holds for any face of S'. Hence, if x is in the boundary of S', the segment $xg(x)$ lies in $S - \mathring{S}'$.

The map g already has been defined as a map of S' onto S. It will now be extended to a map of S onto itself. If $x \in S - \mathring{S}'$, the segment bx meets $\dot{S}'$ in a point x' and, if extended, meets $\dot{S}$ in a point x''. Note that $x'' = g(x')$ and that, as remarked previously, the whole segment $x'x''$ is in $S - \mathring{S}'$. Now define $g(x) = x''$. It is easy to see that the extended map g is continuous on the whole of S. Also, it has the property that it maps S' onto S and $S - \mathring{S}'$ onto $\dot{S}$. The most important property of g is that it is homotopic to the identity, points of $\dot{S}$ being fixed throughout the deformation. This will now be proved.

Lemma 2-3. *The map g just defined is homotopic to the identity mapping of S on itself. This homotopy can be defined by a map $G: S \times I \to S$ such that*

(1) $G(x, t) \in S - \mathring{S}'$ for all x in $S - \mathring{S}'$ and all t in I,

(2) $G(x, t) = x$ for all x in $\dot{S}$ and all t in I.

Proof. To define G it is convenient to write $y = (y^1, y^2, \ldots, y^n) = g(x)$. Then $g(x, t)$ is the point whose jth coordinates is $(1 - t)x^j + ty^j$. In other words, it is the point dividing the segment xy in the ratio t to $1 - t$. G is certainly continuous. Also, $g(x, 0) = x$ for all x and $G(x, 1) = g(x)$. Therefore, G defines a homotopy between g and the identity. If x is in $S - \mathring{S}'$, then by the remarks preceding this lemma, the whole segment $xy = xg(x)$ lies in $S - S'$. So $G(x, t)$ is in $S - \mathring{S}'$ for all t. Finally, if x is in $\dot{S}$, $g(x) = x$ so $G(x, t) = x$ for all t. Thus, G has all the properties stated in the lemma. ∎

Return now to the complex K and to the notation introduced at the beginning of this section.

Lemma 2-4. *$K_r' - V$ is a disjoint union of r simplexes, one contained in each r simplex of $K - L$.*

Proof. This is simply a matter of applying Lemma 2-2 in turn to each r simplex of $K - L$. ∎

Clearly a similar result holds for $K_r' - W$.

Lemma 2-5. *There is a map $g: K_r' \to K_r'$ such that*

(1) *g maps each simplex of $K_r' - V$ linearly on the r simplex of K_r' that contains it,*

(2) $g(V) \subset K_{r-1}'$,

(3) *g is homotopic to the identity, the homotopy being defined by a map $G: K_r' \times I \to K_r'$ such that*

(a) *$G(x, t)$ is in V for all x in V and all t in I,*

(b) *$G(x, t) = x$ for all x in K_{r-1}' and all t in I.*

Proof. The map G will be defined first.

(1) If x is in S, an r-dimensional simplex of $K - L$, define $G(x, t)$ as in Lemma 2-3.

(2) If x is in K_{r-1}' define $G(x, t) = x$ for all t in I.

According to Lemma 2-3, if x is in $\dot{S}$, the first part of the definition of G says that $G(x, t) = x$; this agrees with the second part of the definition since x is in this case in K_{r-1}'. Thus the two parts of the definition fit together to give G as a continuous map of $K_r' \times I$ into K_r'.

If x is in K_{r-1}', $G(x, 0) = x$ by part (2) of the definition of G. Otherwise, using Lemma 2-3, $G(x, 0) = x$. Hence, restricted to $t = 0$, G becomes the identity of K_{r-1}' on itself. Define $g(x) = G(x, 1)$. Then g is the map described in this lemma, and it is certainly homotopic to the identity. Lemma 2-3 and part (2) of the definition of G imply that $g(V) \subset K_{r-1}'$, and Lemma 2-3 implies that g maps each simplex of $K_r' - V$ linearly on the r simplex of K_r' that contains it. Property (b) of G is part of its definition, while property (a) follows from Lemma 2-3. ∎

A similar result holds with V replaced by W.

The homotopy G constructed in the last lemma can be applied to a number of different situations. First, consider the inclusion

$$j: (K_r', K_{r-1}') \to (K_r', V)$$

along with the map g of Lemma 2-5, which is considered now as a map of pairs of spaces:

$$g: (K_r', V) \to (K_r', K_{r-1}')$$

The composed map gj is the same as g on K_r', but it is considered now as a map of the pair (K_r', K_{r-1}') into itself. Similarly, jg is g considered as a map of (K_r', V) into itself. The same homotopy G of Lemma 2-5 shows that both gj and jg are homotopic to the appropriate identity maps. This can be expressed as follows:

Lemma 2-6. *The inclusion map*

$$j: (K'_r, K'_{r-1}) \to (K'_r, V)$$

has a homotopy inverse.

Lemmas 2-5 and 2-6 can also be applied with the complex $K_r - W$ replacing K'_r and with the subset $V - W$ replacing V. The situation is essentially the same, except that a different value of the constant c of Lemma 2-2 is used. The result can be stated most conveniently if an auxiliary space X_r is introduced.

Let X_r be a disjoint union of r simplexes Σ_i, one for each r simplex S_i in $K - L$, with the vertexes of the Σ_i and the S_i paired up by some fixed ordering. Let X_{r-1} be the $(r - 1)$ skeleton of X_r. Let

$$h: (X_r, X_{r-1}) \to (K_r - W, V - W)$$

be the map which maps each Σ_i linearly on the corresponding $S_i - W$ (which is a simplex by Lemma 2-2). There is a natural correspondence (cf. Lemma 2-2) of the vertexes of $S_i - W$ with those of S, consequently with those of Σ_i, and this defines the required linear map. Note that h is essentially a map of the pair (X_r, X_{r-1}) into $K_r - W$ modulo its $(r - 1)$ skeleton, composed with inclusion. When applied to this situation Lemma 2-6 gives the following result.

Lemma 2-7. *The map*

$$h: (X_r, X_{r-1}) \to (K_r - W, V - W)$$

which was just introduced, has a homotopy inverse.

Continuing with the notation of Lemma 2-7, let

$$f: (X_r, X_{r-1}) \to (K'_r, K'_{r-1})$$

be the map that maps each Σ_i of X_r linearly on the corresponding S_i of K_r, with the same pairing of vertexes that was used for h. Consider the following diagram.

$$\begin{array}{ccc} (X_r, X_{r-1}) & \xrightarrow{f} & (K'_r, K'_{r-1}) \\ {\scriptstyle h}\downarrow & & \downarrow{\scriptstyle j} \\ (K_r - W, V - W) & \xrightarrow{i} & (K'_r, V) \end{array}$$

Here f and h are as described above, while i and j are inclusions. The diagram is *not* commutative, but the following lemma implies that the corresponding diagram of induced homomorphisms on homology groups is commutative.

Lemma 2-8. *In the preceding diagram ih and jf are homotopic.*

Proof. Let g' be the restriction of g (Lemma 2-5) to $K_r - W$. Then the appropriate restriction of the map G of Lemma 2-5 gives a homotopy between jg' and i. It follows that $jg'h \simeq ih$ and $g'h$ is homotopic to f. This gives the required result. ∎

The direct sum theorem for complexes can now be stated and proved.

Theorem 2-9. *For each r simplex S_i of $K - L$, the inclusion $(S_i, \dot{S}_i) \to (K'_r, K'_{r-1})$ induces an isomorphism of $H_p(S_i, \dot{S}_i)$ into $H_p(K'_r, K'_{r-1})$ and the group $H_p(K'_r, K'_{r-1})$ is the direct sum of the images of all such isomorphisms as S_i runs over all the r simplexes of $K - L$ for all values of p.*

Proof. The diagram used in Lemma 2-8 induces the following diagram, which is commutative by Lemma 2-8.

$$\begin{array}{ccc} H_p(X_r, X_{r-1}) & \xrightarrow{f_*} & H_p(K'_r, K'_{r-1}) \\ {\scriptstyle h_*}\downarrow & & \downarrow{\scriptstyle j_*} \\ H_p(K_r - W, V - W) & \xrightarrow{i_*} & H_p(K'_r, V) \end{array}$$

By Lemmas 2-6 and 2-7, h and j have homotopy inverses; therefore, h_* and j_* are isomorphisms onto (by the homotopy theorem, Theorem 1-10). By the excision theorem (Theorem 1-22), i_* is an isomorphism onto. Hence, by the commutativity of the diagram, f_* is an isomorphism onto. By Theorem 2-1, $H_p(X_r, X_{r-1})$ is the direct sum of the images of the homomorphisms

$$f_i\colon H_p(\Sigma_i, \dot{\Sigma}_i) \to H_p(X_r, X_{r-1})$$

which are induced by inclusion, and each f_i is an isomorphism. It follows that $H_p(K'_r, K'_{r-1})$ is the direct sum of the images of the homomorphisms $f_* f_i$ and each such homomorphism is an isomorphism. $f_* f_i$ is induced by a linear map of Σ_i on S_i in K'_r, but if Σ_i is identified with S_i this map becomes inclusion, as stated in the theorem. ∎

2-5. HOMOLOGY GROUPS OF CELLS AND SPHERES

The homology groups associated with cells and spheres are worked out in *Introduction*, Chapter VIII. An alternative discussion is given here, based only on properties (1) through (6) of Section 2-1. Here advantage is taken of the facts that an r simplex is an r cell and that its boundary is an $(r - 1)$ sphere, so that the results of the last section can be applied to this situation.

Let S_r be an r simplex and $\dot{S}_r$ its boundary. Let S_{r-1} be one of the $(r - 1)$ faces of S_r and $\dot{S}_{r-1}$ its boundary. Here r is assumed to be greater than 0, and if $r = 1$, $\dot{S}_{r-1}$ is the empty set.

Apply the direct sum theorem, Theorem 2-9, to the case in which $K = \dot{S}_r$, and L is taken to be the union of all faces of S_r except S_{r-1}. If the theorem is applied to the $(r - 1)$ skeleton and the $(r - 2)$ skeleton, it follows that the homomorphism

$$H_p(S_{r-1}, \dot{S}_{r-1}) \to H_p(\dot{S}_r, L) \tag{2-1}$$

which is induced by inclusion, is an isomorphism onto for all p (here $r > 0$). The direct sum of Theorem 2-9 reduces here to one term. Note that this result plays the part of an excision in the present context.

If P is the vertex of S_r opposite the face S_{r-1}, then S_r is the union of segments joining P to the points of S_{r-1}, while L is the union of those segments ending at points of $\dot{S}_{r-1}$. It follows that the inclusion map

$$(P, P) \to (S_r, L)$$

and the map

$$(S_r, L) \to (P, P)$$

which maps all of S_r on P, are homotopy inverse to each other. This can be seen by constructing a shrinking of S_r onto P along line segments radiating from P. Clearly, throughout such an operation L is mapped into itself. It follows, by the homotopy theorem, that

$$H_p(S_r, L) = H_p(P, P)$$

for all p. By Exercise 1-11 this is zero.

Hence, in the homology sequence

$$\to H_p(S_r, L) \to H_p(S_r, \dot{S}_r) \to H_{p-1}(\dot{S}_r, L) \to H_{p-1}(S_r, L) \to$$

the first and last terms are zero. Theorem 1-9 implies that this sequence is exact, so

$$\partial: H_p(S_r, \dot{S}_r) \to H_{p-1}(\dot{S}_r, L) \tag{2-2}$$

is an isomorphism onto for all p, and for $r > 0$. If the composition of the two isomorphisms (2-1) and (2-2) is formed, it follows that

$$H_p(S_r, \dot{S}_r) \cong H_{p-1}(S_{r-1}, \dot{S}_{r-1}) \tag{2-3}$$

for all p and for $r > 0$. Repeated application of this result will now be used to find $H_p(S_r, \dot{S}_r)$. Three cases will be treated separately, according as $p < r$, $p > r$, or $p = r$.

First, let $p < r$. Taking (2-3) with p replaced by $p - 1$, $p - 2$, and so on, combine the resulting isomorphisms. It follows that

$$H_p(S_r, \dot{S}_r) \cong H_0(S_{r-p}, \dot{S}_{r-p}) \tag{2-4}$$

with $r - p > 0$. On the other hand, for any $r > 0$ the lower end of the homology sequence of $(S_r, \dot{S}_r, L)$ is

$$H_0(S_r, L) \to H_0(S_r, \dot{S}_r) \to 0$$

The first term here, as noted before, is zero. Hence, $H_0(S_r, \dot{S}_r)$ is zero for $r > 0$. If r is replaced by $r - p$ and the result is combined with (2-4), it follows that

$$H_p(S_r, \dot{S}_r) \cong H_0(S_{r-p}, \dot{S}_{r-p}) = 0$$

when $r > p$.

Second, let $p > r$. It follows from (2-3), as in the last case, that

$$H_p(S_r, \dot{S}_r) \cong H_{p-r}(S_0, \dot{S}_0) = H_{p-r}(S_0) \tag{2-5}$$

Since $p - r > 0$ and S_0 is one point, $H_{p-r}(S_0) = 0$, by the dimension theorem, Theorem 1-7. Hence $H_p(S_r, \dot{S}_r) = 0$ for $p > r$.

Finally, let $p = r$. In this case, arguing as above, it follows that

$$H_r(S_r, \dot{S}_r) \cong H_0(S_0, \dot{S}_0) = H_0(S_0)$$

Here S_0 is one point, so by Theorem 1-7, $H_0(S_0) = \mathscr{G}$, the coefficient group.

Thus, the combination of the three cases yields the following result.

Theorem 2-10.

$$H_p(S_r, \dot{S}_r) = 0 \qquad \text{if } p \neq r$$
$$H_r(S_r, \dot{S}_r) = \mathcal{G} \qquad \text{the coefficient group}$$

The results of this section can now be combined with the direct sum theorem, Theorem 2-9, to give more information on the homology groups of a simplicial pair. The notation introduced at the beginning of Section 2-4 will be used again.

Theorem 2-11. *In the notation of Theorem* 2-9,

$$H_p(K_r', K_{r-1}') = 0 \qquad p \neq r$$

and $H_r(K_r', K_{r-1}')$ is the direct sum of a number of copies of the coefficient group, one copy for each r simplex in $K - L$.

There is an immediate consequence of this theorem which will be of importance later: it will be shown that it is only necessary to consider the homology groups of a complex up to the dimension of the complex, because all the higher dimensional groups are zero.

Theorem 2-12. *Let (K, L) be a simplicial pair and let the dimension of $K - L$ be n (i.e., the maximum dimension of simplexes in $K - L$ is n). Then $H_p(K, L) = 0$ for $p > n$.*

Proof. Using the notation of Section 2-4, consider the exact homology sequence

$$\to H_{p+1}(K_r', K_{r-1}') \to H_p(K_{r-1}', L) \to H_p(K_r', L) \to H_p(K_r', K_{r-1}') \to$$

If $p > r$, Theorem 2-11 implies that the first and last terms appearing here are zero. Hence

$$H_p(K_r', L) \cong H_p(K_{r-1}', L)$$

for $p > r$.

Apply this isomorphism with r replaced by $n, n-1, \ldots, 1$ (remember that $p > n$ so that the condition $p > r$ will always be satisfied), and combine all the resulting isomorphisms. It follows that

$$H_p(K, L) = H_p(K_n', L) \cong H_p(K_0', L)$$

Here K_0' is the union of L with a finite set of points, $P_1, P_2, \ldots, P_m$. L is open and closed in this union, so the excision theorem implies that $H_p(K_0', L) = H_p(\cup P_i)$. By the direct sum theorem, Theorem 2-9, this is isomorphic to the direct sum of the groups $H_p(P_i)$. Since $p > 0$, these groups are all zero (Theorem 1-7), so $H_p(K_0', L) = 0$. In other words, $H_p(K, L) = 0$, as required. ∎

Exercises. 2-11. Theorem 2-10 says that, if E^n is an n cell and S^{n-1} its boundary, $H_r(E^n, S^{n-1}) = 0$ except when $r = n$, and then $H_n(E^n, S^{n-1}) \cong \mathscr{G}$. Use the homology sequence of (E^n, S^{n-1}) to show that $H_r(S^{n-1}) = 0$ except for $r = 0$ and $r = n - 1$, $H_{n-1}(S^{n-1}) \cong \mathscr{G}(n \neq 1)$, $H_0(S^{n-1}) \cong \mathscr{G}(n \neq 1)$.

There is also a geometrical argument to show that $H_0(S^{n-1}) \cong \mathscr{G}(n \neq 1)$. Show that, in singular homology, if E is an arcwise connected space, $H_0(E) \cong \mathscr{G}$.

2-12. For a Euclidean r simplex S_r, show that $H_r(S_r, \dot{S}_r)$ has a generator represented by a relative cycle which is in fact a singular simplex on S_r, namely, a linear map of Δ_r onto S_r.

2-13. The object of this exercise is to show that, if f is a continuous map of the n cell E^n into itself, f has a fixed point; that is, there is a point x in E^n such that $f(x) = x$. This is the Brouwer fixed point theorem. Prove this result by contradiction. Suppose that f has no fixed point, so that $f(x) \neq x$ for all $x \in E$. Join $f(x)$ to x and extend this segment to meet the boundary S^{n-1} of E^n in $g(x)$. Verify that g is a continuous map of E^n into S^{n-1} which acts as the identity on S^{n-1}. Show that this implies that the identity map of S^{n-1} on itself is homotopic to a constant map, and by examining the homology groups, show that this leads to a contradiction.

2-14. In the excision theorem the condition $\bar{A} \subset \dot{F}$ cannot be dropped. Show this by first taking E as a closed line interval, F as the two end points, and $A = F$, then checking that $H_1(E, F) \neq H_1(E - F)$.

2-6. ORIENTATION

At various points in the discussion of Section 2-5 it was necessary to fix in some way the order of vertexes of a simplex, and it is natural to ask what is the effect of a change of order. In particular, consider $H_r(S, \dot{S})$, where S is an r simplex, and the coefficient group $\mathscr{G}$ is arbitrary. This homology group was shown (Theorem 2-10) to be isomorphic to $\mathscr{G}$. A permutation of the vertexes, however, induces a homeomorphism (a linear map, in fact) of S onto itself, and so it induces an isomorphism of $H_r(S, \dot{S})$ onto itself. In the particular case where $\mathscr{G}$ is the group of integers, $H_r(S, \dot{S})$ is infinite cyclic, and an isomorphism of this group onto itself must correspond to a change of generator. Here there is only a choice of two generators, oppositely signed, so the permutations of the vertexes must fall into two classes corresponding to the two possible choices of generator. It will be shown that these classes coincide with the even and odd permutations. The choice of one of these classes is called orientation of S. The situation for a general coefficient group is a little more troublesome, but it can be described in somewhat similar terms. This is the object of the present section.

To begin with, the one-dimensional case will be examined. The coefficient group is an arbitrary group $\mathcal{G}$.

Let S be a one-dimensional simplex, that is, a line segment. The boundary $\dot{S}$ consists of two points P and Q. Consider the homology sequence

$$\to H_1(S) \to H_1(S, \dot{S}) \xrightarrow{\partial} H_0(\dot{S}) \xrightarrow{i_*} H_0(S) \to H_0(S, \dot{S})$$

The idea is to show that ∂ is an isomorphism into, so that the effect on $H_1(S, \dot{S})$ of reversing S can be deduced from the effect on $H_0(\dot{S})$.

S is a cell, so by the homotopy theorem, it has the same homology groups as a point. Hence $H_1(S) = 0$ and $H_0(S) = \mathcal{G}$. By Theorem 2-10 $H_0(S, \dot{S}) = 0$ and $H_1(S, \dot{S}) = \mathcal{G}$. By Theorem 2-1 $H_0(\dot{S}) = H_0(P) \oplus H_0(Q)$; that is, it is the direct sum of two isomorphic copies of $\mathcal{G}$. To make this more precise, let P_0 be any point and identify $\mathcal{G}$ with $H_0(P_0)$. Let f and g be the two maps of P_0 into $\dot{S}$ defined by $f(P_0) = P$ and $g(P_0) = Q$. Then $H_0(\dot{S})$ is the direct sum of the images of the induced homomorphisms f_* and g_*, both of which are isomorphisms.

The exactness of the above homology sequence implies that i_* is onto and that ∂ has zero kernel. The kernel of i_* will now be computed.

In the first place, note that, since i is the inclusion map of $\dot{S}$ into S, the maps if and ig are homotopic, so by the homotopy theorem and Theorem 1-5, $i_*f_* = i_*g_*$. It follows that any element of $H_0(\dot{S})$ of the form $f_*(\alpha) - g_*(\alpha)$ for any α in $\mathcal{G} = H_0(P_0)$ is in the kernel of i_*. It will now be shown that every element of the kernel of i_* is of this form. Certainly, every element of $H_0(\dot{S})$ is of the form $f_*(\alpha) - g_*(\beta)$ for elements α and β in $\mathcal{G}$. If this element is in the kernel of i_* then $i_*f_*(\alpha) - i_*g_*(\beta) = 0$, but $i_*f_* = i_*g_*$ and so this implies that $i_*f_*(\alpha - \beta) = 0$. Here if is the map of P_0 on P in S, and this has a homotopy inverse, namely, the map of all of S onto P_0. Hence i_*f_* is an isomorphism, so the last equation implies that $\alpha - \beta = 0$. In other words, $\beta = \alpha$. Hence, as asserted, every element of the kernel of i_* is of the form

$$f_*(\alpha) - g_*(\alpha)$$

with α in G.

By the exactness theorem, the kernel of i_* is the image of ∂. Consequently, the latter consists of elements of the form $f_*(\alpha) - g_*(\alpha)$ with α in $\mathcal{G}$.

Now let ϕ be the linear map of S on itself which maps P on Q and Q on P. The restriction of ϕ to $\dot{S}$ can also be denoted by ϕ. Since ϕ interchanges P and Q it is clear that $\phi f = g$ and $\phi g = f$.

Since the boundary homomorphism commutes with induced homomorphisms (Theorem 1-8), the following diagram is commutative:

$$\begin{array}{ccc} H_1(S, \dot{S}) & \xrightarrow{\partial} & H_0(\dot{S}) \\ \phi_* \downarrow & & \phi_* \downarrow \\ H_1(S, \dot{S}) & \xrightarrow{\partial} & H_0(\dot{S}) \end{array}$$

Thus, if α is in $H_1(S, \dot{S})$ then

$$\partial \phi_*(\alpha) = \phi_* \partial \alpha$$

and by the above discussion $\partial \alpha = f_*(\beta) - g_*(\beta)$ for some β in $\mathscr{G}$. Thus

$$\partial \phi_*(\alpha) = \phi_* \partial \alpha = \phi_* f_*(\beta) - \phi_* g_*(\beta) = g_*(\beta) - f_*(\beta) \qquad (2\text{-}6)$$

The last step follows from the equations $\phi f = g$, $\phi g = f$. From (2-6)

$$\partial \phi_*(\alpha) = -\partial \alpha$$

but, because ∂ has zero kernel, it follows that

$$\phi_*(\alpha) = -\alpha$$

The discussion just completed will be taken as the starting point of an induction that will give the general result. It will be shown that the homomorphism induced by the interchange of two vertexes of S changes the sign of elements of $H_r(S, \dot{S})$ where S is an r simplex. Thus, the effect of any permutation depends on whether it is odd or even.

Theorem 2-13. *Let ϕ be the linear map of the r simplex S onto itself which interchanges two vertexes P_1 and P_2 and maps each of the other vertexes on itself. Then, for any γ in $H_r(S, \dot{S})$, $\phi_*(\gamma) = -\gamma$.*

Proof. Assume that the result is true for simplexes of dimension less than r. Note that the above discussion proves the result for $r = 1$. It will now be proved for an r simplex.

Let S_1 be an $(r - 1)$-dimensional face of S that contains the vertexes P_1 and P_2, and let L be the union of the faces of S other than S_1. Note that ϕ maps S_1 onto itself and also maps L onto itself. In particular, it is clear that when ϕ is restricted to S_1 it is related to S_1 in the same way as ϕ itself is related to S.

In the diagram

$$\begin{array}{ccccc} H_r(S, \dot{S}) & \xrightarrow{\partial} & H_{r-1}(\dot{S}, L) & \xleftarrow{h} & H_{r-1}(S_1, \dot{S}_1) \\ \downarrow & & \downarrow & & \downarrow \\ H_r(S, \dot{S}) & \xrightarrow{\partial} & H_{r-1}(\dot{S}, L) & \xleftarrow{h} & H_{r-1}(S_1, \dot{S}_1) \end{array} \qquad (2\text{-}7)$$

the vertical maps are all induced by ϕ or its restrictions and are all denoted by ϕ_*, ∂ is the boundary homomorphism, and h is induced by inclusion. The left square in the diagram is commutative because it is part of the homology sequence of the triple $(S, \dot{S}, L)$ mapped into itself by the induced homomorphism ϕ_*. The right square is commutative since it is induced by a commutative diagram of continuous maps, namely, ϕ and the inclusion. By Section 2-5 (mapping 2-2), ∂ has kernel zero. Also, h is precisely the map shown in Section 2-5 to be an isomorphism onto. Let γ be in $H_r(S, \dot{S})$ and let γ_1 be the element of $H_{r-1}(S_1, \dot{S}_1)$ defined by $\gamma_1 = h^{-1}\partial\gamma$. The induction hypothesis says that $\phi_*(\gamma_1) = -\gamma_1$, that is,

$$\phi_* h^{-1}\partial\gamma = -h^{-1}\partial\gamma$$

The commutativity of the diagram (2-7) then implies that

$$h^{-1}\partial\phi_*\gamma = -h^{-1}\partial\gamma$$

Since h^{-1} and ∂ both have zero kernels, it follows that $\phi_*\gamma = -\gamma$, as was to be shown. ∎

Theorem 2-14. *Let S be an r simplex and ϕ a linear homomorphism of S onto itself. Then for any γ in $H_r(S, \dot{S})$, $\phi_*(\gamma) = \pm\gamma$, the sign being $+$ or $-$ depending on whether the permutation induced by ϕ on the vertexes of S is even or odd.*

Proof. A linear map is uniquely defined by its effect on the vertexes, so ϕ can be expressed as the composition of linear homeomorphisms, each of which interchanges just two vertexes. The required result then follows by repeated application of Theorem 2-14. ∎

Theorem 2-14 suggests that the choice of an isomorphism between $\mathscr{G}$ and $H_r(S_r, \dot{S}_r)$, where S_r is an r simplex, is connected with the choice of one of the permutation classes of vertexes of S. This choice has a simple meaning in the low dimensional cases. If $r = 1$, S_1 is a segment $y_0 y_1$. There are just two permutations of the vertexes, namely, the identity and the interchange of y_0 and y_1, and the choice of one of these can be specified by marking a direction along the segment S_1. If $r = 2$, S_2 is a triangle $y_0 y_1 y_2$. The three even permutations on the vertexes all name them in the same cyclic order, while the odd permutations name them in the opposite cyclic order. Thus, the choice in this case is that of a direction of rotation around the triangle.

These two geometric notions, namely, fixing a direction on a segment and fixing a direction of rotation on a triangle, will now be unified in the general concept of orientation of a simplex.

Definition 2-7. Let the $(r + 1)!$ orderings of the vertexes of an r simplex S be divided into two classes such that the orderings in any one class differ from each other by an even permutation. Each of these classes is called an *orientation of S*.

Thus the choice of an orientation means the choice of an order of vertexes defined up to an even permutation.

As pointed out previously, Theorem 2-14 links the notion of orientation with the choice of an isomorphism between $H_r(S_r, \dot{S}_r)$ and $\mathcal{G}$. To make this definite, some sort of standardization must be introduced, and the following is a convenient way of doing this.

The proof of Theorem 2-10 gives a sequence of isomorphisms

$$H_r(\Delta_r, \dot{\Delta}_r) \xrightarrow{h_r} H_{r-1}(\Delta_{r-1}, \dot{\Delta}_{r-1}) \xrightarrow{h_{r-1}} \cdots \longrightarrow H_1(\Delta_1, \dot{\Delta}_1) \xrightarrow{h_i} H_0(\Delta_0) = \mathcal{G}$$

where the Δ_i are the standard Euclidean simplexes. It will be remembered that Δ_{i-1} is a face of Δ_i. Let ϕ_r be an isomorphism of $H_r(\Delta_r, \dot{\Delta}_r)$ on $\mathcal{G}$ defined as follows.

$$\phi_1 : H_1(\Delta_1, \dot{\Delta}_1) \to \mathcal{G}$$

is the last isomorphism in the above sequence, namely, h_1. Then, if

$$\phi_{r-1} : H_{r-1}(\Delta_{r-1}, \dot{\Delta}_{r-1}) \to \mathcal{G}$$

has already been defined, ϕ_r is $(-1)^r \, \phi_{r-1} h_r$. This alternation in sign is introduced to make the boundary formula work out properly.

Now take any r simplex S_r with vertexes $y_0, \hat{y}_1, \ldots, y_r$. Corresponding to that order of vertexes there is a linear map of Δ_r onto S_r (namely $(y_0 y_1 \cdots y_r)$ in the notation of Definition 1-10) which induces an isomorphism

$$\theta_r : H_r(\Delta_r, \dot{\Delta}_r) \to H_r(S_r, \dot{S}_r)$$

This can be composed with ϕ_r^{-1} to get an isomorphism $\theta_r \phi_r^{-1}$ of $\mathcal{G}$ onto $H_r(S_r, S_r)$. It is convenient to use a notation for this map which will draw attention to the order of vertexes chosen in S_r: for any α in $\mathcal{G}$, write

$$\theta_r \phi_r^{-1}(\alpha) = \alpha((y_0 y_1 \cdots y_r))$$

Definition 2-8. The isomorphism $((y_0 y_1 \cdots y_r))$ of $H_r(S_r, \dot{S}_r)$ and $\mathcal{G}$ is called an *oriented simplex* on S_r.

By Theorem 2-14 there are two oriented simplexes on S_r with opposite signs, according to the permutation class of the vertexes. A comparison of Definitions 2-7 and 2-8 shows that each oriented simplex on S_r corresponds to the choice of an orientation on S_r.

$((y_0 y_1 \cdots y_r))$ is a map, but the notation has been chosen so that the elements of $H_r(S_r, \dot{S}_r)$ can be thought of as multiples of $((y_0 y_1 \cdots y_r))$ by elements of $\mathscr{G}$. In particular, if $\mathscr{G} = Z$, the group of integers, then the choice of an isomorphism of $H_r(S_r, \dot{S}_r)$ and $\mathscr{G}$ is equivalent to the choice of a generator of $H_r(S_r, \dot{S}_r)$, so $((y_0 y_1 \cdots y_r))$ can be identified with one such generator.

If K is a complex, L a subcomplex, and the notation of Theorem 2-9 is used, then $H_r(K'_r, K'_{r-1})$ is the direct sum of the isomorphic images (induced by inclusion) of the groups $H_r(S, \dot{S})$ where S runs over the simplexes of $K - L$. It is convenient to identify these images with the $H_r(S, \dot{S})$ themselves. Now associate with each S an oriented simplex σ. As above, $H_r(S, \dot{S})$ can be written as the group of multiples of σ by elements of $\mathscr{G}$. Hence $H_r(K'_r, K'_{r-1})$, being the direct sum of such groups as S varies over the r simplexes $K - L$, has elements that can be written uniquely as linear combinations of the σ (for various S) with coefficients in $\mathscr{G}$. In other words, $H_r(K'_r, K'_{r-1})$ is the free module generated over $\mathscr{G}$ by the oriented simplexes associated with the r simplexes of $K - L$.

2-7. HOMOLOGY GROUPS OF A SIMPLICIAL PAIR

It will now be shown that the homology groups of a simplicial pair (K, L) can be computed in terms of the oriented simplexes associated, as in the last section, with the simplexes of K which are not in L. It will first be proved that the r-dimensional homology is carried by the r skeleton. This is geometrically reasonable, for if an r cycle is thought of as an r-dimensional surface in K, it should be possible to push it out of the interiors of simplexes of dimension greater than r onto their boundaries, and so on until it reaches the r skeleton.

According to the notation of Section 2-4, K_r is the r skeleton of K, and $K'_r = K_r \cup L$. Throughout the discussion the coefficient group is an arbitrary group $\mathscr{G}$.

Theorem 2-15. *The homomorphism* $i\colon H_r(K'_r, L) \to H_r(K, L)$ *induced by inclusion is onto.*

Proof. Consider the exact homology sequence

$$\to H_{r+1}(K'_{q+1}, K'_q) \to H_r(K'_q, L) \to H_r(K'_{q+1}, L) \to H_r(K'_{q+1}, K'_q) \to \qquad (2\text{-}8)$$

If $r \leqq q$, the last term is zero (Theorem 2-11), so the map in the middle is onto. The map i of this theorem can be expressed as the composition of the following sequence of homomorphisms, all of which are induced by inclusions (using the composition theorem for induced homomorphisms, Theorem 1-5).

$$\to H_r(K'_r, L) \xrightarrow{i_1} H_r(K'_{r+1}, L) \xrightarrow{i_2} H_r(K'_{r+2}, L) \to \cdots \to H_r(K, L) \quad (2\text{-}9)$$

Each of these is of the type just shown to be onto. Hence, the composition i is onto, as required. ∎

In fact, more can be said about (2-9). In (2-8), if $r < q$, both first and last terms are zero, so the map in the middle is an isomorphism onto. This means that, all the maps in (2-9) except i_1 are isomorphisms onto. It follows, in particular, that the kernel of i is the same as the kernel of i_1.

Theorem 2-15 shows that $H_r(K, L)$ is a quotient group of $H_r(K'_r. L)$. The next step is to identify the latter with a subgroup of the group $H_r(K'_r, K'_{r-1})$. This automatically implies that $H_r(K'_r, L)$ is finitely generated over $\mathscr{G}$, hence $H_r(K, L)$ is also.

Theorem 2-16. *The kernel of the homomorphism*

$$j: H_r(K'_r, L) \to H_r(K'_r, K'_{r-1})$$

is zero.

Proof. Consider the sequence

$$\to H_r(K'_{r-1}, L) \to H_r(K'_r, L) \xrightarrow{j} H_r(K'_r, K_{r-1}) \to$$

This is exact because it is part of the homology sequence of the triple (K'_r, K'_{r-1}, L). By Theorem 2-12, $H_r(K'_{r-1}, L) = 0$, since the dimensions of all simplexes of K'_{r-1} which are not in L are less than r. Hence, the kernel of j is zero, as was to be shown. ∎

The final stage of the computation has now been reached. The map i in Theorem 2-15 is onto, therefore, $H_r(K, L) = H_r(K'_r, L)/(\text{kernel of } i)$. The isomorphism j can now be used to embed the numerator and denominator of this quotient in $H_r(K'_r, K'_{r-1})$. The obvious advantage of this step is that $H_r(K, L)$ is thus expressed in terms of subgroups of a free module over $\mathscr{G}$ with known generators, namely the oriented simplexes. Thus

$$H_r(K, L) = \frac{jH_r(K'_r, L)}{j(\text{kernel of } i)}$$

To identify the numerator and denominator in a more satisfactory way, consider the following diagram:

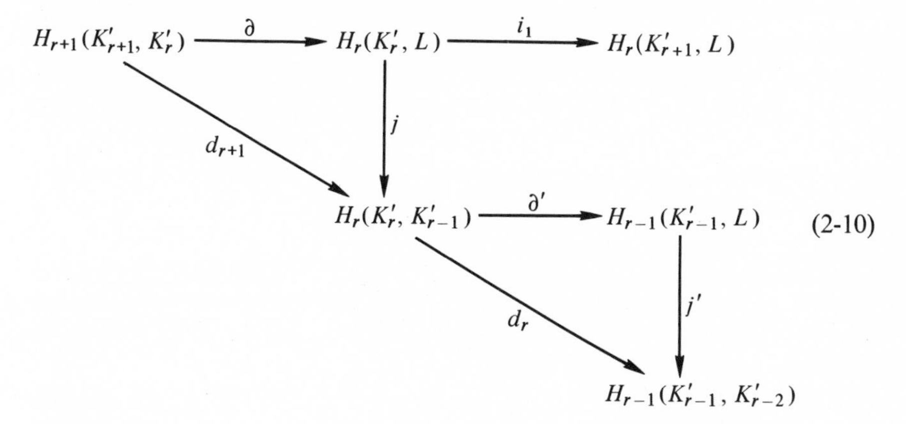

Note that the two triangles are essentially the same with a change of 1 in the dimension. d_{r+1} and d_r are defined as $j\partial$ and $j'\partial'$, respectively, so the diagram is automatically commutative. On the other hand, d_{r+1} is the boundary homomorphism in the homology sequence of the triple $(K'_{r-1'}, K'_r, K'_{r-1})$. To see this consider the diagram

$$\begin{array}{ccc} H_{r+1}(K'_{r+1}, K'_r) & \xrightarrow{\partial} & H_r(K'_r, L) \\ \downarrow & & \downarrow j \\ H_{r+1}(K'_{r+1}, K'_r) & \xrightarrow{\partial''} & H_r(K'_r, K'_{r-1}) \end{array}$$

The horizontal lines are parts of the appropriate homology sequences, and the horizontal maps are boundary homomorphisms. In particular, the top one is the same as ∂ in (2-10). The vertical maps are induced by inclusion. The one on the left is the identity, while that on the right is j in (2-10). This diagram is commutative (Theorem 1-8), so $j\partial$ is equal to ∂'' composed with the identity. In other words, $\partial'' = d_{r+1}$, as was asserted.

The image of j is the same as the kernel of ∂', by the exactness of the homology sequence of the triple (K'_r, K'_{r-1}, L), and this is the same as the kernel of d_r, since j' is an isomorphism (Theorem 2-16). On the other hand, the kernel of i is the kernel of i_1 (cf. the remark following Theorem 2-15), which is the same as the image of ∂, by the exactness of the homology sequence of (K'_{r+1}, K'_r, L). Thus j(kernel of i) $= j$(image of ∂) $=$ image of $j\partial =$ image of d_{r+1}. The quotient formula for $H_r(K, L)$ now can be expressed as follows.

Theorem 2-17. *$H_r(K, L) \cong$ (kernel of d_r)/(image of d_{r+1}).*

By now, some algebraic similarity between the statement of Theorem 2-17 and the original definition of singular homology groups can be noticed. For the groups $H_r(K'_r, K'_{r-1})$ form, as do the $C_r(E)$ or the $C_r(E, F)$, a sequence of Abelian groups with a homomorphism from the rth to the $(r-1)$th for each r, such that the composition of any two successive homomorphisms is zero. This is expressed in the singular case by the equation $d^2 = 0$, where d is the boundary operator, and here by the evident fact that the image of d_{r+1} is contained in the kernel of d_r, so that $d_r d_{r+1} = 0$. The following definition brings out this analogy more clearly.

Definition 2-9. $H_r(K'_r, K'_{r-1})$ is written as $\mathscr{C}_r(K, L)$ and is called the *group of oriented r chains of* K modulo L. d_r is called the *oriented boundary operator.* Its kernel is called the *group of oriented r cycles* $\mathscr{Z}_r(K, L)$ and its image the *group of oriented* $(r-1)$ *boundaries* $\mathscr{B}_{r-1}(K, L)$. The quotient group $\mathscr{Z}_r(K, L)/\mathscr{B}_r(K, L)$ is written as $\mathscr{H}_r(K, L)$ and is called the *oriented simplicial r-dimensional homology group of K modulo L.*

Note that, by Theorem 2-9, $\mathscr{C}_r(K, L)$ is generated over $\mathscr{G}$ by the oriented simplexes (cf. Definition 2-8) associated with the r simplexes in K but not in L. Note also that the same notation is being used for the oriented boundary operator as for the boundary operator on singular chains, but the context always shows which is meant at any given time. As usual, the subscript on d_r will be omitted if the dimension is clear from the context.

In terms of the notation just introduced, Theorem 2-17 states that

$$H_r(K, L) = \mathscr{H}_r(K, L)$$

For actual computation, the operator d must be expressed in terms of the oriented simplexes.

Theorem 2-18. *In the notation of Definition* 2-8

$$d\alpha((y_0 y_1 \cdots y_r)) = \sum (-1)^i \alpha((y_0 y_i \cdots \hat{y}_i \cdots y_r))$$

where α *is any element of* $\mathscr{G}$.

Proof. As noted in the proof of Theorem 2-16, d is simply the boundary homomorphism in the homology sequence of the triple $(K'_r, K'_{r-1}, K'_{r-2})$. Let S be the simplex with vertexes $y_0, y_1, \ldots, y_r$, $\dot{S}$ its boundary, and $\ddot{S}$ its $(r-2)$ skeleton, and consider the simplex itself as a complex. The fact that the boundary homomorphisms commute with the homomorphisms induced by the inclusion $(S, \dot{S}, \ddot{S}) \to (K'_r, K'_{r-1}, K'_{r-2})$ implies that it is sufficient to prove that

$$\partial\alpha((y_0 y_1 \cdots y_r)) = \sum (-1)^i \alpha((y_0 y_1 \cdots \hat{y}_i \cdots y_r))$$

where ∂ is the boundary homomorphism in the homology sequence of the triple $(S, \dot{S}, \ddot{S})$. In fact, replacing S by Δ_r, it is sufficient to prove that

$$\partial\alpha((x_0 x_i \cdots x_r)) = \sum (-1)^i \alpha((x_0 x_1 \cdots \hat{x}_i \cdots x_r))$$

where ∂ is the boundary homomorphism of the homology sequence of $(\Delta_r, \dot{\Delta}_r, \ddot{\Delta}_r)$. The result for S can then be obtained by applying the linear map $(y_0 y_1 \cdots y_r)$ of Δ_r onto S.

Now consider the isomorphism (2-3) in Section 2-5. If Λ is the union of the faces of Δ_r other than Δ_{r-1}, then h_r is the composition of the isomorphisms ∂' and j^{-1} in the diagram

$$H_r(\Delta_r, \dot{\Delta}_r) \xrightarrow{\partial'} H_{r-1}(\dot{\Delta}_r, \Lambda) \xleftarrow{j} H_{r-1}(\Delta_{r-1}, \dot{\Delta}_{r-1})$$

where j is induced by inclusion.

In the diagram

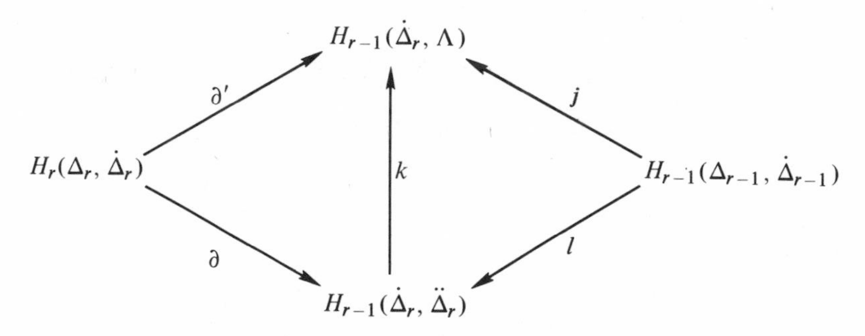

j, k, l are induced by the appropriate inclusion maps, while ∂ and ∂' are the boundary homomorphisms of the appropriate triples. By Theorems 1-5 and 1-8 the diagram commutes.

Since $kl = j$ is an isomorphism onto, it follows that $H_{r-1}(\dot{\Delta}_r, \ddot{\Delta}_r)$ is the direct sum of the image of l and the kernel of k. Hence $h_r = j^{-1}\partial'$ can be obtained by applying ∂, which maps into $H_{r-1}(\dot{\Delta}_r, \ddot{\Delta}_r)$, and then taking the component in the direct sum decomposition, which lies in the image of l. In particular, when $\alpha((x_0 x_1 \cdots x_r))$ is in $H_r(\Delta_r, \dot{\Delta}_r)$, its image in $H_{r-1}(\dot{\Delta}_r, \ddot{\Delta}_r)$ under ∂ is of the form $\sum \alpha_i((x_0 x_1 \cdots \hat{x}_i \cdots x_r))$. The component in the image of l is $\alpha_r((x_0 x_1 \cdots x_{r-1}))$, so

$$h_r(\alpha((x_0 x_1 \cdots x_r))) = \alpha_r((x_0 x_1 \cdots x_{r-1}))$$

Now

$$\phi_r(\alpha((x_0 x_1 \cdots x_r))) = \alpha$$
$$\phi_{r-1}(\alpha_r((x_0 x_1 \cdots x_{r-1}))) = \alpha_r$$

and, by definition of h_r,

$$\phi_r = (-1)^r \phi_{r-1} h_r$$

Hence

$$\begin{aligned}\alpha = \phi_r(\alpha((x_0 x_1 \cdots x_r))) &= (-1)^r \phi_{r-1} h_r(\alpha((x_0 x_1 \cdots x_r))) \\ &= (-1)^{r-1} \phi_{r-1}(\alpha_r((x_0 x_1 \cdots x_{r-1}))) \\ &= (-1)^i \alpha_r\end{aligned}$$

If we put all this together it follows that

$$\partial\alpha((x_0 x_1 \cdots x_r)) = \sum \alpha_i((x_0 x_1 \cdots \hat{x}_i \cdots x_r))$$

with $\alpha_r = (-1)^r \alpha$.

This gives the coefficient of the term corresponding to the omission of the last vertex. For computing α_i, a permutation of the vertexes is made so that x_i becomes the last vertex. Thus

$$\partial\alpha((x_0 x_1 \cdots \hat{x}_i \cdots x_r x_i)) = \sum \alpha_j'((x_0 x_1 \cdots \hat{x}_i \cdots \hat{x}_j \cdots x_r x_i))$$

The last coefficient is α_i', given by the above calculation as

$$\alpha_i' = (-1)^r \alpha$$

However,

$$((x_0 x_1 \cdots \hat{x}_i \cdots x_r x_i)) = (-1)^{r-1}((x_0 x_1 \cdots x_r))$$

by Theorem 2-14, so $\alpha_i' = (-1)^{r-1} \alpha_i$. Hence

$$\alpha_i = (-1)^i \alpha$$

The formula

$$\partial\alpha((x_0 x_1 \cdots x_r)) = \sum (-1)^i \alpha((x_0 x_1 \cdots \hat{x}_i \cdots x_r))$$

is thus proved, and as pointed out, the general formula for any simplex follows, by the use of the appropriate linear map. ∎

2-8. FORMAL DESCRIPTION OF SIMPLICIAL HOMOLOGY

Definitions of oriented simplexes, chains, boundaries, and so on can be formulated in a more abstract way, from the knowledge only of the simplexes of K, independent of any meaning that symbols $((y_0 y_1 \cdots y_r))$ had when originally introduced. As this point is of some importance in the next chapter, the formal description is given in full.

If K is a simplicial complex, oriented simplexes and chains are first to be defined on K. Throughout this discussion a fixed coefficient group $\mathscr{G}$ will be used. Let $\overline{\mathscr{C}}_r(K)$ be the free group over $\mathscr{G}$ generated by the symbols $[y_0 y_1 \cdots y_r]$, where this symbol runs over all the r simplexes of K. Let $R_r(K)$ be the subgroup of $\overline{\mathscr{C}}_r(K)$ generated by elements of the form

$$\alpha[y_0 y_1 \cdots y_r] - \varepsilon\alpha[z_0 z_1 \cdots z_r]$$

where α is in $\mathscr{G}$, the z_i are a permutation of the y_j and ε is $+1$ or -1 according as the permutation is even or odd. Let $\mathscr{C}_r(K)$ be the quotient group $\overline{\mathscr{C}}_r(K)/R_r(K)$. Write the image of $[y_0 y_1 \cdots y_r]$ in this quotient as $((y_0 y_1 \cdots y_r))$. Note that this means that $\mathscr{C}_r(K)$ is generated over $\mathscr{G}$ by the symbols $((y_0 y_1 \cdots y_r))$ with the relations

$$((y_0 y_1 \cdots y_r)) = \pm((z_0 z_1 \cdots z_r))$$

whenever the z_i are a permutation of the y_j, the sign being $+$ or $-$ according as the permutation is even or odd.

Definition 2-10. The symbol $((y_0 y_1 \cdots y_r))$ is called an *oriented r simplex* on $[y_0 y_1 \cdots y_r]$. There are two such oriented simplexes on an r simplex with opposite signs, obtained from each other by permutation of the vertexes.

Definition 2-11. If L is a subcomplex of K, $\mathscr{C}_r(K, L)$ is defined as $\mathscr{C}_r(K)/\mathscr{C}_r(L)$. Clearly $\mathscr{C}_r(K, L)$ can be identified with the group generated over $\mathscr{G}$ by the oriented simplexes on $K - L$. $\mathscr{C}_r(K, L)$ is called the *r-dimensional oriented chain group* of the simplicial pair (K, L).

Definition 2-12. The *oriented boundary operator* d is defined by the formula

$$d\alpha((y_0 y_1 \cdots y_r)) = \sum (-1)^i \alpha((y_0 y_1 \cdots \hat{y}_i \cdots y_r))$$

α being any element of $\mathscr{G}$. This defines d as a homomorphism of $\mathscr{C}_r(K, L)$ into $\mathscr{C}_{r-1}(K, L)$.

Definition 2-13. The kernel of $d: \mathscr{C}_r(K, L) \to \mathscr{C}_{r-1}(K, L)$ is called the *group of oriented r cycles* of K modulo L and is denoted by $\mathscr{Z}_r(K, L)$.

Definition 2-14. The image of $d: \mathscr{C}_{r+1}(K, L) \to \mathscr{C}_{r+1}(K, L)$ is called the *group of oriented r boundaries* of K modulo L and is denoted by $\mathscr{B}_r(K, L)$.

It is easy to check that the oriented boundary operator satisfies the relation $d^2 = 0$, so $\mathscr{B}_r(K, L)$ is a subgroup of $\mathscr{Z}_r(K, L)$.

Definition 2-15. The quotient group $\mathscr{Z}_r(K, L)/\mathscr{B}_r(K, L)$ is called the *r-dimensional oriented simplicial homology group* of K modulo L and is denoted by $\mathscr{H}_r(K, L)$.

It can be seen at once that this reproduces all the algebraic features of the oriented chains, cycles, and so on, as introduced in Definition 2-9, but simply avoids indentifying them with relative singular homology classes. Of course, the result

$$H_r(K, L) \cong \mathscr{H}_r(K, L)$$

still stands with this new definition of the right-hand side.

Certain kinds of maps between simplicial complexes induce homomorphisms between the corresponding simplicial homology groups, expressed entirely in terms of the symbols $((y_0 y_1 \cdots y_r))$. These are the simplicial maps that are defined as follows.

Definition 2-16. Let K and X be two simplicial complexes and let $f\colon K \to X$ be a map with the property that its restriction to any simplex S of K is a linear map onto some simplex of X. f is called a *simplicial map* of K into X.

It is obvious that f is a continuous map. Also given that f is a simplicial map, clearly it is completely defined as soon as its values are given on the vertexes of K, since this property is satisfied by linear maps. In fact, if a map f is defined on the vertexes of K with values in X such that, whenever $y_0, y_1, \ldots, y_r$ are vertexes of a simplex in K, $f(y_0), f(y_1), \ldots, f(y_r)$ are vertexes of a simplex in X, then f can be extended in just one way to a simplicial map of K into X.

Definition 2-17. If $f\colon K \to X$ is a simplicial map and L, Y are subcomplexes of K and X respectively, such that $f(L) \subset Y$, then f is called a *simplicial map of the simplicial pair* (K, L) *into the pair* (X, Y).

A simplicial map of simplicial pairs induces in a natural way a homomorphism of $\mathscr{C}_r(K, L)$ into $\mathscr{C}_r(X, Y)$. For if $y_0, y_1, \ldots, y_r$ are vertexes of a simplex in K, not in L, and if $z_i = f(y_i)$, then

$$\hat{f}\alpha((y_0 y_1 \cdots y_r)) = \alpha((z_0 z_1 \cdots z_r)) \tag{2-11}$$

for any α in $\mathscr{G}$. If the z_i are all different, then an even or odd permutation of the y_j induces a similar permutation of the z_i, so $\hat{f}$ is compatible with the relations on the oriented chain groups. Of course, if the z_i are not all different, the right-hand side of Eq. (2-11) defining $\hat{f}$ is to be zero. Thus, $\hat{f}$ can be extended by linearity to a homomorphism $\hat{f}\colon \mathscr{C}_r(K, L) \to \mathscr{C}_r(X, Y)$. By the boundary formula (Definition 2-12), it follows that

$$d\hat{f}\alpha((y_0 y_1 \cdots y_r)) = \hat{f}d\alpha((y_0 y_1 \cdots y_r))$$

Extending this by linearity,

$$d\hat{f}\lambda = \hat{f}d\lambda$$

for any λ in $\mathscr{C}_r(K, L)$. It follows from this that if λ is a cycle so is $\hat{f}(\lambda)$ and if λ is a boundary so is $\hat{f}(\lambda)$. In other words, $\hat{f}$ maps $\mathscr{Z}_r(K, L)$ and $\mathscr{B}_r(K, L)$ into $\mathscr{Z}_r(X, Y)$ and $\mathscr{B}_r(X, Y)$ respectively. Hence, $\hat{f}$ induces a homomorphism $\hat{f}_*\colon \mathscr{H}_r(K, L) \to \mathscr{H}_r(X, Y)$.

It is natural to ask whether the induced homomorphism $\hat{f}_*$ just defined agrees with that which would have resulted if the oriented chains had been obtained by way of singular homology theory, as in Section 2-7; that is, does the diagram

$$\begin{array}{ccc} H_r(K, L) & \to & \mathscr{H}_r(K, L) \\ {\scriptstyle f_*}\downarrow & & {\scriptstyle \hat{f}_*}\downarrow \\ H_r(X, Y) & \to & \mathscr{H}_r(X, Y) \end{array} \tag{2-12}$$

commute, where the horizontal maps are isomorphisms as in Theorem 2-17.

To examine this question, return to the notation of Section 2-4. Thus K_r is the r skeleton of K and $K_r' = K_r \cup L$, X_r is the r skeleton of X, and $X_r' = X_r \cup Y$. In the diagram

$$\begin{array}{ccccc} H_r(K_r', K_{r-1}') & \overset{j}{\leftarrow} & H_r(K_r', L) & \overset{i}{\rightarrow} & H_r(K, L) \\ \downarrow{\scriptstyle f_*'} & & \downarrow{\scriptstyle f_*'} & & \downarrow{\scriptstyle f_*} \\ H_r(X_r', X_{r-1}') & \overset{j'}{\leftarrow} & H_r(X_r', Y) & \overset{i'}{\rightarrow} & H_r(X, Y) \end{array} \tag{2-13}$$

i and j are as in Theorems 2-15 and 2-16, while i' and j' play the same parts for X and Y. f_* is the induced homomorphism for singular homology groups corresponding to the continuous map f. Since f is simplicial it is linear on any r simplex of K and so maps it into a simplex of dimension at most r in X. Thus, f maps K_r into X_r, and since it maps L into Y, it also maps K_r' into X_r', for each r. If f' is the restriction of f to K_r', f_*' on the left of the diagram is the corresponding induced homomorphism on the singular homology groups, while the f_*' in the middle is the restriction to the group $H_r(K_r', L)$ which is identified by j with a subgroup of $H_r(K_r', K_{r-1}')$. The whole diagram commutes since all the maps are induced homomorphisms corresponding to a commutative diagram of continuous maps.

Consider in detail the operation of f'_* on the left side of the above diagram. Take a generator of $H_r(K'_r, K'_{r-1})$. According to Section 2-6, this can be written as $((y_0 y_1 \cdots y_r))$, where $[y_0 y_1 \cdots y_r]$ is an r simplex of $K - L$. Remember that in this context $((y_0 y_1 \cdots y_r))$ stands for the homology class represented by the singular simplex $(y_0 y_1 \cdots y_r)$ (a linear map of Δ_r into K). It is thus clear that f'_* maps $((y_0 y_1 \cdots y_r))$ on $((z_0 z_1 \cdots z_r))$ where $z_i = f(y_i)$ for each i. This means that the action of f'_* on the oriented simplexes coincides with that of $\hat{f}$, which is defined in a purely formal way by (2-11) Hence, identifying $H_r(K'_r, L)$ with $\mathscr{Z}_r(K, L)$ by means of j, and similarly $H_r(X'_r, Y)$ with $\mathscr{Z}_r(X, Y)$, diagram (2-13) becomes the commutative diagram

$$\begin{array}{ccc} \mathscr{Z}_r(K, L) & \xrightarrow{h} & H_r(K, L) \\ \downarrow \hat{f} & & \downarrow f_* \\ \mathscr{Z}_r(X, Y) & \xrightarrow{h'} & H_r(X, Y) \end{array}$$

where $h = ij^{-1}$, $h' = i'j'^{-1}$ (restricted to the images of j and j', respectively). By Theorem 2-17, the kernels of h and h' are $\mathscr{B}_r(K, L)$ and $\mathscr{B}_r(X, Y)$ respectively, and $\hat{f}$ carries the first of these into the second. It follows by taking quotient groups that diagram (2-12) is commutative as required.

Exercises. 2-15. Let $K \cup L$ be a simplicial complex, K and L being subcomplexes. Prove that the inclusion maps

$$(K, K \cap L) \to (K \cup L, L)$$

$$(L, K \cap L) \to (K \cup L, K)$$

(both simplicial maps) induce isomorphisms of the simplicial homology groups.

Note that this is a form of excision theorem for simplicial homology.

2-16. Let K be a simplicial complex of dimension n. K is called an n circuit if every $(n-1)$ simplex of K is a face of exactly two n simplexes, and if every pair S and T of n simplexes can be connected by a chain $S = S_1, S_2, \ldots, S_r = T$ of n simplexes such that each pair S_i, S_{i+1} has a common $(n-1)$ face. K is called orientable if all the n simplexes can be oriented so that whenever two of them have a common $(n-1)$ face they induce opposite orientations on it. In this case the n simplexes are said to be coherently oriented. If this is impossible the circuit is called nonorientable. For example, when the compact 2 manifolds (cf. *Introduction*, p. 192) are triangulated they become 2 circuits and the definition of orientability here coincides with that given in *Introduction*.

Prove that, if K is an orientable n circuit, then $\mathscr{H}_n(K) \cong \mathscr{G}$, and that a generator of $\mathscr{H}_n(K)$ is represented by a cycle that is the sum of all the coherently oriented n simplexes. If K is nonorientable show that $\mathscr{H}_n(K) = 0$. In fact, show that in this case K carries no nonzero n cycle.

2-17. The last exercise can be generalized. Let (K, L) be a simplicial pair with K of dimenion n. Suppose that every $(n-1)$ simplex in $K - L$ is a face of exactly two n simplexes in $K - L$ and that any two n simplexes of $K - L$ can be connected by a chain of n simplexes in $K - L$ such that consecutive simplexes have a common $(n-1)$ face. Then (K, L) is called a relative n circuit. It is orientable if the n simplexes of $K - L$ can be coherently oriented, otherwise it is nonorientable.

If (K, L) is an orientable relative n circuit prove that $H_n(K, L) \cong \mathscr{G}$; if it is nonorientable show that $\mathscr{H}_n(K, L) = 0$.

An n cell modulo its boundary, to give an example, is an orientable relative n circuit. A Möbius strip (a rectangle that has one pair of opposite sides identified after being given a half twist) modulo its edge is a nonorientable relative 2 circuit.

If the pair (E^n, S^{n-1}) is triangulated in any way the result will be a relative n circuit. This is intuitively clear but requires some proof, which is not entirely trivial. The idea is that if, in some triangulation, an $(n - 1)$ simplex is a face of r n simplexes with $r \neq 2$, then a point of that $(n - 1)$ simplex would fail to have spherical neighborhoods, as it should. For any triangulation the n circuit (E^n, S^{n-1}) must be orientable. This follows from the fact that $\mathscr{H}_n(E^n, S^{n-1})$ is independent of the triangulation and is in fact equal to $H_n(E^n, S^{n-1}) \cong \mathscr{G}$. More generally, it can be shown that if (K, L) is any simplicial pair with $K - L$ homeomorphic to an open n cell, then (K, L) is an orientable relative n circuit.

2-9. CELL COMPLEXES

Apparently, we now have an algorithm for computing the homology groups of triangulable spaces. But even quite simple spaces when triangulated may contain a very large number of simplexes so that the computation of the cycle and boundary groups involves the solution of a large number of linear equations in many variables. The idea of this section is to work out a method of calculating the homology groups by using chain groups with fewer generators. If the number of generators needed can be cut down sufficiently, then the computation may be quite a simple matter. The object is to group the simplexes together to form cells so that the boundary of a cell will consist of cells of lower dimension. Chain groups will then be constructed from the cells instead of from the simplexes, and cycle and boundary groups can be defined whose quotients are shown to be the simplicial homology groups.

The notion of the simplexes of a complex K has already been introduced. Here it is more convenient to speak of open p simplexes, that is, p simplexes from which all the $(p - 1)$-dimensional faces have been removed. In this terminology a simplicial complex K is the disjoint union of all the open simplexes of K.

Definition 2-18. Suppose that a complex K is expressed as the disjoint union of open cells of various dimensions, called the cells of K, that satisfy the following conditions.

(1) Each open p cell of K is the union of open simplexes of K of dimensions $\leqq p$, and is homeomorphic to an open solid p sphere.

(2) Each open p cell of K is of the form $X - Y$, where X and Y are subcomplexes of K and Y is the union of cells of K of dimensions $<p$.

(3) If $X - Y$ is a p cell of K (as in (2)) the pair (X, Y) has the same homology groups as a p cell (closed solid p sphere) modulo its boundary. In particular, if a generator of $\mathscr{H}_p(X, Y)$ is represented by the relative cycle α, α is called an oriented p cell on K associated with $X - Y$.

(4) If α is an oriented p cell on K (as in (3)) $d\alpha$ is a linear combination of oriented $(p - 1)$ cells on K. Here d is the oriented simplicial boundary operator.

If these conditions are satisfied, K is said to be expressed as a *cell complex*.

Note that the conditions given are somewhat redundant, since it can be shown that, in the notation of condition (2), the fact that $X - Y$ is topologically a p cell is sufficient to guarantee that the pair (X, Y) has the homology groups of a p cell modulo its boundary (cf. Exercise 2-17 and the note following it). This is a little troublesome to prove, however, and in most cases it is less troublesome to see that all the conditions stated above are satisfied. It is also possible to show that the p simplexes in X can be oriented so that α, representing a generator of $\mathscr{H}_p(X, Y)$, is sum of all oriented p simplexes on X. In fact, each $(p - 1)$ simplex on X, not in Y, is a face of exactly two p simplexes, and when suitable orientations are given and the boundary is taken, the common face will cancel out. Again, all this is a bit of a nuisance to prove, and it is not really needed in the computation procedure to be worked out here.

Examples

2-6. The open simplexes of a complex K can themselves be taken as cells and all the above conditions can easily be seen to hold.

2-7. Consider the torus K with the two circles A and B marked on it (cf. Fig. 7) intersecting at the point P. K can be triangulated so that the union of A and B is a subcomplex L, the point P being a vertex. Clearly $K - (A \cup B)$ is an open 2 cell, $A - P$ and $B - P$ are open 1 cells and P itself is a 0 cell, so K is expressed as the union of open cells, each of which is itself a union of open simplexes of K. It has already been seen (cf. Exercise 2-17) that the pairs (K, L), (A, P), (B, P) all have the homology of cells of the appropriate dimensions modulo their boundaries. Also, if c, a, b, p are oriented cells associated with $K - L$, $A - P$, $B - P$, P, respectively, it is known (cf. Exercise 2-17) that

$$dc = da = db = dp = 0$$

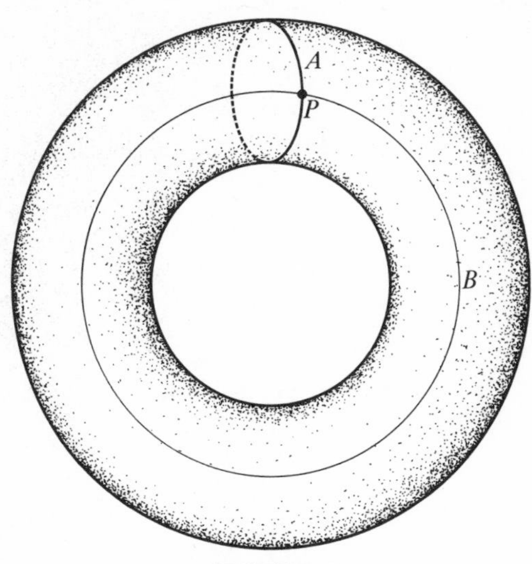

FIGURE 7

so that condition (4) of Definition 2-18 is satisfied. Thus K has been made into a cell complex.

Note that the argument used in Exercise 2-17 to show that the homology condition (3) of Definition 2-18 is satisfied (i.e., by combining homotopy and excision) is precisely what is needed in general to show that if $X - Y$ is a p cell then (X, Y) has the correct homology groups. The catch is that, in general, the homotopy and excision are rather complicated to describe, while in special cases such as this, the argument is fairly easy to construct.

2-8. It is worthwhile to note the kind of decomposition that is not admissible by Definition 2-18. Consider the one-dimensional complex K in Fig. 8. The diagram shows K decomposed into a disjoint union of segments A, B (supposed open), and the three points P, Q, and R. However, the boundary of an oriented cell corresponding to B is not a linear combination of oriented cells associated with P, Q, and R so condition (4) of Definition 2-18 fails to hold in this case. In Definition 2-18, condition (4) is the most important one, for it is this condition that allows the definition of chain groups from which the homology groups can be computed.

2-9. A cell decomposition will now be obtained for P^n, the real n-dimensional projective space. This is the space obtained from the n sphere S^n by identifying diametrically opposite points. (The case $n = 2$, the projective plane, is studied in *Introduction*.) There is a continuous map

$$f\colon S^n \to P^n$$

such that $f^{-1}(p)$ for each p of P^n is a pair of diametrically opposite points of S^n.

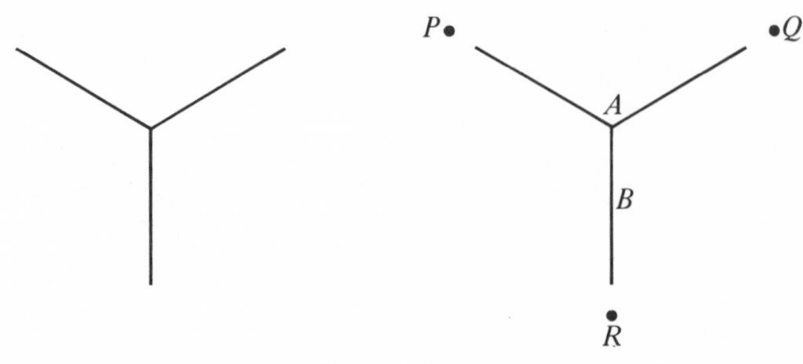

FIGURE 8

For the present purpose, S^n and P^n must be made into simplicial complexes. So in $(n+1)$ space let x_i^1 and x_i^{-1} denote the points on the ith axis $(i = 0, 1, \ldots, n)$ at unit distance from the origin in the positive and negative directions respectively (cf. Fig. 9 for the case $n = 2$). Take the union of all the Euclidean simplexes $[x_0 x_1 \cdots x_n]$, where the indices are given the values 1 and -1 in all possible ways. The result is a simplicial complex homeomorphic to S^n. The symbol S^n throughout this example denotes this simplicial complex. Note that $S^r (r < n)$ is obtained as a subcomplex of S^n by restriction to the space spanned by the first $(r+1)$ axes.

The images of the simplexes of S^n under f, however, do not, form a simplicial decomposition of P^n. For example, with $n = 2$, the images of $[x_0^1 x_1^1]$ and $[x_0^1 x_1^{-1}]$ will have two points in common, contradicting the definition of a simplicial complex. On the other hand, it is not hard to see that the images under f of the simplexes of BS^n (the barycentric subdivision of S^n) give a simplicial decomposition of P^n.

The simplexes of this decomposition of P^n are now grouped together to form cells. Going back to the simplicial complex S^n, define C_r^1 and C_r^{-1} as the open stars (cf. Exercise 2-7) in the subcomplex S^r of the vertexes x_r^1 and x_r^{-1}, respectively. These are both open r cells. Figure 9 shows these cells for the case $n = 2$. It is easily seen that $f(C_r^1) = f(C_r^{-1}) = C_r$ is an open r cell in P^n and that P^n is the union of $C_0, C_1, \ldots, C_n$. In the simplicial decomposition of P^n obtained from the complexes BS^n, each C_r is a union of open simplexes of P^n. Also, each $\bar{C}_r$ is a subcomplex of P^n and $C_r = \bar{C}_r - \bar{C}_{r-1}$. $\bar{C}_{r-1}$ is the union of the C_i for $i \leqq r-1$ and so condition (2) of Definition 2-18 is satisfied. To check condition (3) it must be shown that f induces an isomorphism of $H_i(\bar{C}_r^1, \bar{C}_{r-1}^1 \cup \bar{C}_{r-1}^{-1})$ onto $H_i(\bar{C}_r, \bar{C}_{r-1})$ for each i. This is proved by using the simplicial homology groups, which are known to depend only on the simplexes in the interiors of $\bar{C}_r^1$ and $\bar{C}_r$.

Finally, it is easy to see that the C_r^1 and C_r^{-1} form a cell decomposition of S^n. In particular, on oriented cell c_r^1 can be defined on C_r^1 by taking the sum of oriented simplexes on C_r^1 in a coherent orientation. Choose the coherent orientation that corresponds to the order $x_0^1, x_1^1, \ldots, x_r^1$ of the vertexes of $[x_0{}^1 x_1{}^1 \cdots x_r{}^1]$. This determines the appropriate orientations of the other r simplexes. Thus $c_r{}^1$ contains the term $((x_0{}^1\, x_1{}^1 \cdots x_r{}^1))$. Similarly, an oriented cell c_r^{-1} is taken on C_r^{-1} so that it contains the term $((x_0^{-1} x_1^{-1} \cdots x_r^{-1}))$. At this stage the object is to map $c_r{}^1$ and c_r^{-1} over into P^n by the homomorphism induced by f. The results should be equal and should be an oriented cell on P^n associated with C_r. The catch is that, strictly speaking, an oriented cell associated with C_r should be a sum of oriented simplexes on P^n associated with the simplicial decomposition of P^n, which is obtained not from S^n but from BS^n. So let f_1 be the homomorphism induced by f on oriented simplicial chains of P^n. B can be defined on oriented simplexes by the same inductive procedure used in Definition 1-33 for singular simplexes. So in particular, $Bc_r{}^1$ and Bc_r^{-1} become oriented cells on BS^n and $f_1 Bc_r{}^1 = f_1 Bc_r^{-1}$ can be defined as c_r, an oriented cell on C_r. The boundary condition (4) of Definition 2-18 will now be checked.

$$dc_r = df_1 Bc_r{}^1 = f_1 dBc_r{}^1 = f_1 Bdc_r{}^1 \tag{2-14}$$

by the analog of Theorem 1-18, for the operator B on simplicial chains. It is clear, from previous knowledge of the homology of cells and spheres, that $dc_r{}^1 = \pm c_{r-1}^1 \pm c_{r-1}^{-1}$. It follows that dc_r will be 0 or $\pm 2c_{r-1}$. This verifies condition (4) of Definition 2-18.

An explicit formula for dc_r will be needed later. To compute this note that $c_r{}^1$ contains, in addition to the term $((x_0{}^1 x_1{}^1 \cdots x_r{}^1))$, the term

$$(-1)^r((x_0^{-1} x_1^{-1} \cdots x_{r-1}^{-1}\, x_r{}^1))$$

This is so since a change in one index corresponds to crossing one $(r-1)$ face, on which opposite orientations are to be induced, and r indices are to be changed. Thus $dc_r{}^1$ contains the terms

$$(-1)^r((x_0{}^1 x_1{}^1 \cdots x_{r-1}^1)) + ((x_0^{-1} x_1^{-1} \cdots x_{r-1}^{-1}))$$

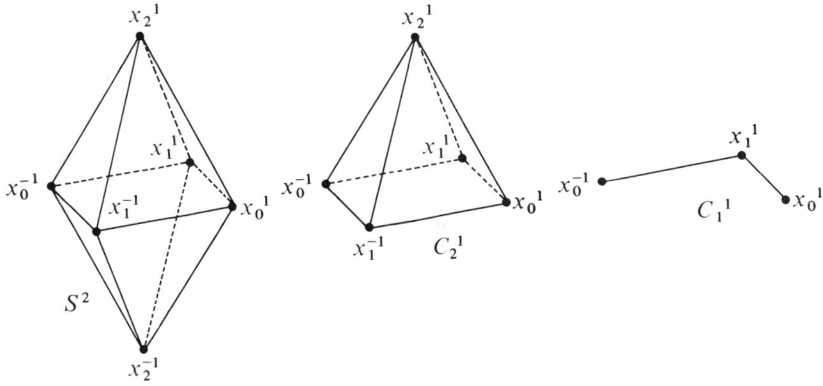

FIGURE 9

All other terms obtained from $((x_0^1 x_1^1 \cdots x_r^1))$ and $((x_0^{-1} x_1^{-1} \cdots x_{r-1}^{-1} x_r^1))$ contain the vertex x_r^1. These cancel, while the boundaries of other simplexes give terms whose indices are mixed $+1$ and -1. It follows that

$$dc_r^1 = (-1)^r c_{r-1}^1 + c_{r-1}^{-1}$$

Combining this with (2-14) yields the formula

$$dc_r = [(-1)^r + 1]c_{r-1}$$

Take the odd and even values of r separately, remembering that there is no c_{-1}. The boundary relations thus have the following form:

$$dc_0 = 0$$

$$dc_r = 0 \qquad \text{(if } r \text{ is odd)}$$

$$dc_r = 2c_{r-1} \qquad \text{(if } r \text{ is even and not 0)}$$

Chain, cycle, and boundary groups will now be associated with a cell decomposition of a complex K. Let $\bar{\mathscr{C}}_r(K)$ be the free group over the coefficient group $\mathscr{G}$ generated by the oriented p cells of K that correspond to the given decomposition.

Definition 2-19. $\bar{\mathscr{C}}_p(K)$ is called the group of *cellular p chains on K*, corresponding to the given cell decomposition. Condition (4) of Definition 2-18 implies that d is a homomorphism of $\bar{\mathscr{C}}_p(K)$ into $\bar{\mathscr{C}}_{p-1}(K)$ for each p. The kernel of $d: \bar{\mathscr{C}}_p(K) \to \bar{\mathscr{C}}_{p-1}(K)$ is denoted by $\bar{\mathscr{Z}}_p(K)$ and is called the group of *cellular p cycles on K*. The image of $d: \bar{\mathscr{C}}_{p+1}(K) \to \bar{\mathscr{C}}_p(K)$ is written as $\bar{\mathscr{B}}_p(K)$ and called the group of *cellular p boundaries of K*.

The following is the basic theorem on cell decompositions of a complex.

Theorem 2-19. *In the notation used above*

$$\mathscr{H}_p(K) = \frac{\bar{\mathscr{Z}}_p(K)}{\bar{\mathscr{B}}_p(K)}$$

Proof. This can all be expressed in a more algebraic way as will be shown later. In the meantime, however, the proof will be carried out by a straightforward verification in two steps. First, it will be shown that every oriented simplicial cycle α on K is homologous to a linear combination of oriented cells (which are, after all, simplicial chains on K). Secondly, it will be shown that if $\alpha \in \bar{\mathscr{Z}}_p(K)$ and if $\alpha = d\beta$ for some simplicial chain β, then in fact

$\alpha = d\beta'$ where β' is in $\bar{\mathscr{C}}_{p+1}(K)$. These two steps together imply that the inclusion maps $\bar{\mathscr{C}}_p(K) \to \mathscr{C}_p(K)$, for each p, induce isomorphisms onto of the quotient groups $\bar{\mathscr{Z}}_p(K)/\bar{\mathscr{B}}_p(K)$ and $\mathscr{Z}_p(K)/\mathscr{B}_p(K)$.

Let α be a simplicial p cycle on K and let $X - Y$ be a q cell of the cell decomposition of K (using the terminology and notation of Definition 2-18), and suppose that there is a singular p simplex on X appearing with nonzero coefficient in α. Write $\alpha = \alpha_1 + \alpha_2$ where α_1 is a linear combination of singular simplexes on X while the singular simplexes of α_2 are all carried by p simplexes not in X. It is clear that α_1 is a relative cycle (in the sense of simplicial homology) of X modulo Y. It follows that, if $q > p$, $\alpha_1 = d\beta + \gamma$, where β is a simplicial chain on X and γ is a simplicial chain on Y. Hence, if $q > p$

$$\alpha = \gamma + \alpha_2 + d\beta$$

In other words, α is homologous to a cycle $\gamma + \alpha_2$ which has no singular simplexes with nonzero coefficient on the interior of the q cell $X - Y$. If this argument is repeated, it follows that α is homologous to a chain on the union of cells of dimensions $\leqq p$ (cf. Theorem 2-15). On the other hand, suppose $q = p$. Then α_1 must be homologous modulo Y to a multiple of the oriented cell c associated with $X - Y$. Repetition of this argument, shows that α is homologous to a cycle

$$\alpha' + \alpha''$$

where α' is a linear combination of oriented p cells on K; α'', a simplicial p chain on the union of cells of K of dimensions $<p$, must in fact be 0. This completes the first step of the proof.

For the second step of the proof, let α be a cellular p cycle and suppose that there is a simplicial $(p + 1)$ chain β such that $\alpha = d\beta$. Again let $X - Y$ be a q cell of K with $q > p + 1$ and suppose that there are oriented simplexes on X with nonzero coefficients in β. Write $\beta = \beta_1 + \beta_2$, where β_1 is a chain in X and β_2 is a linear combination of oriented simplexes not on X. The simplexes of which $d\beta$ is a linear combination are certainly not on the interior of $X - Y$ (the union of cells of dimension $\leqq p$). Thus, as before, β_1 is a relative cycle of X modulo Y, so since $q > p + 1$, $\beta_1 = \beta_3 + d\gamma$, where β_3 is a chain on Y and γ a chain on X. Then

$$\alpha = d(\beta_3 + \beta_2)$$

In this way, step by step, a chain β'' is obtained on the union of cells of dimensions $\leqq p + 1$ such that

$$\alpha = d\beta''$$

Repeat the argument now with β''. Let $X - Y$ be a $(p + 1)$ cell and write $\beta'' = \beta_1 + \beta_2$, where β_1 is a chain on X and β_2 is a linear combination of simplexes not on X. This time β_1, a relative $(p + 1)$ cycle of X modulo Y, is homologous to a multiple of c, the oriented cell associated with $X - Y$. Thus

$$\beta_1 = kc + \beta_3 + d\gamma$$

where $k \in \mathscr{G}$, β_3 is a $(p + 1)$ chain on Y, and γ is a chain on X. Here the dimension of Y is $\leq p$, so $\beta_3 = 0$. Repetition of this argument gives a cellular $(p + 1)$ chain β' such that $\alpha = d\beta'$, as required. ∎

The use of this theorem is illustrated in the following examples and exercises.

Examples

2-10. Consider the torus K with the cell decomposition described in Example 2-7. The cellular chain groups are generated as follows over the coefficient group $\mathscr{G}$:

$\bar{\mathscr{C}}_2$ has one generator c

$\bar{\mathscr{C}}_1$ has two generators a and b

$\bar{\mathscr{C}}_0$ has one generator p

The boundary relations $dc = da = db = dp = 0$ ensure that these same elements generate the cellular cycle groups of the appropriate dimensions and that the boundary groups are all zero. Hence, by Theorem 2-19, $\mathscr{H}_2(K) = \mathscr{G}$

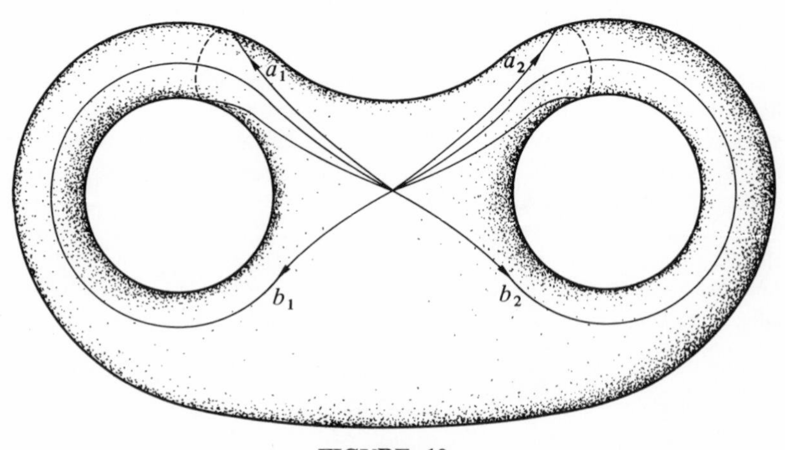

FIGURE 10

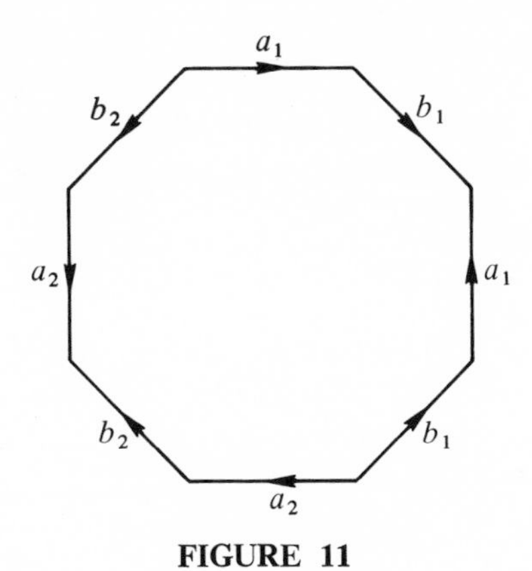

FIGURE 11

(generated by c), $\mathscr{H}_1(K) = \mathscr{G} \oplus \mathscr{G}$ (generated by a and b), and $\mathscr{H}_0(K) = \mathscr{G}$ (generated by p).

Note that the principle involved in Theorem 2-19 is essentially that used in the computation of various homology groups of surfaces in *Introduction* (Chapter IX, p. 192), but Theorem 2-19 formulates this principle in a more systematic way.

Exercises. 2-18. Working as in Example 2-10, show that if K is the projective plane then $\mathscr{H}_2(K) = 0$, $\mathscr{H}_0(K) \cong \mathscr{G}$, and $\mathscr{H}_1(K) = \mathscr{G}/2\mathscr{G}$ (provided that $\mathscr{G}$ has no elements of order 2).

2-19. Exercise 2-18 and Example 2-10 are special cases of the computation of the homology groups of 2 manifolds, spaces such that every point has a neighborhood that is homeomorphic to an open 2 cell. The compact 2 manifolds are known to be triangulable and fall into two groups, the orientable and the nonorientable. The orientable 2 manifolds have the property that all the 2 simplexes can be oriented so that when two of them have a common side they induce opposite orientations on this side. For the nonorientable type this cannot be done. A compact orientable 2 manifold is homeomorphic to a sphere with p handles attached for some p (*Introduction*, p. 192). Note that the sphere and the torus are the simplest cases of this with p equal to 0 and 1, respectively. A compact nonorientable 2 manifold can always be obtained by cutting k holes in the surface of a sphere, for some k, and then identifying diametrically opposite points on the circumference of each hole. The projective plane is the simplest case of this with $k = 1$.

Let K be a sphere with p handles (cf. Fig. 10, and *Introduction*, p. 158). For each handle, mark paths a_i, b_i as shown, starting at a point p. If K is cut along the a_i and b_i a polygon with $4p$ sides is obtained with identifications like those marked in Fig. 11. It can be shown that K can be triangulated so that the union of the a_i and b_i forms a subcomplex and P is a vertex. Show that a cell decomposition of K exists which has one 2 cell, $2p$ 1 cells and one 0 cell, hence, $\mathscr{H}_2(K) = \mathscr{H}_0(K) \cong \mathscr{G}$ while $\mathscr{H}_1(K)$ is free with $2p$ generators.

In the nonorientable case where K is a sphere with k holes, in each of which the diametrically opposite points of the circumference are identified, show in the same way that $\mathscr{H}_2(K)=0$, $\mathscr{H}_0(K)\cong\mathscr{G}$, and $\mathscr{H}_1(K)$ has k generators $\alpha_1, \alpha_2, \ldots, \alpha_k$ satisfying the relation $2c(\alpha_1+\alpha_2+\cdots+\alpha_k)=0$ for any c in the coefficient group $\mathscr{G}$.

2-20. Use the cell decomposition described in Example 2-9 to show that (provided $\mathscr{G}$ has no elements of order 2)

$$H_0(P^n)=\mathscr{G}$$

$$H_r(P^n)=0 \qquad \text{(if } r \text{ is even and not 0)}$$

$$H_r(P^n)=\frac{\mathscr{G}}{2\mathscr{G}} \qquad \text{(if } r \text{ is odd)}$$

2-10. CANONICAL BASES

Despite the possible simplification effected by the introduction of cell complexes a systematic procedure for the computation of homology groups is sometimes desirable. Clearly the algebra involved depends on the nature of the coefficient group, and the discussion here will be confined to the case in which this group is the group of integers.

Suppose that K is a simplicial complex and let $\mathscr{C}_p(K)$ be generated by $c_1{}^p, c_2{}^p, \ldots, c_k{}^p$, which initially are oriented simplexes (if K were a cell complex they would be oriented cells). A linear change of basis will be made for each $\mathscr{C}_p(K)$ so that, in terms of the new generators, the action of the boundary operator can be written in a specially simple form. For each i, $dc_i{}^p$ is a $(p-1)$ chain, so

$$dc_i{}^p=\sum a_{ij}^p c_j^{p-1}$$

Here the integers a_{ij}^p form a matrix A^p, called the p-dimensional incidence matrix. The relation $d^2=0$ means that

$$0=d^2c_i{}^p=d\sum a_{ij}^p c_j^{p-1}=\sum a_{ij}^p a_{jh}^{p-1} c_h^{p-2}$$

Here the c_h^{p-1} are linearly independent over Z, so in matrix notation,

$$A^pA^{p-1}=0 \tag{2-15}$$

By suitable choice of generators of the groups $\mathscr{C}_p(K)$, all the incidence matrices can be made into diagonal matrices. Remember that if A is any matrix of integers there are matrices H and K of integers, both with determinant 1, such that HAK is a matrix of the form

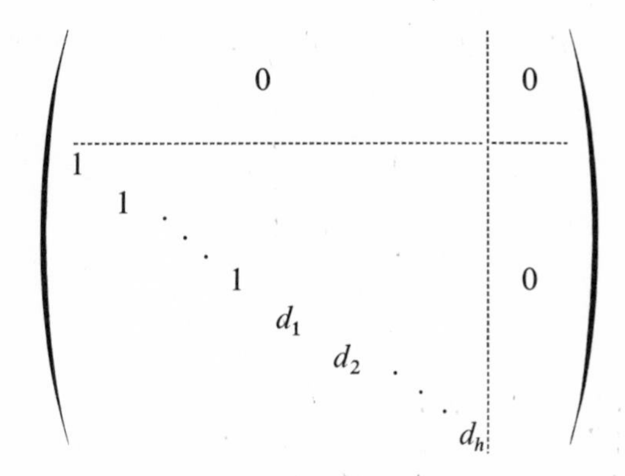

where the submatrices in the top left, top right, and bottom right and all unmarked entries are filled by zeros. In addition, d_i is a factor of d_{i+1} for each i. This diagonalization process will be applied to each A^p in turn.

Starting with A^1 find matrices of integers H^1 and K^1 with unit determinant such that

$$H^1 A^1 K^1 = \left(\begin{array}{c|c} 0 & 0 \\ \hline \begin{matrix} * & & \\ & * & \\ & & * \end{matrix} & 0 \end{array}\right) = D^1$$

where the stars denote the nonzero part of the matrix on the right. Replace the bases $c^0 = \begin{pmatrix} c_1{}^0 \\ c_2{}^0 \\ \vdots \\ c_p{}^0 \end{pmatrix}$ and $c^1 = \begin{pmatrix} c_1{}^1 \\ c_2{}^1 \\ \vdots \\ c_q{}^1 \end{pmatrix}$ of $\mathscr{C}_0(K)$ and $\mathscr{C}_1(K)$ respectively by $(K^1)^{-1}c^0$ and H^1c^1 (note that $(K^1)^{-1}$ is also a matrix of integers here). In terms of these new bases for the chain groups, the boundary relations become

$$dH^1c^1 = H^1A^1c^0 = H^1A^1K^1(K^1)^{-1}c^0 = D^1(K^1)^{-1}c^0$$

Thus the new one-dimensional incidence matrix is D^1. If the new two-dimensional incidence matrix is $\bar{A}^2$ the relation (2-15) becomes

$$\bar{A}^2 D^1 = 0$$

so $\bar{A}^2$ is of the form $(* \mid 0)$, the star denoting the nonzero part. There are as many columns of zeros as there are nonzero elements in D^1. Making a further change of basis for $\mathscr{C}_2(K)$ and $\mathscr{C}_1(K)$, diagonalize $\bar{A}^2$ to give a matrix of the form

$$D^2 = \left(\begin{array}{ccc|c} & 0 & & 0 \\ \hline * & & & \\ & * & & 0 \\ & & * & \end{array}\right)$$

which will have at least as many columns of zeros as $\bar{A}^2$ has. Here the change of basis for $\mathscr{C}_1(K)$ can certainly be made so that the incidence matrix of dimension 1 is still D^1. At this stage then the one- and two-dimensional incidence matrices are D^1 and D^2. And now this process is repeated step by step until all the incidence matrices are brought into diagonal form.

The final result is that, for each p, a new basis $g^p = \begin{pmatrix} g_1{}^p \\ g_2{}^p \\ \vdots \\ g_k{}^p \end{pmatrix}$ has been found for $\mathscr{C}_p(K)$ so that the boundary relations, in matrix notation, become

$$dg^p = D^p g^{p-1}$$

where

$$D^p = \left(\begin{array}{cccccc|c} & & & 0 & & & 0 \\ \hline 1 & & & & & & \\ & 1 & & & & & \\ & & \ddots & & & & \\ & & & 1 & & & 0 \\ & & & & t_1^p & & \\ & & & & & \ddots & \\ & & & & & t_{k(p)}^p & \end{array}\right)$$

and of course the relation $D^p D^{p-1} = 0$ holds for each p.

It is now convenient to partition the single-column matrix g^p into a number of subcolumns, for each p, writing

$$g^p = \begin{pmatrix} \alpha^p \\ \beta^p \\ \gamma^p \\ \delta^p \\ \varepsilon^p \end{pmatrix}$$

Here the total number of elements in α^p, β^p, and γ^p combined is the number of zero rows in D^p, so that the relations

$$d\alpha_i^p = d\beta_i^p = d\gamma_i^p = 0$$

hold for all the entries in these subcolumns. The number of entries in δ^p is equal to the number of units in D^p and this in turn is the number of entries in α^{p-1}; thus, the relations

$$d\delta_i^p = \alpha_i^{p-1}$$

hold. The number of entries in ε^p is $k(p)$ (the number of t_i^p in D^p) and this will be the number of entries in β^{p-1}. Thus the relations

$$d\varepsilon_i^p = t_i^p \beta^{p-1}$$

hold. The fact that the total number of entries in δ^p, ε^p does not exceed the number in α^{p-1}, β^{p-1}, γ^{p-1} follows from the relation $D^p D^{p-1} = 0$.

In summary, a new basis has been found for $\mathscr{C}_p(K)$ which consists of elements denoted by α_i^p, β_i^p, γ_i^p, δ_i^p, ε_i^p with boundary relations

$$d\alpha_i^p = d\beta_i^p = d\gamma_i^p = 0,\ d\delta_i^p = \alpha_i^{p-1},\ d\varepsilon_i^p = t_i^p \beta_i^{p-1}$$

Definition 2-20. Bases of the $\mathscr{C}_p(K)$ with these properties are called *canonical bases.*

From the boundary relations, it can be seen at once that the cycle group $\mathscr{Z}_p(K)$ is generated by the α_i^p, β_i^p, γ_i^p. Of these, the α_i^p are all boundaries while, for each i, $t_i^{p+1}\beta_i^p$ is a boundary. No linear combination of the γ_i^p is a boundary. It follows that $\mathscr{H}_p(K)$ is isomorphic to a direct sum $B_p + T_p$, where B_p is a free Abelian group generated by the γ_i^p and T_p is a direct sum of finite cyclic groups of orders $t_1^{p+1}, t_2^{p+1}, \ldots, t_{k(p+1)}^{p+1}$. Thus, the structure of the homology groups of K over Z can be read off at once when the canonical bases have been constructed.

Note that exactly the same discussion could have been carried out for a cell decomposition of K, using the groups $\mathscr{C}_p(K)$ instead of $\mathscr{C}_p(K)$ and taking the original generators c_i^p as oriented cells.

The following exercises involve certain arithmetical properties of the homology groups of a complex.

First a definition is needed.

Definition 2-21. The number of γ_i^p in a canonical basis, that is, the number of independent generators of the free part of $\mathscr{H}_p(K)$, is called the pth *Betti number of* K and is written as $b_p(K)$. B_p is the pth *Betti group of* K. The order of T_p is called the pth *torsion coefficient of* K and is written as $t_p(K)$. T_p is the pth *torsion group of* K.

Exercises. 2-21. Returning to the incidence matrices A^p, let n_p be the number of generators of $\mathscr{C}_p(K)$, or what is the same thing, the number of rows in A^p. This is the number of p simplexes in K, if K is a simplicial complex, or the number of p cells if it is represented as a cell complex. Prove that the Betti number b_p is given by

$$b_p = n_p - \text{rank of } A^p - \text{rank of } A^{p+1}$$

2-22. Use Exercise 2-19 to show that

$$\sum (-1)^p b_p = \sum (-1)^p n_p$$

2-23. The number $\sum (-1)^p b_p$ is called the Euler characteristic of K. Using a cell decomposition, show that the Euler characteristic of a sphere with p handles is $2 - 2p$. Also check this by using the known structure of the homology groups. For a 2 sphere, $p = 0$ so Exercise 2-21 implies that if a sphere is represented in any way as a cell complex then

$$n_2 - n_1 + n_0 = 2$$

where n_i is the number of i cells. This is the Euler formula.

2-24. If rational numbers instead of the integers are used as the coefficient group, prove that $\mathscr{H}_p(K)$ is a vector space over the rationals of dimension $b_p(K)$.

The remaining exercises outline an interesting application of homology theory. The object is to obtain a proof of the so-called "fundamental theorem of algebra" which states that the field of complex numbers is algebraically closed. More explicitly, this theorem states that, if $f(z) = a_0 + a_1 z + \cdots + a_n z^n$ is a polynomial in z with complex coefficients, then it has a complex root; that is, there is a complex number c such that $f(c) = 0$. The complex plane can be completed to a 2 sphere S^2 by adding a point at infinity, denoted by ∞, and a map $f: S^2 \to S^2$ can be defined by setting $f(z) = a_0 + a_1 z + \cdots + a_n z^n$ and $f(\infty) = \infty$. Note that if the fundamental theorem of algebra were false, the map f could not be onto, for no point would be mapped on 0. So to prove the theorem it is sufficient to show that f is onto.

2-25. In the notation just introduced show that f is continuous. Also, if f is not onto prove that it is homotopic to a constant map.

(*Hint*: Essentially, if f is not onto, it is a map into an open disc which can be shrunk to a point.)

2-26. Prove that f is homotopic to the map g defined by $g(z) = z^n$, $q(\infty) = \infty$.

(*Hint*: Change the coefficients of f continuously to those of the polynomial g.)

2-27. Combining the last two exercises and the preceding remarks proves the fundamental theorem of algebra if it can be shown that g cannot be homotopic to a constant map. Prove this by showing that $g_* : H_2(S^2) \to H_2(S^2)$ is not the zero homomorphism. In fact, if $\alpha \in H_2(S^2)$, show that $g_*(\alpha) = n\alpha$.

(*Hint*: Carry this out in two stages. First note that g induces a map of the unit disc E^2 onto itself and also a map of the unit circle S^1 onto itself. Then show that the above assertion on g_* is true if and only if the corresponding result is true for $g_*: H_2(E^2, S^1) \to H_2(E^2, S^1)$ (use an excision argument) and, by combining with ∂, for $g_*: H_1(S^1) \to H_1(S^1)$. Finally, compute explicitly the homomorphism $g_*: H_1(S^1) \to H_1(S^1)$.)

3

Chain Complexes—Homology and Cohomology

The underlying algebra of homology theory will be taken up in this chapter. The study is motivated by the ideas of Chapters 1 and 2 as well as by new illustrations that will be introduced here.

3-1. A PAUSE FOR MOTIVATION

It is clear from the last two chapters that there is a fundamental type of algebraic structure underlying homology theory, namely, a sequence of groups K_p with homomorphisms $\phi_p: K_p \to K_{p-1}$ for each p such that $\phi_{p-1}\phi_p = 0$. This has been illustrated by the groups of singular chains on a topological space E (with $K_p = C_p(E)$) and by the groups of oriented chains on a simplicial complex E (with $K_p = \mathscr{C}(E)$), the ϕ_p in each case being the appropriate boundary operators. It is desirable to examine the algebraic properties of such structures in themselves. Topological results already obtained and others to be derived later will then appear as special cases or consequences of the algebraic results.

Indeed, this policy of pausing to take algebraic stock of the situation becomes even more advisable when the notion of cohomology is introduced. This notion will prove to have algebraic features very similar to the results just mentioned, and in addition, it is linked to homology by a purely algebraic procedure, which should be studied first, independently of any topological application.

The idea of cohomology is perhaps best motivated by the following geometric considerations. A brief description of one of the typical problems will be given which leads to an introduction of cohomology. This description will be given in a fairly informal way, without any attempt at setting up things rigorously.

Consider a smooth surface E, smooth meaning that the notion of a tangent vector to E at any point makes sense, and at the same time, let E be homeomorphic to a simplicial complex. The triangles forming E are assumed to be oriented coherently. In other words, a direction of rotation is marked on each triangle such that, for each pair of adjacent triangles, the orientations induced on the common side are opposite. A surface E whose triangles can be oriented in this way is called an orientable surface.

A unit tangent vector field on E is the assignment to each point of E of a unit tangent vector which varies continuously with its point of contact. The problem of constructing such a field can be approached in the following way. First, attach a unit tangent vector to each vertex of the complex E. If a 1 simplex of E has ends p and q, a continuous rotation will carry the tangent vector assigned to p into that assigned to q. In this way a continuous unit tangent vector field is given over the 1 skeleton of E. The essential step in the construction is the extension of this field over the triangles.

Let S be a triangle of the complex E. The set of unit tangent vectors at a point can be parametrized by their end points; that is, this set can be identified with the set of points of a circle S^1. It follows that the set of unit tangent vectors at points of S can be identified with $S \times S^1$. Since there is already a continuous unit tangent vector field on the 1 skeleton, there is a unit tangent vector $(p, v(p))$ attached to each point p of the boundary of S, so that v, mapping p on $v(p)$, is a continuous map of the boundary of S into S^1. The extension of the vector field over S is equivalent to the extension of this map to a map of all of S into S^1. The boundary of S is itself a circle, and up to homotopy, it can be identified in two ways with a standard copy $S_0{}^1$ of the circle, according to the direction of rotation. The orientation given on S fixes this direction, so v determines a map $S_0{}^1 \to S^1$ up to homotopy. In other words, an element of $\pi_1(S^1)$ (cf. Appendix A) is determined. Note that if the opposite orientation on S had been chosen, the resulting element of $\pi_1(S^1)$ would be oppositely signed. Hence this procedure assigns to the oriented simplex σ on S an element $f(\sigma)$ of $\pi_1(S^1)$ such that $f(-\sigma) = -f(\sigma)$. Repeating this for each triangle of E generates a homomorphism $f\colon \mathscr{C}_2(E)$

$\to \pi_1(S^1)$. $\pi_1(S^1)$ can be identified with Z so f can be thought of as a homomorphism of $\mathscr{C}_2(E)$ into Z.

It has been pointed out that the tangent vector field over the boundary of S can be extended over S if and only if the map v can be extended over S, that is, if and only if the corresponding element $f(\sigma)$ in $\pi_1(S^1)$ is zero. Thus the unit tangent vector field over the 1 skeleton of E can be extended over the whole of E if and only if the homomorphism f is the zero homomorphism.

At this stage it is conceivable that the extension of the unit tangent vector field to E may fail not because there is no such field possible, but simply because of the unfortunate choice of the field over the 1 skeleton. So it is natural to ask whether the field over the 1 skeleton can be adjusted so that the integer $f(\sigma)$ obtained above becomes zero for each triangle of E. Such an adjustment will now be examined, leaving the initial assignment on the vertexes unchanged (in fact, when working on a surface no generality is lost by making this assumption on the vertexes).

Suppose then that T is a 1 simplex of E. The argument already used shows that the assignment of a unit tangent vector field along T is equivalent to giving a map of T into S^1. If this field is changed on T without being changed at the end points, two maps of T into S^1 are obtained which agree at the end points; these may be put together to give a map of a circle into S^1. This is done by identifying T in turn with the two semicircles of the standard circle $S_0{}^1$. Of course the results depends on the orientation given to T. If τ is an oriented simplex on T, the map of $S_0{}^1$ into S^1 just described represents an element $g(\tau)$ of $\pi_1(S^1)$ such that $g(-\tau) = -g(\tau)$. In this way, if an adjustment of the vector field is made over the 1 skeleton of E, a homomorphism g of $\mathscr{C}_1(E)$ into $\pi_1(S^1) = Z$ is obtained. Then if σ is an oriented 2 simplex of E and if τ_1, τ_2, τ_3 are its sides with the induced orientations, it can be checked that the integer $f(\sigma)$ obtained from the first vector field over the 1 skeleton is replaced, for the new vector field, by $f(\sigma) + g(\tau_1) + g(\tau_2) + g(\tau_3)$, that is, by $f(\sigma) + g(d\sigma)$.

There is a more convenient algebraic way of expressing this: for any p, the set of homomorphisms $\mathscr{C}_p(E) \to Z$ is an additive group $\mathscr{C}^p(E)$. Thus in the above discussion, f is in $\mathscr{C}^2(E)$ and g is in $\mathscr{C}^1(E)$. Any homomorphism ϕ from $\mathscr{C}_p(E)$ to $\mathscr{C}_q(E)$ has a dual ϕ', which is defined as follows. Map each homomorphism $h: \mathscr{C}_q(E) \to Z$ on the homomorphism $h\phi: \mathscr{C}_p(E) \to Z$, and call the latter $\phi'(h)$. Thus the element h of $\mathscr{C}^q(E)$ is mapped on the element $\phi'(h)$ of $\mathscr{C}^p(E)$. It is easy to see that ϕ' is a homomorphism.

In this terminology let the dual of d be δ, a homomorphism of $\mathscr{C}^1(E)$ into $\mathscr{C}^2(E)$. Then $f(\sigma) + g(d\sigma)$ can be written as $(f + \delta g)(\sigma)$. Thus the original assignment of vectors on the 1 skeleton leads to the function f, while the new assignment leads to $f + \delta g$ as the obstruction to extending the unit tangent vector field over E. Then, the condition for the existence of such an extension is that it must be possible to choose g so that $f + \delta g = 0$, that is, $f = -\delta g$.

The last equation is expressed in words by saying that f is the coboundary of g, or that f is cohomologous to zero. f itself is called an obstruction cocycle. Thus a unit tangent vector field exists over E if and only if the obstruction cocycle is cohomologous to zero.

Algebraically the above situation is as follows, a complex K of any dimension being substituted for E. For each chain group $\mathscr{C}_p(K)$ define the corresponding cochain group $\mathscr{C}^p(K)$ as the group of homomorphisms of $\mathscr{C}_p(K) \to Z$. The boundary homomorphism $d\colon \mathscr{C}_p(K) — \mathscr{C}_{p-1}(K)$ has a dual $\delta\colon \mathscr{C}^{p-1}(K) \to \mathscr{C}^p(K)$ and it is not hard to see that $\delta^2 = 0$. Thus, algebraically, the sequence of cochain groups is somewhat similar to that of the chain groups, except that the operator δ raises dimension. It is natural to define cocycles, coboundaries, and cohomology from the cochain sequence, imitating the definitions of cycles, boundaries, and homology.

Clearly a similar construction is possible given any sequence of chain groups, for example, the singular chain groups on a space. The use of the integers as coefficients is accidental. Any coefficient group could be used instead.

The algebraic concepts that have been suggested here will now be studied.

3-2. CHAIN COMPLEXES

The discussion at the beginning of Section 3-1 suggests the following definition.

Definition 3-1. A *chain complex* K is a sequence of additive Abelian groups and homomorphisms

$$\to K_p \xrightarrow{d_p} K_{p-1} \to$$

such that $d_{p-1}\, d_p = 0$.

Note that the condition on the d_p implies that the image of d_p is contained in the kernel of d_{p-1}. It is assumed that d_0 maps K_0 on 0, and the sequence is assumed to terminate at this point. This will be true in all the applications here.

Examples

3-1. If E is a topological space let $K_p = C_p(E)$ and let d_p be the boundary operator as defined in Chapter 1. The chain complex so obtained is called the *singular chain complex* on E over the coefficient group $\mathscr{G}$.

3-2. If E is a simplicial complex, take $K_p = \mathscr{C}_p(E)$ and let d_p be the oriented boundary operator as in Chapter 2. The resulting chain complex is the *oriented simplicial chain complex* on E over $\mathscr{G}$.

3-3. Let E be a simplicial complex with a cell decomposition. Let $K_p = \bar{\mathscr{C}}_p(E)$, in the notation of Definition 2-19. K is then the *cellular chain complex* of E corresponding to its cell decomposition.

These examples motivate the following definition.

Definition 3-2. If K is a chain complex as in Definition 3-1, K_p is called the *group of p chains of K*. d_p is called the *boundary operator* on K_p. The elements of K_p also are said to be of *dimension p*.

If the context makes it clear what dimensions are involved, the subscripts on the d_p are sometimes dropped. The condition $d_{p-1} d_p = 0$ then is written as $d^2 = 0$.

To follow up the remarks in Section 3-1, the notion of duality will now be introduced. Let $\mathscr{G}$ be a fixed additive Abelian group.

Definition 3-3. If H is an additive Abelian group then the *dual of H with respect to $\mathscr{G}$* is the set of homomorphisms of H into $\mathscr{G}$. The dual of H is denoted by H'.

If ϕ and ψ are in H' then $\phi + \psi$ can be defined as the homomorphism of H into $\mathscr{G}$ such that $(\phi + \psi)(x) = \phi(x) + \psi(x)$ for all x in H. This operation makes H' into an additive Abelian group.

Definition 3-4. Let K and L be two additive Abelian groups and let $f\colon K \to L$ be a homomorphism. If ϕ is in L' the composition ϕf is denoted by $f'(\phi)$ and is a homomorphism of K into $\mathscr{G}$, that is, an element of K'. Thus f' is a map of L' into K', called the *dual of f with respect to $\mathscr{G}$*.

Lemma 3-1. *In the notation of the above definitions, f' is a homomorphism of L' into K'.*

Proof. Let ϕ, ψ be in L', so that ϕ, ψ and $\phi + \psi$ are all homomorphisms of L into $\mathscr{G}$. By definition, $f'(\phi + \psi)$ is the composition $(\phi + \psi)f$. So if x is any element of K,

$$\begin{aligned} f'(\phi + \psi)(x) &= (\phi + \psi)(f(x)) \\ &= \phi(f(x)) + \psi(f(x)) \\ &= (\phi f + \psi f)(x) \\ &= (f'(\phi) + f'(\psi))(x) \end{aligned}$$

Hence

$$f'(\phi + \psi) = f'(\phi) + f'(\psi)$$

as was to be shown. ∎

Lemma 3-2. *Let K, L, M be additive Abelian groups and let $f: K \to L$ and $g: L \to M$ be homomorphisms. Then the relation*

$$(gf)' = f'g'$$

holds between the duals.

Proof. This is a matter of simple verification and is left as an exercise. ∎

Note that it follows from this lemma that, if a commutative diagram is given, then the dual diagram obtained by replacing all the groups and homomorphisms by their duals, and by reversing the arrows, is also commutative.

It should be noted in particular that if $f: K \to L$ is the zero homomorphism, then the dual $f': L' \to K'$ is also zero. Also, the dual of the identity is the identity.

If K is a chain complex, the corresponding cochain complex over a given coefficient group $\mathscr{G}$ is constructed by dualizing all the groups and homomorphisms of K.

Definition 3-5. Let K be a chain complex consisting of the groups and homomorphisms

$$\to K_p \xrightarrow{d_p} K_{p-1} \to$$

Denote by K^p the dual of K_p with respect to the coefficient group $\mathscr{G}$. K^p is called *the group of p cochains of K over $\mathscr{G}$*. Denote by δ^p the dual of d_p. δ^p is called the *coboundary operator* on K^{p-1}. The system of groups and homomorphisms

$$\leftarrow K^p \xleftarrow{\delta^p} K^{p-1} \leftarrow$$

is called the *cochain complex over $\mathscr{G}$ associated with* K, and is denoted by K^*.

It is clear, by Lemma 3-2, and the fact that the dual of the zero homomorphism is zero, that $\delta^p\delta^{p-1} = 0$, or dropping the subscripts, $\delta^2 = 0$. This means that the image of δ^{p-1} is contained in the kernel of δ^p.

Both homology and cohomology groups can be associated with a chain complex, the definitions being modeled on those appearing in singular and simplicial homology theory. Let K be the chain complex

$$\to K_{p+1} \xrightarrow{d_{p+1}} K_p \xrightarrow{d_p} K_{p-1} \to$$

Definition 3-6. The kernel of d_p is denoted by $Z_p(K)$ and is called *the group of p cycles of K.* The image of d_{p+1} is denoted by $B_p(K)$ and is called *the group of p boundaries of K.*

The relation $d_p d_{p+1} = 0$ implies that $B_p(K) \subset Z_p(K)$, so the quotient group can be constructed.

Definition 3-7. The quotient group $Z_p(K)/B_p(K)$ is denoted by $H_p(K)$ and is called *the* pth *homology group of K.*

Cohomology is defined similarly, starting instead with the dual complex of K with respect to a coefficient group $\mathscr{G}$, namely

$$\to K^{p-1} \xrightarrow{\delta^p} K^p \xrightarrow{\delta^{p+1}} K^{p+1} \to$$

Definition 3-8. The kernel of δ^{p+1} is denoted by $Z^p(K)$ and is called *the group of p cocycles of K over* $\mathscr{G}$. The image of δ^p is denoted by $B^p(K)$ and is called *the group of p coboundaries of K over* $\mathscr{G}$.

The relation $\delta^{p+1}\delta^p = 0$ implies that $B^p(K) \subset Z^p(K)$, so the quotient group can be defined.

Definition 3-9. The quotient group $Z^p(K)/B^p(K)$ is denoted by $H^p(K)$ and is called the pth *cohomology group of K over* $\mathscr{G}$.

In the definitions associated with cohomology, the coefficient group could be inserted in the notation, but this is usually unnecessary as the context makes the coefficient group clear.

If α and β are cycles (or cocycles) representing the same element of $H_p(K)$ (or $H^p(K)$) they are called homologous (or cohomologous). In either case the relation is written $\alpha \sim \beta$.

Examples

3-4. K can be taken as the singular chain complex on a topological space E, that is, $K_p = C_p(E)$ for each p. This can be assumed to be over an arbitrary coefficient group. The homology groups of K are then as in Chapter 1, namely, the singular homology groups of the space E.

In particular, let K be the singular chain complex of E over the group of integers as coefficient group. Let $\mathscr{G}$ be any additive Abelian group. Let K^* be the cochain complex of K with respect to $\mathscr{G}$. Then K^p is *the group of singular p cochains of E over* $\mathscr{G}$ and the corresponding cohomology groups are *the singular cohomology groups of E over* $\mathscr{G}$.

3-5. Similarly, K can be taken as the oriented simplicial chain complex on a simplicial complex E over a coefficient group $\mathscr{G}$. The homology groups of K are then the simplicial homology groups $H_p(E)$ of E.

If K is the oriented simplicial chain complex of E over Z then the dual K^* with respect to $\mathscr{G}$ is *the oriented simplicial cochain complex of* E *over* $\mathscr{G}$, and the corresponding cohomology groups are called *the simplicial cohomology groups of* E *over* $\mathscr{G}$.

3-3. CHAIN HOMORPHISMS

In this section the notion of homomorphisms between chain complexes will be introduced. These are, in fact, families of homomorphisms on groups. Let K and L be chain complexes with boundary operators d and d' respectively. Let a family of homomorphisms $f_p: K_p \to L_p$ be given, one for each p, such that the diagram

$$\begin{array}{ccc} K_p & \xrightarrow{d} & K_{p-1} \\ {\scriptstyle f_p}\downarrow & & \downarrow{\scriptstyle f_{p-1}} \\ L_p & \xrightarrow{d'} & L_{p-1} \end{array} \tag{3-1}$$

is commutative for each p.

Definition 3-10. A system of f_p like the one just described is called a *chain homomorphism from* K *to* L. It is denoted by the notation $f: K \to L$.

Examples

3-6. Let E and F be two topological spaces, and let K and L be their singular chain complexes; that is, $K_p = C_p(E)$ and $L_p = C_p(F)$. Let $f: E \to F$ be a continuous map. The induced homomorphisms on chain groups, previously denoted by f_1 (Definition 1-9), form a family of homomorphisms $f_{1p}: K_p \to L_p$ that satisfies the above commutativity condition, so they constitute a chain homomorphism of K into L.

3-7. Let E and F be simplicial complexes, K and L the associated oriented chain complexes, whose p-dimensional groups are, respectively, $\mathscr{C}_p(E)$ and $\mathscr{C}_p(F)$, and let f be a simplicial map from E into F. As shown in Section 2-8, f induces a homomorphism $\hat{f}: \mathscr{C}_p(E) \to \mathscr{C}_p(F)$ for each p, and these homomorphisms commute with the oriented boundary operators. Hence they constitute a chain homomorphism from K into L.

A chain homomorphism $f: K \to L$ induces by duality a similar operation between the corresponding cochain complexes for any coefficient group $\mathscr{G}$;

as explained in the remark following Lemma 3-2, the commutative diagram (3-1) leads to a commutative diagram

$$\begin{array}{ccc} K^p & \xleftarrow{\delta} & K^{p-1} \\ {\scriptstyle f^p}\uparrow & & \uparrow{\scriptstyle f^{p-1}} \\ L^p & \xleftarrow{\delta'} & L^{p-1} \end{array} \tag{3-2}$$

where f^p is the dual of f_p with respect to $\mathcal{G}$.

Definition 3-11. A family of homomorphisms $f' = \{f^p\}$ that makes the diagram (3-2) commute for all p is called a *cochain homomorphism of the cochain complex L into the cochain complex K*. Here f' is called the *cochain homomorphism dual to f.*

3-4. INDUCED HOMOMORPHISMS ON HOMOLOGY AND COHOMOLOGY GROUPS

The notion of induced homomorphisms corresponding to continuous or simplicial maps has already been introduced (Sections 1-6, 2-8). A similar concept will now be described algebraically for chain and cochain complexes.

Let K and L be chain complexes and let $f: K \to L$ be a chain homomorphism. The commutativity of the diagram

$$\begin{array}{ccc} K_p & \xrightarrow{d} & K_{p-1} \\ {\scriptstyle f_p}\downarrow & & \downarrow{\scriptstyle f_{p-1}} \\ L_p & \xrightarrow{d} & L_{p-1} \end{array}$$

implies that f_p maps $Z_p(K)$ into $Z_p(L)$ and $B_p(K)$ into $B_p(L)$. Hence it induces a homomorphism of the corresponding quotient groups, namely

$$f_*: H_p(K) \to H_p(L)$$

Here, if the element $\bar{\alpha}$ of $H_p(K)$ is represented by α in $Z_p(K)$, then $f_*(\bar{\alpha})$ is the element of $H_p(L)$ represented by $f_p(\alpha)$.

Definition 3-12. f_* is called the *induced homomorphism on homology groups associated with f.*

Examples

3-8. If K and L are the singular chain complexes on topological spaces E and F, and if the chain homomorphism f is induced by a continuous map of E into F, then f_* coincides with the induced homomorphism on homology groups as described in Section 1-6.

3-9. Another example is furnished by taking K and L as the oriented simplicial chain complexes on simplicial complexes E and F and f as the chain homomorphism induced by a simplicial map (cf. Section 2-8). Again f_* coincides with the induced homomorphism on homology groups as described in Section 2-8.

If f is a chain homomorphism from K to L, let f' be the dual cochain homomorphism from L^* to K^*, with respect to some coefficient group. As in the case of chain homomorphisms, the commutativity of the diagram

$$\begin{array}{ccc} K_p & \xrightarrow{\delta} & K^{p+1} \\ {\scriptstyle f'^p}\uparrow & & \uparrow{\scriptstyle f'^{p+1}} \\ L^p & \xrightarrow{\delta} & L^{p+1} \end{array}$$

implies that f'^p carries $Z^p(L)$ and $B^p(L)$ into $Z^p(K)$ and $B^p(K)$, respectively, and so induces a homomorphism

$$f^*: H^p(L) \to H^p(K)$$

Definition 3-13. f^* is called the *induced homomorphism on cohomology groups associated with* f.

In discussing singular homology theory, it was shown that the homomorphism on homology groups induced by a composition of maps is the composition of the homomorphisms induced by the maps separately. The algebraic counterpart of this, for both homology and cohomology groups, will now be proved.

Theorem 3-3. *Let* K, L, M *be chain complexes and let* $f: K \to L$ *and* $g: L \to M$ *be chain homomorphisms. Let* f' *and* g' *be the dual cochain homomorphisms with respect to some coefficient group. Then* gf *is a chain homomorphism of* K *into* M, $f'g'$ *is its dual cochain homomorphism, and*

$$(gf)_* = g_* f_*$$

$$(gf)^* = f^* g^*$$

Also, if $f: K \to K$ *is the identity, then* f_* *and* f^* *are the identity.*

Proof. If the diagrams showing that f and g commute with the appropriate boundary operators are put together, the commutativity of gf with boundary operators can be seen at once. Hence gf is a chain homomorphism. Similarly, $f'g'$ is a cochain homomorphism. The fact that $f'g'$ is the dual of gf follows at once from Lemma 3-2.

Now let $\bar{\alpha}$ be an element of $H_p(K)$, represented by $\bar{\alpha}$ in $Z_p(K)$. By the definition of induced homomorphisms, $(gf)_*(\bar{\alpha})$ is the class modulo $B_p(M)$ of $(gf)(\alpha) = g(f(\alpha))$. But $f_*(\bar{\alpha})$ is the class modulo $B_p(L)$ of $f(\alpha)$, so $(g_* f_*)(\bar{\alpha}) = g_*(f_*(\bar{\alpha}))$ is the class modulo $B_p(M)$ of $g(f(\alpha))$, the definition of induced homomorphisms being used twice here. Hence $(gf)_*(\bar{\alpha}) = (g_* f_*)(\bar{\alpha})$ for all $\bar{\alpha}$ in $H_p(K)$, so $(gf)_* = g_* f_*$, as was to be shown. Similarly $(gf)^* = f^* g^*$.

The last statement in the theorem is almost trivial. ∎

3-5. CHAIN HOMOTOPY

The idea here is to reproduce the algebraic features of the notion of homotopic maps. The motivation for the following definition comes from the formula of Theorem 1-16 which connects the induced homomorphisms associated with homotopic maps.

Let K and L be chain complexes, where both boundary operators are denoted by d, and let f and g be two chain homomorphisms from K into L.

Definition 3-14. For each p let $h_p\colon K_{p-1} \to L_p$ be a homomorphism such that

$$f_p - g_p = h_p\, d_p + d_{p+1} h_{p+1} \tag{3-3}$$

for all p. Then f and g are said to be *algebraically homotopic.*

Note that the h_p do *not* constitute a chain homomorphism. If the homomorphisms are arranged (as is often convenient) in a diagram like the following this diagram is not commutative.

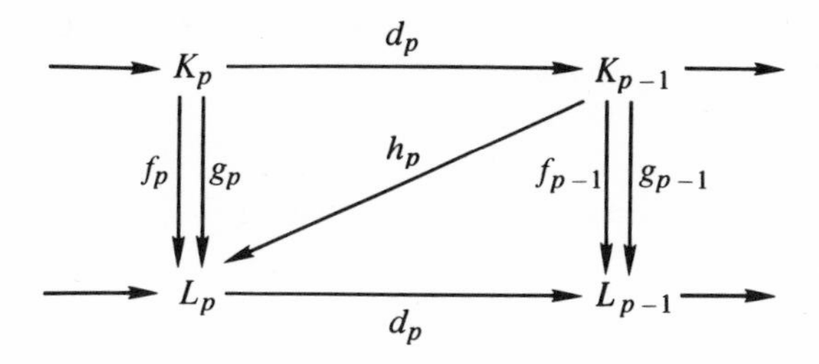

Examples

3-10. Consider first singular homology. Let K and L be the singular complexes on the topological spaces E and F, respectively, and let f and g be two homotopic continuous maps of E into F. f and g induce chain homomorphisms f and g, respectively, of K into L. Then if h_p is taken to be the prism operator P (for the appropriate dimension) introduced in Section

1-10, the condition (3-3) is satisfied. Thus the chain homomorphisms f and g are algebraically homotopic. As already pointed out, this example is the motivation for the use of the term "homotopy" in this algebraic sense.

3-11. Let K be the singular chain complex on a space E. Let B be the barycentric subdivision operator introduced in Section 1-11. B is a homomorphism of $C_p(E)$ into itself for each p, and it commutes with the boundary operator d. In other words, B is a chain homomorphism of K into itself. Now compare B with the identity map K onto itself. In Section 1-11, an operator H was constructed such that $B -$ identity $= dH + Hd$. This equation says that B and the identity are algebraically homotopic.

3-12. Let E be a simplicial complex and let $E \times I$ be triangulated as it was in Exercise 2-2. Let f and g be simplicial maps of E into $E \times I$ which are defined by setting $f(x) = (x, 0)$ and $g(x) = (x, 1)$. Then the induced homomorphisms on simplicial chain groups corresponding to f and g are algebraically homotopic. The point is that an operator P similar to that of Section 1-10 can be constructed and a formula similar to that of Theorem 1-14 can be obtained entirely in terms of oriented simplexes on E and $E \times I$. This example is, in effect, the homotopy theorem for simplicial homology theory.

The notion of algebraic homotopy can be applied to cochain homomorphisms by dualizing Definition 3-14. Using the notation of that definition, let $\mathcal{G}$ be a coefficient group and let K^* and L^* be the duals of K and L, respectively, with respect to $\mathcal{G}$. Also let the duals of the homomorphisms f_p, g_p, h_p, d be f^p, g^p, h^p, δ, respectively. Equation (3-3) can be dualized, giving

$$f^p - g^p = \delta h^p + h^{p+1}\delta \tag{3-4}$$

Definition 3-15. The cochain homomorphism $f' = \{f^p\}$, $g' = \{g^p\}$ are said to be *algebraically homotopic* if Eq. (3-4) is satisfied for a family of homomorphisms h^p.

Note that this definition is formally similar to Definition 3-14.

3-6. THE ALGEBRAIC HOMOTOPY THEOREM

Algebraically homotopic linear maps will now be shown to induce the same homomorphisms on the homology and cohomology groups. In particular, if K and L are the singular complexes on two topological spaces, and if the algebraically homotopic maps are induced by homotopic maps in the topological sense, this result reduces to the homotopy theorem for singular homology theory. It is also the basis of the proof of the excision theorem with the help of barycentric subdivision.

Theorem 3-4. *Let f and g be algebraically homotopic chain homomorphisms of a chain complex K into a chain complex L. Then, in the notation of* Definitions 3-12, *and* 3-13, $f_* = g_*$ *and* $f^* = g^*$.

Proof. Let $\bar{\alpha} \in H_p(K)$ and suppose that $\bar{\alpha}$ is represented by α in $Z_p(K)$. Apply the defining equation of algebraic homotopy to α. Then, in the notation of Definition 3-14,

$$\begin{aligned} f_p(\alpha) - g_p(\alpha) &= h_p\, d(\alpha) + dh_{p+1}(\alpha) \\ &= dh_{p+1}(\alpha) \end{aligned}$$

acknowledging that $d\alpha = 0$. But $f_p(\alpha)$ and $g_p(\alpha)$ represent $f_*(\bar{\alpha})$ and $g_*(\bar{\alpha})$, respectively (cf. Definition 3-12), and the last equation says that they are in the same homology class modulo $B_p(L)$. Hence $f_* = g_*$, as was to be shown.
Similarly it can be shown that $f^* = g^*$. ∎

The notion of homotopy inverse was introduced in singular homology theory. This concept can now be formulated algebraically.

Definition 3-16. Let $f\colon K \to L$ and $g\colon L \to K$ be chain homomorphisms such that the compositions fg and gf are algebraically homotopic to the appropriate identity homomorphisms. f and g are called *algebraic homotopy inverses.*

Theorem 3-5. *If $f\colon K \to L$ has an algebraic homotopy inverse, the induced homomorphisms f_* and f^* are both isomorphisms onto.*

Proof. Consider f_*. According to the hypotheses and Definition 3-16, there is a chain homomorphism $g\colon L \to K$ such that the compositions fg and gf are algebraically homotopic to the appropriate identities. Hence $(fg)_* =$ (identity)$_*$, by Theorem 3-4; this is the identity homomorphism. Theorem 3-3 says, however, that $(fg)_* = f_* g_*$, so $f_* g_*$ is the identity homomorphism. Similarly, $g_* f_*$ is the identity, so f_* is an isomorphism onto for each dimension, as required.
Similarly f^* is an isomorphism onto. ∎

The following is a partial converse to the previous theorem.

Theorem 3-6. *Let K and L be chain complexes such that all the K_p and L_p are free Abelian groups, and let $f\colon K \to L$ be a chain homomorphism such that f_* is an isomorphism onto. Then f has a homotopy inverse.*

Proof: Before the proof of this theorem is tackled, its hypotheses will be converted into a more usable form. Since K_{p-1} is free Abelian, its subgroup $B_{p-1}(K)$ is also free. The map $d\colon K_p \to K_{p-1}$ has image $B_{p-1}(K)$ and kernel $Z_p(K)$, so $B_{p-1}(K) \cong K_p/Z_p(K)$. In other words, the quotient group $K_p/Z_p(K)$ is free and so is K_p, from which it follows that $Z_p(K)$ is a direct summand. Thus K_p can be written as

$$K_p = Z_p(K) \oplus W_p$$

where W_p is isomorphic to $B_{p-1}(K)$ under $d_p\colon K_p \to K_{p-1}$.

Another preliminary step consists in proving the theorem in a very special case, namely, that in which $L = 0$. Here f is the zero map, and the hypothesis that f_* is an isomorphism onto means that $H_p(K) = 0$ for all p. The conclusion of the theorem is that there is a g that is a homotopy inverse to f. g can only map $L = 0$ on the zero element of K, so fg is automatically the identity map of L on itself. Also, gf is simply the zero map of K into itself, that is, the map carrying all of K onto the zero element of K. Hence what is to be proved is that the identity map of K on itself is algebraically homotopic to the zero map. This will be done by constructing the homomorphisms $h_p\colon K_{p-1} \to K_p$ corresponding to the definition of algebraic homotopy.

$K_{p-1} = Z_{p-1}(K) \oplus W_{p-1}$ is a direct sum so h_p can be defined separately on the summands. Let h_p be zero on W_{p-1}. Since $H_{p-1}(K) = 0$, $Z_{p-1}(K) = B_{p-1}(K)$ and $d_p\colon W_p \to B_{p-1}(K)$ is an isomorphism onto. Now h_p will be defined on $Z_{p-1}(K) = B_{p-1}(K)$ as the inverse of this isomorphism.

It must be shown that the h_p so defined give the algebraic homotopy between the identity map of K on itself and the zero map. To do this take x in K_p and write it as $y + z$ with $y \in Z_p(K)$ and $z \in W_p$. Then

$$d_p x = d_p y + d_p z = d_p z$$

Since h_p acts as the inverse of d_p restricted to W_p,

$$h_p d_p x = h_p d_p z = z$$

Also, $h_{p+1}x = h_{p+1}y + h_{p+1}z = h_{p+1}y$, so

$$d_{p+1}h_{p+1}x = d_{p+1}h_{p+1}y = y$$

again acknowledging that h_{p+1} and d_{p+1} restricted to W_{p+1}, are inverses. Hence

$$x = y + z = d_{p+1}h_{p+1}x + h_p d_p x$$

This equation says that the identity is algebraically homotopic to the zero map as was to be shown.

The proof of Theorem 3-6 in the general case will now be motivated. The method is to construct a new complex whose homology groups are all zero and to apply to it the result of the special case. To see the geometric meaning of such a construction, think for a moment of the elements of K and L as cells in the topological sense, elements of K_p and L_p corresponding to cells of dimension p. The boundary of a cell is thought of as a linear combination of cells on its geometric boundary. Note that no attempt at rigor is being made here, only an attempt to make things geometrically plausible.

Referring to Fig. 12, construct a new complex K' whose cells are all those of L and also cells obtained in the following manner. For each cell x of K form the product $x \times I$ and identify $\{p\} \times \{1\}$ with $f(p)$ in L for each p in x. Call the resulting cell x' (it may be degenerate!). Geometrically speaking, K' is the mapping cylinder of f. Since f induces isomorphisms of the homology groups so does the inclusion map of K into K'. From the exactness of the homology sequence of K' modulu K, it follows that $H_p(K', K) = 0$ for all p. Thus the complex to which the special case of the theorem will be applied is the quotient complex K'/K. So far the reasoning has been heuristic, but now the algebraic definitions will be set up following the lines indicated by the geometry.

Note first that an element of K', above, is a linear combination of cells of L and cells of the type x', the latter being in one-to-one correspondence with the cells of K. Thus, an element of K' can be thought of as a pair (x, y) with x in K and y in L; this notation can also be used to represent an element of K'/K.

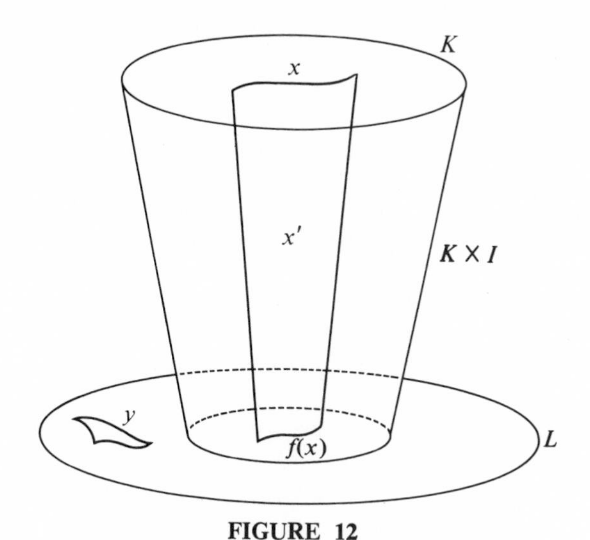

FIGURE 12

Following up this observation algebraically, define K^* to be the chain complex whose p-dimensional chain groups K_p^* is the direct sum of K_{p-1} and L_p. The elements of K_p^* are written as pairs (x, y) with $x \in K_{p-1}$ and $y \in L_p$.

The boundary of x' can be thought of (see Fig. 12) as $f(x) - x - (dx)'$, where $(dx)'$ represents a linear combination of cells of K' corresponding to the cells of dx in the same way as x' is related to x. Reduction modulo K gives $f(x) - (dx)'$, now to be represented algebraically as the pair $(-dx, f(x))$. If y is in L its boundary is dy, which is written in K as the pair $(0, dy)$.

These remarks suggest the appropriate definition for the boundary operator in K^*: if (x, y) is in K^*, define

$$d^*(x, y) = [-dx, dy + f(x)]$$

It is easy to see that this is a homomorphism from K_p^* into K_{p-1}^*, and it can be verified that $d^{*2} = 0$.

The foregoing heuristic discussion suggests that $H_p(K)^*$ is zero for all p. To check this algebraically, consider what it means for (x, y) to be a cycle in K^*. Let $(x, y) \in K_p^*$ and let $d^*(x, y) = 0$. Then $dx = 0$ and $f(x) = -dy$. In other words, x is a cycle in K_{p-1} and $f(x)$ is homologous to zero in L. Since f_* is an isomorphism, however, this implies that x is homologous to zero in K; that is, $x = dz$ for some z in K. Note that $f(x) = f(dz) = df(z)$ and also $f(x) = -dy$, so $d[f(z) + y] = 0$; that is to say $f(z) + y$ is a cycle on L_p. Since f_* is onto, it follows that there is a cycle u on K and an element w of K such that $f(u) = f(z) + y + dw$. Hence $y = f(u - z) - dw$. On the other hand, $x = dz = -d(u - z)$ (u is a cycle), so

$$d^*(u - z, -w) = [x, -dw + f(u - z)] = (x, y)$$

In other words, the assumption that $d^*(x, y) = 0$ implies that (x, y) is in the image of d^*. Hence $H_p(K_p^*) = 0$ for all p, as was to be shown.

The special case of the theorem will now be applied to the complex K^*. For each p there is a homomorphism

$$h\colon K_p^* \to K_{p-1}^*$$

(in order to simplify the notation the subscript is dropped) such that

$$(x, y) = (d^*h + hd^*)(x, y)$$

for all (x, y) in K_p^*. Since the K_p^* are direct sums it is possible to write the homomorphisms h in the form

$$h(x, y) = (h_{11}x + h_{12}y, h_{21}x + h_{22}y)$$

in terms of the four homomorphisms

$$h_{11}: K_{p-1} \to K_p$$
$$h_{12}: L_p \to K_p$$
$$h_{21}: K_{p-1} \to L_{p+1}$$
$$h_{22}: L_p \to L_{p+1}$$

The equation $d^*h + hd^* =$ identity, when expressed in terms of the h_{ij}, becomes

$$\begin{aligned}(x, y) &= d^*(h_{11}x + h_{12}y, h_{21}x + h_{22}y) + h(-dx, dy + f(x)) \\ &= (-dh_{11}x - dh_{12}y, dh_{21}x + dh_{22}y + f(h_{11}x + h_{12}y)) \\ &\quad + (-h_{11}\,dx + h_{12}\,dy + h_{12}f(x), -h_{21}\,dx + h_{22}\,dy + h_{22}f(x))\end{aligned}$$

Equating the components on each side of this yields the relations

$$x = -dh_{11}x - h_{11}\,dx - dh_{12}y + h_{12}\,dy + h_{12}f(x) \tag{3-5}$$

$$y = dh_{21}x - h_{21}\,dx + dh_{22}y + h_{22}\,dy + fh_{11}x + fh_{12}y + h_{22}fx \tag{3-6}$$

These are identities in x and y, that is, they hold for all values of x and y, so various bits of information can be obtained by giving x and y particular values. Setting $x = 0$ in (3-5) yields

$$-dh_{12}y + h_{12}\,dy = 0$$

for all y in L. In other words, h_{12} commutes with d, so it is a chain homomorphism of L into K. If y is set equal to 0 in (3-5) it follows that

$$x - h_{12}f(x) = -dh_{11}x - h_{11}\,dx$$

for all x in K_{p-1}. This equation says that the composition $h_{12}f$ is algebraically homotopic to the identity of K onto itself. On the other hand, if x is set equal to 0 in (3-6) it follows that

$$y - fh_{12}(y) = dh_{22}y + h_{22}\,dy$$

for all y in L_p, and this means that fh_{12} is algebraically homotopic to the identity homomorphism of L onto itself. Thus, an algebraic homotopy inverse to f, namely, h_{12}, has been constructed to complete the proof of the theorem. ∎

3-7. SOME APPLICATIONS OF ALGEBRAIC HOMOTOPY

To illustrate Theorems 3-5 and 3-6 simplicial and singular homology and cohomology will first be compared.

Let E be a simplicial complex, let $K = \mathscr{C}(E)$ be the oriented chain complex on E, and let $L = C(E)$ be the singular chain complex on E. Theorem 2-17 says that $H_p(K) \cong H_p(L)$ for all p and any coefficient group. In particular, this isomorphism holds for the groups over the integers, and for the moment, this will be taken as the coefficient group. In this case, the chain groups $L_p = C_p(E)$ are free Abelian groups. K_p will be shown to be isomorphic to a subgroup of L_p, and hence to be also free Abelian. In addition, it will be shown that the isomorphisms $H_p(K) \cong H_p(L)$ are induced by inclusion. Then, by Theorem 3-6, this inclusion will be shown to have an algebraic homotopy inverse.

First, let all the vertexes of E be ordered in some way, arbitrary but fixed once and for all, to be called the standard order. If $((x_0 x_1 \cdots x_p))$ is an oriented simplex on E (in the notation of Section 2-8) and if $y_0, y_1, \ldots, y_p$ are the x_i arranged in standard order, then $((x_0 x_1 \cdots x_p)) = \pm((y_0 y_1 \cdots y_p))$, and there are no nontrivial relations between the symbols $((y_0 y_1 \cdots y_p))$ with the vertexes in standard order. Thus the oriented chain group $\mathscr{C}_p(E)$ over Z is a free Abelian group generated by the symbols $((x_0 x_1 \cdots x_p))$ with the vertexes x_i appearing in standard order.

Define the map $h\colon K_p \to L_p$ for each p by setting

$$h((x_0 x_1 \cdots x_p)) = (x_0 x_1 \cdots x_p)$$

where the symbol on the right is a singular simplex defined by a linear map, and the x_i are understood to be in standard order. The extension of h by linearity becomes a homomorphism for each p. The boundary operators in K and L are given by the formulas

$$d((x_0 x_1 \cdots x_p)) = \sum (-1)^i ((x_0 x_1 \cdots \hat{x}_i \cdots x_p))$$
$$d(x_0 x_1 \cdots x_p) = \sum (-1)^i (x_0 x_1 \cdots \hat{x}_i \cdots x_p)$$

From these it is easy to see that h commutes with d and so is a chain homomorphism. It is also clearly an isomorphism into for each p.

The isomorphism between $H_p(K)$ and $H_p(L)$ will now be analyzed. As before, let E_p be the p skeleton E. Then, since the map $H_p(E_p) \to H_p(E)$ induced by the inclusion map i is onto (Theorem 2-15), any p-dimensional singular homology class $\bar{\alpha}$ on E can be represented by a singular p cycle α on E_p. On the other hand, α is homologous in E_p modulo E_{p-1} to a linear combination $\sum a_i(y_0 y_1 \ldots y_p)$ of singular simplexes defined by linear maps (cf. Theorem 2-9 and Exercise 2-12), and by the map j: $H_p(E_p) \to H_p(E_p, E_{p-1})$ of Theorem 2-16, the homology class of α goes into $\sum a_i((y_0 y_1 \cdots y_p))$. Since this is in the image of j, it is a cycle under the oriented boundary operator (Definition 2-9). The similarity in form between the oriented and singular boundary operators implies that $\sum a_i(y_0 y_1 \cdots y_p)$ is a singular cycle α_0.

Thus $\alpha = \alpha_0 + \beta + d\gamma$, with β on E_{p-1} and γ on E_p. α and α_0 are cycles, hence β is too. Then β is a p cycle on the $(p-1)$ complex E_{p-1}, so β is homologous to 0. Hence, if γ is changed appropriately it follows that $\alpha = \alpha_0 + d\gamma$. In other words, $\bar{\alpha}$ in $H_p(E)$ is represented by $\alpha_0 = \sum a_i(y_0 y_1 \cdots y_p)$; this is the image under h of $\sum a_i((y_0 y_1 \cdots y_p))$.

Consider now the homomorphisms

$$Z_p(K) \xrightarrow{h} Z_p(L) \xrightarrow{f} H_p(L) = H_p(E)$$

where h is the restriction of the h defined previously and f is the natural homomorphism of $Z_p(L)$ on the quotient group $Z_p(L)/B_p(L)$. The composition fh maps the oriented cycle $\sum a_i((y_0 y_1 \cdots y_p))$ on the homology class of $\sum a_i(y_0 y_1 \cdots y_p)$, so it coincides with the homomorphism ij^{-1} (where i and j are as in Theorems 2-15, 2-16) restricted to the image of j (which is $Z_p(K)$). The kernel of $fh = ij^{-1}$ is $B_p(K)$, so the composition fh induces the isomorphism of $H_p(K) = \mathscr{H}_p(E)$ onto $H_p(E)$ which was obtained in Theorem 2-17. On the other hand, h carries $B_p(K)$ into $B_p(L)$ and so induces a homomorphism of $H_p(K)$ into $Z_p(L)/B_p(L) = H_p(L)$, while f induces the identity map of $Z_p(L)/B_p(L)$ onto $H_p(L)$. Hence the isomorphism $H_p(K) \to H_p(L)$ of Theorem 2-17, namely, that induced by fh, is the same as the homomorphism h_*: $H_p(K) \to H_p(L)$ induced by the chain homomorphism h. Thus h induces isomorphisms of the homology groups and, as previously pointed out, the groups K_p and L_p are all free. Hence Theorem 3-6 implies that the map h has an algebraic homotopy inverse.

An immediate consequence of this result is the following theorem.

Theorem 3-7. *The oriented simplicial chain complex and the singular chain complex of any simplicial complex have the same cohomology groups over any coefficient group. In addition, these isomorphisms are induced by the map h (or rather its dual) introduced in this section.*

As an exercise, check that this theorem also holds for relative simplicial and singular homology and cohomology.

As a further illustration of the use of algebraic homotopy the oriented chain complex of a simplicial complex E will now be compared with its ordered chain complex. The latter complex $\overline{\mathscr{C}}(E)$ is defined as follows. First, let S be the set of all finite sequences (possibly with repeated terms) of vertexes of E with the additional restriction that the sequence $\{x_0 x_1 \cdots x_p\}$ is in S if and only if the x_i are all in the same simplex of E. Then $\bar{\mathscr{C}}_p(E)$ is the free Abelian group generated by all the elements of S containing $p+1$ terms (not necessarily distinct). To make the $\bar{C}_p(E)$ into a complex $\bar{\mathscr{C}}(E)$ define the boundary operator d: $\bar{C}_p(E) \to \bar{\mathscr{C}}_{p-1}(E)$ by the formula

$$d\{x_0 x_1 \cdots x_p\} = \sum (-1)^i \{x_0 x_1 \cdots \hat{x}_i \cdots x_p\}$$

then extending by linearity. It is easy to check that $d^2 = 0$.

Note that the definition of $\bar{\mathscr{C}}(E)$ is similar to that of the oriented chain complex $\mathscr{C}(E)$ on E in terms of the symbols $((x_0 x_1 \cdots x_p))$. In that case, however, no repetitions of vertexes were allowed, and permutations of the vertexes were defined either to leave such a symbol unchanged or to change its sign.

Also note that, although $\bar{\mathscr{C}}(E)$ is being defined here with the integers as the coefficient group, a similar definition could be made for any other coefficient group.

Definition 3-17. The chain complex $\bar{\mathscr{C}}(E)$ just constructed is called the *ordered chain complex on E*. Its homology groups are called the *ordered simplicial groups on* E and are denoted by $\bar{\mathscr{H}}_p(E)$. Again, it should be noted that the coefficient group is Z, but other coefficient groups could be used which might have to be indicated in the notation.

Presently, it will be shown, by an application of the algebraic homotopy theorem, that $\bar{\mathscr{H}}_p(E) \cong \mathscr{H}_p(E)$ for all p. In the meantime, this theorem will be used to show that, if S is a simplex, then $\bar{\mathscr{H}}_p(S) = 0$ for all $p > 0$, while $\bar{\mathscr{H}}_0(S) = Z$ (the coefficient group). The motivation for this is that the ordered homology groups of a single point are zero in the positive dimensions and Z in dimension 0 (this can easily be checked by explicit computation) and that a simplex S can be shrunk to a point, one of its vertexes x_0, for instance. As S shrinks, the path of a chain can be thought of as the join of that chain to x_0, defined in a manner similar to that of Definition 1-31. This suggests constructing the homomorphisms h of the definition of algebraic homotopy as follows.

Let $\{y_0 y_1 \cdots y_p\}$ be a generator of the ordered chain group $\bar{\mathscr{C}}_p(S)$ of the simplex S, and define

$$h\{y_0 y_1 \cdots y_p\} = \{x_0 y_0 y_1 \cdots y_p\}$$

where x_0 is a vertex of S. Then h is extended by linearity to become a homomorphism of $\bar{\mathscr{C}}_p(S)$ into $\bar{\mathscr{C}}_{p+1}(S)$ for each p. It is easy to see that

$$d\{x_0 y_0 y_1 \cdots y_p\} = \{y_0 y_1 \cdots y_p\} - \sum (-1)^i \{x_0 y_0 y_1 \cdots \hat{y}_i \cdots y_p\}$$

If $p > 0$, the last set of terms is equal to $hd\{y_0 y_1 \cdots y_p\}$, so for any $\alpha \in \bar{\mathscr{C}}_p(S)$ with $p > 0$,

$$dh\alpha = \alpha - hd\alpha$$

That is,

$$\alpha = dh\alpha + hd\alpha \tag{3-7}$$

If $p = 0$,

$$d\{x_0 y_0\} = \{y_0\} - \{x_0\}$$

Now let $\alpha \in \bar{\mathscr{C}}_0(S)$, $\alpha = \sum a_i\{y_i\}$, where the y are vertexes of S and the a_i are in Z, and write $n(\alpha) = \sum a_i$. By the last equation

$$dh\alpha = \alpha - n(\alpha)\{x_0\}$$

This can be written as

$$\alpha - n(\alpha)\{x_0\} = dh\alpha + hd\alpha \tag{3-8}$$

for $d\alpha = 0$ if $\alpha \in \bar{\mathscr{C}}_0(S)$.

Equations (3-7) and (3-8) together state that the identity map of the complex $\bar{\mathscr{C}}(S)$ onto itself is algebraically homotopic to the map f defined by setting

$$f(\alpha) = 0 \qquad \text{if } \alpha \in \bar{\mathscr{C}}_p(S), p > 0$$

$$f(\alpha) = n(\alpha)\{x_0\} \qquad \text{if } \alpha \in \bar{\mathscr{C}}_0(S)$$

It is easy to compute that the homomorphism f_* induced by f on the homology groups $\bar{\mathscr{H}}_p(S)$ is as follows: f_* maps each $\bar{\mathscr{H}}_p(S)$ on 0 if $p > 0$, and f_* maps $\bar{\mathscr{H}}_0(S)$ isomorphically on Z. The algebraic homotopy theorem (Theorem 3-4) says, however, that f_* is equal to the homomorphism induced by the identity, that is, the identity homomorphism (Theorem 3-3). Thus $\bar{\mathscr{H}}_p(S) = 0$, for $p > 0$, and $\bar{\mathscr{H}}_0(S) = Z$, as asserted.

If E is now any simplicial complex, the oriented chain complex $\mathscr{C}(E)$ can be shown to be isomorphic to a subcomplex of the ordered chain complex $\bar{\mathscr{C}}(E)$, by a procedure similar to that employed at the beginning of this section for the singular complex.

Fix a standard order for the vertexes of E, and note that $\mathscr{C}_p(E)$ is the free Abelian group generated by the oriented simplexes $((x_0 x_1 \cdots x_p))$ with the x_i in standard order. Define the map f of $\mathscr{C}_p(E)$ into $\bar{\mathscr{C}}_p(E)$, for each p, by setting

$$f((x_0 x_1 \cdots x_p)) = \{x_0 x_1 \cdots x_p\}$$

where the x_i are in standard order, and then extending by linearity. Clearly, f is an isomorphism into for each p. It is also obvious that f commutes with d, so it is a chain homomorphism.

An algebraic homotopy inverse j will now be constructed for f. In the symbol $\{x_0 x_1 \cdots x_p\}$, if the x_i are not all distinct, set $j\{x_0 x_1 \cdots x_p\} = 0$. If the x are all different, set

$$j\{x_0 x_1 \cdots x_p\} = \pm((y_0 y_1 \cdots y_p))$$

where the y_i are the x_i arranged in standard order and the sign is + or − accordingly as the y_i form is an even or an odd permutation of the x_i. Then j is extended by linearity to a homomorphism of $\overline{\mathscr{C}}_p(E)$ into $\mathscr{C}_p(E)$ for each p, and it is easily proved to be a chain homomorphism.

If $x_0, x_1, \ldots, x_p$ are in standard order,

$$jf((x_0 x_1 \cdots x_p)) = j\{x_0 x_1 \cdots x_p\} = ((x_0 x_1 \cdots x_p))$$

Since the $((x_0 x_1 \cdots x_p))$ with the x_i in standard order generate $\mathscr{C}_p(E)$ for each p, it follows that jf is the identity.

On the other hand, if $x_0, x_1, \ldots, x_p$ is any sequence of vertexes of some simplex of E, then

$$fj\{x_0 x_1 \cdots x_p\} = 0$$

if any two of the x_i are equal; if the x_i are all different and $y_0, y_1, \ldots, y_p$ is the set of x_i in standard order,

$$fj\{x_0 x_1 \cdots x_p\} = \pm f((y_0 y_1 \cdots y_p)) = \pm\{y_0 y_1 \cdots y_p\}$$

where the sign is + or − accordingly as the y_i form an even or an odd permutation of the x_i.

Now fj will be proved to be algebraically homotopic to the identity map by constructing inductively the homomorphisms h of the definition of algebraic homotopy (Definition 3-14). Suppose that, for $p < r$, the homomorphisms

$$h_p \colon \overline{\mathscr{C}}_r(E) \to \overline{\mathscr{C}}_{p+1}(E)$$

have been constructed such that

$$fj - i = h_{p-1}\, d + dh_p \tag{3-9}$$

where i is the identity. Also assume that, for $p < r$, h_p maps $\bar{\mathscr{C}}_p(S)$ into $\bar{\mathscr{C}}_{p+1}(S)$ for each simplex S of E. Noting that $fj = i$ on $\bar{\mathscr{C}}_0(E)$, take $h_0 = 0$ to the induction. Now h_r must be defined. (Note the similarity of the following to the construction of H in Section 1-11.)

It is sufficient to define $h_r\{x_0 x_1 \cdots x_r\}$ for each generator of $\bar{\mathscr{C}}_r(E)$ as an element of $\bar{\mathscr{C}}_{r+1}(E)$ with the property that

$$dh_r\{x_0 x_1 \cdots x_r\} = fj\{x_0 x_1 \cdots x_r\} - \{x_0 x_1 \cdots x_r\} - h_{r-1}\, d\{x_0 x_1 \cdots x_r\} \tag{3-10}$$

In fact, if the induction is to be carried on, $h_r\{x_0 x_1 \cdots x_r\}$ must be found as an element of $\bar{\mathscr{C}}_{r+1}(S)$, where S is the simplex whose vertexes are the x_i. By the induction hypothesis $h_{r-1}\, d\{x_0 x_1 \cdots x_r\}$ is defined and is in $\bar{\mathscr{C}}_r(S)$. Thus, the whole right side of the last equation is in $\bar{\mathscr{C}}_r(S)$; it will now be shown to be a cycle.

$$\begin{aligned}
&d(fj\{x_0 x_1 \cdots x_r\} - \{x_0 x_1 \cdots x_r\} - h_{r-1}\, d\{x_0 x_1 \cdots x_r\}) \\
&\quad = dfj\{x_0 x_1 \cdots x_r\} - d\{x_0 x_1 \cdots x_r\} - dh_{r-1}\, d\{x_0 x_1 \cdots x_r\} \\
&\quad = dfj\{x_0 x_1 \cdots x_r\} - d\{x_0 x_1 \cdots x_r\} - fjd\{x_0 x_1 \cdots x_r\} + d\{x_0 x_1 \cdots x_r\} \\
&\qquad + h_{r-2}\, d^2\{x_0 x_1 \cdots x_r\}
\end{aligned}$$

where Eq. (3-9) of the induction hypothesis is used in the last step. Here $dfj = fjd$ and $d^2 = 0$, so everything cancels out and the right-hand side of (3-10) is a cycle in $\bar{\mathscr{C}}_r(S)$. It has already been shown that $\bar{\mathscr{H}}_r(S) = 0$ for $r > 0$, so for $r > 0$, the right-hand side of (3-10) is in the image of d. Hence, there is an element of $\bar{\mathscr{C}}_{r+1}(S)$ which will be written as $h_r\{x_0 x_1 \cdots x_r\}$, such that Eq. (3-10) holds. Now h_r is extended by linearity to a homomorphism of $\bar{\mathscr{C}}_r(E)$ into $\bar{\mathscr{C}}_{r+1}(E)$. Equation (3-9) is satisfied. Also, h_r carries $\bar{\mathscr{C}}_r(S)$ into $\bar{\mathscr{C}}_{r+1}(S)$ for each simplex S of E, so the inductive construction is complete.

Thus, it has been shown that jf is equal to the identity and fj is algebraically homotopic to the identity, so f and j are algebraically homotopy inverse, as was to be shown. It follows that the ordered chain complex and the oriented chain complex on E have the same homology groups and the same cohomology groups over any coefficient group. Note that the same argument would show that the ordered and oriented chain complexes over other coefficient groups have the same homology groups.

The type of argument used here can be adapted to a more general situation. It is evident that the essential point is that both fj and the identity map an element $\{x_0 x_1 \cdots x_p\}$ into elements carried by a simplex, and the essential feature of a simplex is that its homology groups (as defined by the ordered complex) are all zero in the positive dimensions. This situation can be generalized by introducing the notion of acyclic carrier functions.

Definition 3-18. Let E and F be simplicial complexes and let $f\colon \bar{\mathscr{C}}(E)—\bar{\mathscr{C}}(F)$ be a chain homomorphism. An *acyclic carrier function associated with f* is a map K, which assigns to each simplex S of E a subcomplex $K(S)$ of F such that:

(1) If S' is a face of S, $K(S')$ is a subcomplex of $K(S)$;
(2) $\bar{\mathscr{H}}_p[K(S)] = 0$ for all $p > 0$ and all simplexes S in E;
(3) $f\{x_0 x_1 \cdots x_p\}$ is in $\bar{\mathscr{C}}_p(K(S))$ when the x_i are vertexes of S.

In the preceding comparison of the oriented and ordered chain complexes, for example, the map fj of $\bar{\mathscr{C}}(E)$ into itself had an acyclic carrier function attached to it, namely, $K(S) = S$ for each simplex S in E.

Suppose now that g and f are two chain homomorphisms of $\bar{\mathscr{C}}(E)$ into $\bar{\mathscr{C}}(F)$ with the same acyclic carrier function, and suppose that $f = g$ on $\bar{\mathscr{C}}_0(E)$. Then the inductive construction of the h_p just used in comparing the complexes $\bar{\mathscr{C}}(E)$ and $\mathscr{C}(E)$ can be followed through more or less word for word, leading to the result that f and g are algebraically homotopic.

This discussion has been concerned with the ordered chain complexes but of course a similar situation arises in connection with other complexes. For example, acyclic carrier functions could be defined for chain maps between the oriented chain complexes of E and F. The definition would be as in Definition 3-18 with oriented chains and homology substituted for ordered chains and homology. In this case, two chain maps from $\mathscr{C}(E)$ to $\mathscr{C}(F)$, agreeing on the zero-dimensional group and having the same acyclic carrier function, would be algebraically homotopic.

This will now be illustrated by a comparison of the oriented simplicial homology groups of a complex E with those of its barycentric subdivision BE (defined in Exercise 2-4).

Since E and BE have the same underlying topological space they can be expected to have the same homology groups. This will now be proved, entirely in the context of oriented simplicial homology, by setting up a pair of maps between $\mathscr{C}(E)$ and $\mathscr{C}(BE)$ which are algebraic homotopy inverses to each other.

First define a chain homomorphism $B\colon \mathscr{C}(E) \to \mathscr{C}(BE)$ by the following inductive construction.

(1) $B\sigma = \sigma$ if $\sigma \in \mathscr{C}_0(E)$.
(2) Assume that B is already defined on $\mathscr{C}_{p-1}(E)$.
(3) If $((x_0 x_1 \cdots x_p))$ is an oriented simplex on E, set

$$B((x_0 x_1 \cdots x_p)) = \sum (-1)^i ((bx_0 x_1 \cdots \hat{x}_i \cdots x_p))$$

where b is the barycenter of the simplex $[x_0 x_1 \cdots x_p]$.

(4) If $\alpha = \sum a_i \sigma_i$, where the σ_i are oriented p simplexes on E and the a_i are in the coefficient group, set $B\alpha = \sum a_i B\sigma_i$.

Note that this construction parallels the definition of the operator B on the singular chain complex (cf. Section 1-11), the geometric motivation being the same in both cases. It is easy to check that B commutes with the boundary operator and so is, in fact, a chain homomorphism.

Now an algebraic homotopy inverse j to B will be constructed. Geometrically, this can be thought of as the homomorphism induced by a simplicial map of BE into E, which is homotopic to the identity (in the topological sense) and which is obtained by systematically displacing the barycenter of each simplex onto one of its vertexes. To make this process definite suppose that all the vertexes of E have been ordered in some way. Then note that, for any vertex x of BE, there is a unique smallest simplex S of E containing x and, in fact, x is the barycenter of S. Let y be the earliest vertex of S in the ordering of the vertexes of E. Then, if $x_0, x_1, \ldots, x_p$ are vertexes of a simplex of BE, a set of corresponding vertexes $y_0, y_1, \ldots, y_p$ of E is obtained, along with a set of corresponding simplexes $S_0, S_1, \ldots, S_p$ of E. If the x_i are suitably numbered it can be assumed (as in Exercise 2-5) that

$$S_0 \subset S_1 \subset \cdots \subset S_p$$

so the y_i are actually all vertexes of S_p. Thus, the symbol $((y_0 y_1 \cdots y_p))$ is defined: it is either 0, if the y_i are not all different, or it is an oriented simplex on E. Define

$$j((x_0 x_1 \cdots x_p)) = ((y_0 y_1 \cdots y_p))$$

and extend j by linearity to make it a homomorphism of $\mathscr{C}_p(BE)$ into $\mathscr{C}_p(E)$. From the definition of j, $dj((x_0 x_1 \cdots x_p)) = jd((x_0 x_1 \cdots x_p))$, so j is a chain homomorphism.

Now B and j will be shown to be algebraic homotopy inverses. Consider first the composition jB. Let σ be an oriented simplex carried by the simplex S of E. Then the vertexes of the simplexes of BE that carry the terms of the chain $B\sigma$ are all barycenters of S or its faces. This can be seen at once from a comparison of the definition of $B\sigma$ and that of BS (cf. Definition 1-33 and Exercise 2-4). Hence, by the definition of j, all the vertexes of simplexes carrying terms of the chain $jB\sigma$ are vertexes of S. Thus $jB\sigma$ is an oriented chain carried by S. A simplex has zero homology groups in the positive dimensions for the oriented simplicial homology theory, so the assignment of S to S is an acyclic carrier function corresponding to the chain homomorphism jB and also of course for the identity, jB is equal to the identity on 0 chains, so the above remarks imply that jB is algebraically homotopic to the identity.

Now consider Bj. Let σ be an oriented simplex of BE carried by a simplex S of BE and let S be the smallest simplex of E containing S_0. Then it is easy to see that $Bj\sigma$ is an oriented chain on S. S, or strictly speaking BS, is a subcomplex of BE with homology groups zero in the positive dimensions. This should be verified as an exercise, using of course oriented chains, cycles, and so on, and making no appeal to the singular homology theory. So the assignment of BS to σ is an acyclic carrier function for Bj and also for the identity. Hence, it again follows that Bj is algebraically homotopic to the identity, so B and j are algebraic homotopy inverses, as was to be shown.

Exercises. 3-1. Note that in the above discussion the assignment of y_i in E to x_i in BE induces a linear map $f: BE \to E$, which has the property that for any simplex S of BE there is a simplex of E containing both S and $f(S)$. f induces the homomorphisms j above. Prove that the map f is homotopic to the identity. [Define a homotopy by displacing x in BE along the segment joining x and $f(x)$.]

3-2. The preceding exercise can be generalized. Let E and F be simplicial complexes, and let f and g be simplicial maps of E into F. f and g are called *contiguous* if for every simplex S in E there is a simplex $C(S)$ in F containing both $f(S)$ and $g(S)$. Prove that when this condition holds, f and g are homotopic.

Contiguous maps induce the same homomorphisms on homology groups. Prove this result entirely within the framework of simplicial homology. In other words, set up an algebraic homotopy between the homomorphisms induced by f and g on simplicial chains.

Note that this idea can also be extended to simplicial maps f and g of the simplicial pair (E, E') into (F, F'). Here the condition of continuity is that for every simplex S in E there is a simplex $C(S)$ in F containing both $f(S)$ and $g(S)$ and if $S \subset E'$ then $C(S) \subset F'$. Show that two such maps are homotopic, as maps of pairs. Also show, entirely in terms of simplicial homology, that they induce the same homomorphisms on simplicial homology groups.

3-3. Here is another application of Theorem 3-6.

Let K be a simplicial complex with a cell decomposition. Let $\overline{\mathscr{C}}(K)$, $\mathscr{C}(K)$, and $C(K)$ denote the cellular chain complex, the simplicial chain complex, and the singular chain complex on K, respectively, all over Z. Then there are inclusion maps

$$\overline{\mathscr{C}}(K) \underset{i}{\to} \mathscr{C}(K) \underset{j}{\to} C(K)$$

Show that i and j and, consequently, ji have algebraic homotopy inverses.

3-8. SUBCOMPLEXES AND QUOTIENT COMPLEXES

In this section, some ideas implicit in foregoing discussions will be elaborated. Let K and L be chain complexes and suppose that for each dimension K_p is a subgroup of L_p; thus there is a set of inclusion maps $i: K_p \to L_p$.

Definition 3-19. If the i form a chain homomorphism of K into L then K will be called a *subcomplex of* L. Note that the boundary operator of K is simply that of L restricted to K.

Examples

3-13. The oriented simplicial chain complex on a simplicial complex E can be identified with a subcomplex of the ordered simplicial chain complex, which in turn can be identified with a subcomplex of the singular chain complex (cf. Section 3-7).

3-14. If a simplicial complex has a cell decomposition, the cellular chain complex is a subcomplex of the oriented simplicial chain complex.

Let K be a subcomplex of L. Then it is clear that for each p the boundary operator d of L induces a homomorphism $d\colon L_p/K_p \to L_{p-1}/K_{p-1}$. This homomorphism also satisfies $d^2 = 0$. Thus the groups L_p/K_p with d as boundary operator form a chain complex.

Definition 3-20. The chain complex just constructed is denoted by L/K and called the *quotient complex of L modulo K*. Note that the natural map $L_p \to L_p/K_p$ for each p constitutes a chain homomorphism.

Exercises. 3-4. Let K be a subcomplex of L. Imitate the proof of Theorem 1-9 to show that there is an exact homology sequence

$$\to H_p(K) \xrightarrow{i} H_p(L) \xrightarrow{j} H_p\left(\frac{L}{K}\right) \xrightarrow{\partial} H_{p-1}(K) \to$$

where i is induced by inclusion, j by the natural map on the quotient, and ∂ is constructed in a manner similar to that used to construct ∂ in Theorem 1-9. In addiyion, ∂ commutes with homomorphisms induced by chain homomorphisms.

3-5. Reformulate all of this section for cochain complexes.

3-6. Let K and L be chain complexes with K a subcomplex of L. By taking duals with respect to a coefficient group $\mathcal{G}$ and hence forming cohomology groups, construct an exact cohomology sequence

$$\to H^p\left(\frac{K}{L}\right) \xrightarrow{j} H^p(L) \xrightarrow{i} H^p(K) \xrightarrow{\partial} H^{p+1}\left(\frac{L}{K}\right)$$

where i and j are induced by inclusion and the natural map on the quotient, respectively. δ commutes with induced homomorphisms.

3-9. COMPUTATION OF COHOMOLOGY GROUPS

The computation of the cohomology groups of a simplicial complex presents the same sort of problems as the computation of homology groups, and as might be expected, the methods described in Sections 2-9 and 2-10 yield corresponding techniques here.

First suppose that K is a simplicial complex and denote by $\mathscr{C}(K)$ its oriented simplicial chain complex over Z as coefficient group. Also suppose that K is expressed as a cell complex and let $\bar{\mathscr{C}}(K)$ be its cellular chain complex. It has been seen that $\bar{\mathscr{C}}(K)$ is a subcomplex of $\mathscr{C}(K)$. Denoting the inclusion map by i, reformulate Theorem 2-9 as follows.

Theorem 3-8. *$i: \bar{\mathscr{C}}(K) \to \mathscr{C}(K)$ induces isomorphisms of the homology groups over z.*

The chain groups of both $\bar{\mathscr{C}}(K)$ and $\mathscr{C}(K)$ are free Abelian, so Theorem 3-6 can be applied to show that i has an algebraic homotopy inverse (cf. Exercise 3-3), and this in turn implies that it induces isomorphisms of the cohomology groups. The following theorem is a formal statement of this.

Theorem 3-9. *$i: \bar{\mathscr{C}}(K) \to \mathscr{C}(K)$ induces isomorphisms onto of the cohomology groups of these complexes over any coefficient group.*

This means that the cohomology groups of a triangulable space can be computed from a cell decomposition. In the following exercises the computation will be carried out for some of the spaces whose cell decompositions were studied in Section 2-9.

Exercises. 3-7. Let K be a sphere with p handles. Using integers as coefficient group, show that $H^0(K) = Z = H^2(K)$, $H^1(K)$ is a free Abelian group with $2p$ generators.

Make a similar computation where K is a nonorientable surface, that is, a sphere with k holes, on which the diametrically opposite points on the circumference of each hole are identified.

3-8. Let K be the n-dimensional projective space P^n. Use the cell decomposition of Example 2-9 to show that

$$H^0(P^n) = Z$$
$$H^r(P^n) = 0 \qquad \text{(if } r \text{ is odd)}$$
$$H^r(P^n) = Z_2 \qquad \text{(if } r \text{ is even and not 0)}$$

In the case of more complicated spaces the computation can be made systematic by dualizing the notion of canonical bases. Here, for simplicity, the coefficient group is taken as Z. Let K be a cell complex and let $\bar{\mathscr{C}}(K)$ be the corresponding algebraic complex of cellular chains. Let $\bar{\mathscr{C}}_p(K)$ be generated by $c_{p1}, c_{p2}, \ldots, c_{pk}$ and denote by c^{pi} the p cochain whose value is 1 on c_{pi} and 0 on $c_{pj} (j \neq i)$. Then if ϕ is any p cochain on $\bar{\mathscr{C}}(K)$, over Z, it is clear that ϕ and $\sum \phi(c_{pi})c^{pi}$ take the same values on all elements of $\bar{\mathscr{C}}_p(K)$; that is, they are equal. Also, since no linear combination of the c^{pi} is 0, $\bar{\mathscr{C}}^p(K)$ is a free Abelian group generated by $c^{p1}, c^{p2}, \ldots, c^{pk}$. This is then a dual set of generators to the set $c_{p1}, c_{p2}, \ldots, c_{pk}$. It is convenient to call c^{pi} the cocell associated with c_{pi}, or with the cell of K carrying c_{pi}. The coboundary operator can now be expressed in terms of cocells. In matrix notation, the action of the boundary operator is given by

$$dc_p = A^p c_{p-1}$$

where c_p is the single column matrix of c_{pi} and A^p is the incidence matrix introduced in Section 2-10. By definition of the coboundary operator

$$\begin{aligned} \delta c^{pi}(c_{p+1,j}) &= c^{pi}(dc_{p+1,j}) \\ &= c^{pi}\left(\sum a_{jh}^{p+1} c_{ph}\right) \\ &= \sum a_{jh}^{p+1} c^{pi}(c_{ph}) \\ &= a_{ji}^{p+1} \end{aligned}$$

It follows that

$$\delta c^{pi} = \sum a_{ji}^{p+1} c^{p+1,j} \tag{3-11}$$

Now write c^p for the matrix of one row whose elements are the c^{pi}. In matrix notation, (3-11) becomes

$$\delta c^p = c^{p+1} A^{p+1}$$

Thus the same incidence matrices that describe the boundary relations also describe the coboundary relations.

Clearly, the same remarks could be made for any set of independent generators of the $\bar{\mathscr{C}}_p(K)$, provided that the dual sets of generators are used for the $\bar{\mathscr{C}}_p(K)$. In particular, suppose that g_p denotes the single column matrix consisting of a canonical basis of $\bar{\mathscr{C}}_p(K)$ for each p, so that the boundary relations are given by

$$dg_p = D^p g_{p-1}$$

where

$$D^p = \left(\begin{array}{cccccc|c} & & & 0 & & & 0 \\ \hline 1 & & & & & & \\ & 1 & & & & & \\ & & \ddots & & & & \\ & & & 1 & & & 0 \\ & & & & t_1{}^p & & \\ & & & & & \ddots & \\ & & & & & t_{k^p(p)} & \end{array}\right)$$

as in Section 2-10. Let g^p denote the basis of $\mathscr{C}^p(K)$ dual to g_p, written as a one-rowed matrix. Then the coboundary relations take the form

$$\delta g^p = g^{p+1} D^{p+1} \tag{3-12}$$

Just as the structure of the homology groups was read from the boundary formulas along with the form of the D in Section 2-10, so here the structure of the cohomology groups can be read from the formulas (3-12). The details should be checked as an exercise. The result obtained is the following:

Theorem 3-10. *If $H_p(K) \cong B_p \oplus T_p$, where B_p is free and T_p is the p-dimensional torsion group (notation of Section 2-10) then*

$$H^p(K) \cong B_p \oplus T_{p-1}$$

It might seem from this that cohomology gives no more information about a space than homology. But it will be seen that additional structure can be defined on the cohomology groups, making them much more powerful tools than the homology groups.

3-10. ATTACHING CONES AND CELLS

Here, in a sequence of exercises, a technique will be introduced which is sometimes useful in the computation of homology and cohomology groups. The idea to be described is closely related to the notion of a cell complex, and is particularly useful when a sequence of spaces is constructed, each by attaching a cell to the previous one.

Definition 3-21. Let E be a topological space. *The cone over E*, denoted by CE, is the space obtained from $E \times I$ by identifying all points of $E \times \{1\}$ with each other. More precisely, an equivalence relation R is defined on $E \times I$ by making (x, t) equivalent only to itself if $t \neq 1$ and $(x, 1)$ equivalent to $(y, 1)$ for all x and y. CE is the quotient space of $E \times I$ by R.

Examples

3-15. If $E = S^1$, CE is a cone in the ordinary sense. Note that here CS^1 is the 2 cell E^2. More generally, $CS^{n-1} = E^n$.

Note that the given space E can be identified with the subspace $E \times \{0\}$ of CE.

Definition 3-22. Let E and F be topological spaces and let $f: E \to F$ be a continuous map. Then the space formed by *attaching CE to F with the* *attaching map* f is the space formed from the union of F and CE by identifying each point x in $E \subset CE$ with $f(x)$ in F.

Note that in the simplest case E would be mapped by f homeomorphically into F and CE could be thought of as a cone with base E, so the attaching process simply means sticking the base E of CE onto the homeomorphic set $f(E)$. Also, since a cell is a cone over a sphere, any simplicial or cell complex can be thought of as obtained by a sequence of operations of attaching cones. In general, this would give no useful information, but in each of the following exercises the attaching map has a specially simple form, so information on the homology and cohomology groups can be deduced.

Exercises. 3-9. Let CE be attached to F with the attaching map f, as in Definition 3-22, and let the resulting space be F'. Show that there is an isomorphism between $H_p(CE, E)$ and $H_p(F', F)$ for each p (use homotopy and excision). In addition, show that this isomorphism makes the following diagram commutative.

$$\begin{array}{ccc} H_p(CE,E) & \xrightarrow{\cong} & H_p(F',F) \\ \downarrow \partial & & \downarrow \partial \\ H_{p-1}(E) & \xrightarrow[f_*]{} & H_{p-1}(F) \end{array}$$

3-10. A complex n-dimensional projective space PC^n is defined as follows. Take the complex affine space C^{n+1} of dimension $n+1$ from which the origin has been removed and define the equivalence relation $\sim$ by making $(z_0', z_1', \ldots, z_n') \sim (z_0, z_1, \ldots, z_n)$ if and only if there is a complex number $\lambda \neq 0$ such that $z_i' = \lambda z_i$ for each i. The quotient space of C^{n+1} with respect to this relation is PC^n. Clearly, it would be enough to take, instead of the whole of C^{n+1} minus the origin, the subspace for which $\sum z_i \bar{z}_i = 1$, which is a $(2n+1)$ sphere. Then in the equivalence relation λ would satisfy the equation $\lambda\bar{\lambda} = 1$. Thus there is a continuous map f^n: $S^{2n+1} \to PC^n$ such that the inverse image of each point is a circle. Show that PC^{n+1} is obtained from PC^n by attaching the cell $E^{2n+2} = CS^{2n+1}$ with the attaching map f^n. Hence compute the homology and cohomology groups of the complex projective spaces.

3-11. The suspension operation is essentially the operation of attaching two identical cones to each other. Let E be a space and let $E_1 = CE$ be the cone over E. Let E_2 be a second copy of CE and attach it to E_1, using as attaching map the inclusion map $E \subset E_1$. The result is called the suspension of E and is denoted by SE.

Note, for example, that S^n is the suspension of S^{n-1}.

Prove that the inclusion map of E into SE induces an isomorphism of $H_{p-1}(E)$ onto $H_p(SE)$ for $p > 1$.

3-12. Apply the last exercise to compute the homology groups of the spheres.

4

The Cohomology Ring

In this chapter a new topological invariant is introduced, the cohomology ring of a space. The addition is that of cohomology classes, while the idea introduced here is that of multiplication of cohomology classes by the so-called cup product.

4-1. MOTIVATION

In a sense, cochains are generalized functions on a space with values in the coefficient group $\mathscr{G}$. In fact, in singular cohomology, a singular 0 simplex can be identified with a point in the space, so a 0 cochain can be identified with a $\mathscr{G}$-valued function on the space. The cochains are added by summing functional values, and this additive structure then induces the additive structure of the cohomology groups. Now suppose that $\mathscr{G}$ is a ring. In order to multiply $\mathscr{G}$-valued functions on a space their functional values must be multiplied. It is natural to ask whether this multiplication can be extended to higher dimensional cochains and, more important, whether this extension can be made in such a way as to induce a multiplicative structure on the cohomology groups. This extension is made in this chapter, and the collection of cohomology groups, all dimensions taken together, can be made in this way into a ring. This ring is important as a more powerful topological invariant than the cohomology groups; that is, it is possible for two spaces to have the same cohomology groups (taking into account the additive structure only) but to have different cohomology rings. Examples of this situation are given in Section 4-16.

In the body of this chapter singular cohomology will be used. To suggest the formula for defining the product of cochains, however, consider the ordered simplicial cohomology of a simplicial complex (cf. Definition 3-17). Remember that if E is a simplicial complex, the ordered chain complex over Z associated with E is the chain complex whose p-dimensional chain group is the free Abelian group generated by symbols $\{x_0 x_1 \cdots x_p\}$, where the x_i are all vertexes of a simplex in E. The corresponding cochain complex over $\mathscr{G}$ is the ordered cochain complex over $\mathscr{G}$. A 0 cochain in this complex is simply a function on the set of vertexes of E with values in $\mathscr{G}$. Assume now that $\mathscr{G}$ is a ring. Then, if f and g are 0 cochains on E, their product is the 0 cochain (i.e., function on the vertexes) defined by

$$(fg)\{x\} = f\{x\}g\{x\}$$

where the multiplication on the right is ring multiplication in $\mathscr{G}$.

The coboundary of fg is an ordered 1 cochain on E. If $\{xy\}$ is an ordered simplex on E (here x and y are the end points of a 1 simplex of E) then

$$\begin{aligned} \delta(fg)\{xy\} &= (fg)\{y\} - (fg)\{x\} \\ &= f\{y\}g\{y\} - f\{x\}g\{x\} \end{aligned}$$

If this formula is to make sense in terms of cohomology, $\delta(fg)$ should be expressed in terms of f, g, and their coboundaries. With this in view, rewrite the last formula as

$$\begin{aligned} \delta(fg)\{xy\} &= (f\{y\} - f\{x\})g\{y\} + f\{x\}(g\{y\} - g\{x\}) \\ &= \delta f\{xy\}g\{y\} + f\{x\}\delta g\{xy\} \end{aligned} \tag{4-1}$$

So that both of the terms on the right can be thought of as 1 cochains operating on $\{xy\}$, think of the first term as the product of the 1 cochain δf and the 0 cochain g to yield a 1 cochain operating on $\{xy\}$. Similarly, think of the second term as the product of f and δg to yield a 1 cochain operating on $\{xy\}$. For historical reasons, the symbol for this product operation is written as $\cup$, called "cup." The formula suggested here is

$$(\delta f \cup g)\{xy\} = \delta f\{xy\}g\{y\} \tag{4-2}$$

Similarly

$$(f \cup \delta g)\{xy\} = f\{x\}\delta g\{xy\} \tag{4-3}$$

With these notations (4-1) becomes

$$\delta(fg)\{xy\} = (\delta f \cup g)\{xy\} + (f \cup \delta g)\{xy\}$$

or, equating the cochains on the left and right

$$\delta(fg) = \delta f \cup g + f \cup \delta g$$

Formulas (4-2) and (4-3) suggest that, if f and g are p and q cochains respectively, then $f \cup g$ should be a $(p+q)$ cochain defined by the formula

$$(f \cup g)\{x_0 x_1 \cdots x_{p+q}\} = f\{x_0 x_1 \cdots x_p\}g\{x_p x_{p+1} \cdots x_{p+q}\} \tag{4-4}$$

Exercises. **4-1.** Using the defining formula (4-4), prove the associative law

$$(f \cup g) \cup h = f \cup (g \cup h)$$

and the distributive laws

$$f \cup (g + h) = f \cup g + f \cup h$$

$$(f + g) \cup h = f \cup h + g \cup h$$

Also, if e is the 0 cochain that assigns the value 1 to each vertex of E, show that

$$e \cup f = f \cup e = f$$

for any cochain f on E.

4-2. Let f and g be cochains on E of dimensions p and q, respectively. Using the defining formula (4-4), prove that

$$\delta(f \cup g) = \delta f \cup g + (-1)^p f \cup \delta g \tag{4-5}$$

(To do this compute the values of both sides on $\{x_0 x_1 \cdots x_{p+q+1}\}$; a similar computation appears in Theorem 4-3.)

The $\cup$ product just defined is the product of a p cochain with a q cochain which gives a $(p+q)$ cochain. The question now arising is whether this product has any significance in terms of cohomology classes. If f and g are both cocycles, the formula (4-5) in Exercise 4-2 shows that $f \cup g$ is a cocycle. On the other hand, suppose that f is a cocycle and that $g = \delta h$ is a coboundary in which f and h have dimensions p and $q - 1$. Then

$$\begin{aligned}\delta(f \cup h) &= \delta f \cup h + (-1)^p f \cup \delta h \\ &= (-1)^p f \cup g\end{aligned}$$

Thus $f \cup g$ is a coboundary. Similarly, if f is a coboundary and g is a cocycle, then $f \cup g$ is a coboundary. Suppose now that f, f', g, g' are cocycles such that $f - f'$ and $g - g'$ are coboundaries. Then

$$\begin{aligned} f \cup g - f' \cup g' &= f \cup g - f' \cup g + f' \cup g - f' \cup g' \\ &= (f - f') \cup g + f' \cup (g - g') \qquad \text{(by Exercise 4-1)} \end{aligned}$$

Here $f - f'$ and $g - g'$ are coboundaries and g and f' are cocycles, so the remarks just made show that the right-hand side is a coboundary. In other words, $f \cup g$ and $f' \cup g'$ are cohomologous. It follows that if $\bar{f}$ and $\bar{g}$ are in $\mathscr{H}^p(E)$ and $\mathscr{H}^q(E)$, respectively (this is ordered simplicial cohomology), and if f and g are cocycles representing f and g, respectively, then $\bar{f} \cup \bar{g}$ can be defined as the element of $\mathscr{H}^{p+q}(E)$ represented by $f \cup g$; this definition is independent of the representatives f, g chosen for $\bar{f}$ and $\bar{g}$. Thus $\cup$ becomes a product operation on the direct sum $\sum \mathscr{H}^p(E)$. Exercise 4-1 shows that this product operation is associative and satisfies the distributive laws, and the cohomology class of e acts as unit element. It follows that $\sum \mathscr{H}^p(E)$ becomes a ring, namely, the ordered simplicial cohomology ring of E.

The same kind of discussion can be carried out for oriented simplicial cohomology. The appropriate chain groups are now generated by symbols $((x_0 x_1 \cdots x_p))$ where the x_i are all different, are the vertexes of a simplex of E, and are understood to be arranged in some fixed standard order for all the vertexes of E. In this way the oriented simplicial cohomology ring is E is obtained.

4-2. $\cup$ PRODUCT FOR SINGULAR COHOMOLOGY

The construction of the singular cohomology ring of a space will now be discussed. Of course, Section 4-1 is of some independent interest, since it shows how a cohomology ring can be introduced on a simplicial complex by using only its simplicial structure. Even for simplicial complexes, defining a ring structure for singular cohomology has certain advantages. In the first place, this makes clear that the ring structure is topologically invariant; in the second place, considerable freedom and flexibility in the computation of the products is gained (as seen in Sections 4-4 and 4-16).

So let E be a topological space. The notations of Chapters 1 and 3 will be used here for the singular chain and cochain complexes of E. In particular, the coefficient group $\mathscr{G}$ will now be a ring. The first step is to adapt the formula (4-4) to obtain a definition of the product of a singular p cochain with a singular q cochain which gives a singular $(p + q)$ cochain. As usual, let $\Delta_p, \Delta_q, \Delta_{p+q}$ denote the standard Euclidean simplexes of dimensions $p, q, (p + q)$. The vertexes of Δ_{p+q} are denoted by $x_0, x_1, \ldots, x_{p+q}$. Define F_p and L_q as linear maps

$$F_p\colon \Delta_p \to \Delta_{p+q}$$
$$L\colon \Delta_q \to \Delta_{p+q}$$

where F_p maps the vertexes of Δ_p (in standard order) on the vertexes x_0, $x_1, \ldots, x_p$ of Δ_{p+q} (F_p is inclusion) and L_q maps the vertexes of Δ_q (in standard order) on the vertexes $x_p, x_{p+1}, \ldots, x_{p+q}$ of Δ_{p+q}. Thus F_p maps Δ_p on the simplex spanned by the first $p+1$ vertexes of Δ_{p+q}, L_q maps Δ_q on that spanned by the last $q+1$ vertexes; F and L are supposed to stand for first and last in this notation. Now let $f \in C^p(E)$ and $g \in C^q(E)$. Define a singular cochain $f \cup g \in C^{p+q}(E)$ by setting

$$(f \cup g)(\sigma) = f(\sigma F_p) g(\sigma L_q) \tag{4-6}$$

for any singular $(p+q)$ simplex σ on E. The product on the right is ring multiplication in $\mathcal{G}$. Note that, in particular, if E is the space Δ_{p+q} itself and σ is the identity map, then σF_p and σL_q can be identified with the ordered simplexes $\{x_0 x_1 \cdots x_p\}$ and $\{x_p x_{p+1} \cdots x_{p+q}\}$, respectively, and σ can be identified with the ordered simplex $\{x_0 x_1 \cdots x_{p+q}\}$, so that the formula (4-6) reduces to (4-4).

Definition 4-1. The product $\cup$ defined by (4-6) on the direct sum $\sum C^p(E)$ is called the *cup product on singular cochains.*

The following theorem describes the algebraic properties of this $\cup$ product (cf. Exercise 4-1).

Theorem 4-1. *The associative and distributive laws are satisfied by* $\cup$:

$$f \cup (g \cup h) = (f \cup g) \cup h$$

$$f \cup (g + h) = f \cup g + f \cup h$$

$$(f + g) \cup h = f \cup h + g \cup h$$

Also, if e is the zero cochain that takes the value 1 *on each singular zero simplex, then*

$$e \cup f = f \cup e = f$$

for any cochain f.

Proof. Suppose that f, g, h are singular cochains of E of dimensions p, q, r, respectively. Let F_p be the linear map of Δ_p on the face $[x_0 x_1 \cdots x_p]$ of Δ_{p+q+r} and let L_{q+r} be the linear map of Δ_{q+r} on the face $[x_p x_{p+1} \cdots x_{p+q+r}]$, both maps being defined with all the vertexes in standard order. In addition, let F'_q, L'_r be the linear maps taking Δ_q and Δ_r, respectively, on the faces

$[x_0 x_1 \cdots x_q]$, $[x_q x_{q+1} \cdots x_{q+r}]$ of Δ_{q+r}. By definition, if σ is a singular $(p+q+r)$ simplex on E,

$$[f \cup (g \cup h)](\sigma) = f(\sigma F_p)(g \cup h)(\sigma L_{q+r}) \tag{4-7}$$

and

$$(g \cup h)(\sigma L_{q+r}) = g(\sigma L_{q+r} F'_q) h(\sigma L_{q+r} L'_r) \tag{4-8}$$

Now write $F_p = \phi_1, L_{q+r}F'_q = \phi_2, L_{q+r}L'_r = \phi_2$. These are respectively linear maps of Δ_p, Δ_q, and Δ_r on the faces $[x_0 x_1 \cdots x_p]$, $[x_p x_{p+1} \cdots x_{p+q}]$ and $[x_{p+q} x_{p+q+1} \cdots x_{p+q+r}]$ of Δ_{p+q+r} (remember that the composition of linear maps is linear). Hence (4-7) and (4-8) put together become

$$[f \cup (g \cup h)](\sigma) = f(\sigma\phi_1)g(\sigma\phi_2)h(\sigma\phi_3)$$

A similar argument shows that

$$[(f \cup g) \cup h](\sigma) = f(\sigma\phi_1)g(\sigma\phi_2)h(\sigma\phi_3)$$

so $f \cup (g \cup h) = (f \cup g) \cup h$, as required.

The distributive laws are simpler to prove. First, let f be a p cochain on E and let g and h be q cochains. Let F_p and L_q map Δ_p and Δ_q linearly on the faces $[x_0 x_1 \cdots x_p]$, $[x_p x_{p+1} \cdots x_{p+q}]$ of Δ_{p+q}. Let ϕ be a $(p+q)$-dimensional singular simplex on E. Then, by definition

$$\begin{aligned} [f \cup (g + h)](\sigma) &= f(\sigma F_p)(g + h)(\sigma L_q) \\ &= f(\sigma F_p)[g(\sigma L_q) + h(\sigma L_q)] \\ &= f(\sigma F_p)g(\sigma L_q) + f(\sigma F_p)h(\sigma L_q) \\ &= (f \cup g)(\sigma) + (f \cup h)(\sigma) \end{aligned}$$

Hence $f \cup (g + h) = f \cup g + f \cup h$.

A similar proof establishes the other distributive law.

The proof of the last part of the theorem is a trivial exercise. ∎

The last theorem means that the ∪ product makes the direct sum $\sum C^p(E)$ into a ring. Now, if $\phi: E \to F$ is a continuous map, it is known to induce a homomorphism ϕ^1 of the cochain groups. This is in fact a ring homomorphism.

Theorem 4-2. *Let $\phi: E \to F$ be a continuous map and let*

$$\phi^1: C^p(F) \to C^p(E)$$

for each p be the induced homomorphism on the cochain groups. Then, for any cochains g and h on F

$$\phi^1(g \cup h) = \phi^1(g) \cup \phi^1(h)$$

Proof. Suppose that g and h are of dimensions p and q and let F_p and L_q be as before. Also, let ϕ_1 be the homomorphism induced by ϕ on the singular chain groups. Then if σ is a $(p+q)$-dimensional singular simplex on E

$$\begin{aligned}
\phi^1(g \cup h)(\sigma) &= (g \cup h)[\phi_1(\sigma)] \qquad \text{(by duality of } \phi_1 \text{ and } \phi^1)\\
&= (g \cup h)(\phi\sigma) \qquad \text{(Definition 1-9)}\\
&= g(\phi\sigma F_p)h(\phi\sigma L_q) \qquad \text{[by (4-6)]}\\
&= \phi^1(g)(\sigma F_p)\phi^1(h)(\sigma L_q)\\
&= [\phi^1(g) \cup \phi^1(h)](\sigma)
\end{aligned}$$

∎

The next step is to prove a coboundary formula analogous to (4-5).

Theorem 4-3. *Let f and g be cochains in E of dimensions p and q, respectively. Then*

$$\delta(f \cup g) = \delta f \cup g + (-1)^p f \cup \delta g$$

Proof. Let σ be a singular simplex on E of dimension $p + q + 1$. Then

$$\begin{aligned}
\delta(f \cup g)(\sigma) &= (f \cup g)(d\sigma)\\
&= (f \cup g)(\sigma_1\, d(x_0 x_1 \cdots x_{p+q+1})) \qquad \text{(Definition 1-12)}\\
&= \sigma^1(f \cup g)(d(x_0 x_1 \cdots x_{p+q+1}))\\
&= (\sigma^1(f) \cup \sigma^1(g))(d(x_0 x_1 \cdots x_{p+q+1})) \qquad \text{(Theorem 4-2)}\\
&= (\sigma^1(f) \cup \sigma^1(g))\left(\sum (-1)^i (x_0 x_1 \cdots \hat{x}_i \cdots x_{p+q+1})\right)
\end{aligned}$$

Now (4-6) is applied to each term $(x_0 x_1 \cdots \hat{x}_i \cdots x_{p+q+1})$ of the right-hand side. Note that this is already a linear map of Δ_{p+q} into Δ_{p+q+1}, so it can be composed with the F_p and L_q of (4-6) to give linear maps, namely,

$$\begin{aligned}
(x_0 x_1 \cdots \hat{x}_i \cdots x_{p+q+1})F_p &= (x_0 x_1 \cdots \hat{x}_i \cdots x_{p+1}) \qquad \text{if } i < p+1\\
&\text{or } (x_0 x_1 \cdots x_p) \qquad \text{if } i \geqq p+1
\end{aligned}$$

and

$$\begin{aligned}
(x_0 x_1 \cdots \hat{x}_i \cdots x_{p+q+1})L_q &= (x_{p+1} x_{p+2} \cdots x_{p+q+1}) \qquad \text{if } i < p+1\\
&\text{or } (x_p x_{p+1} \cdots \hat{x}_i \cdots x_{p+q+1}) \qquad \text{if } i \geqq p+1
\end{aligned}$$

Hence

$$\delta(f \cup g)(\sigma) = \sum_{i=0}^{p} (-1)^i \sigma^1(f)(x_0 x_1 \cdots \hat{x}_i \cdots x_{p+1}) \sigma^1(g)(x_{p+1} \cdots x_{p+q+1})$$
$$+ \sum_{i=p+1}^{p+q+1} (-1)^i \sigma^1(f)(x_0 x_1 \cdots x_p) \sigma^1(g)(x_p x_{p+1} \cdots \hat{x}_i \cdots x_{p+q+1})$$
$$= \sum_{i=0}^{p+1} (-1)^i \sigma^1(f)(x_0 x_1 \cdots \hat{x}_i \cdots x_{p+1}) \sigma^1(g)(x_{p+1} \cdots x_{p+q+1})$$
$$+ \sum_{p}^{p+q+1} (-1)^i \sigma^1(f)(x_0 x_1 \cdots x_p) \sigma^1(g)(x_p \cdots \hat{x}_i \cdots x_{p+q+1})$$

(here the terms inserted in the two sums are equal but with opposite signs)

$$= \sigma^1(f)[d(x_0 x_1 \cdots x_{p+1})] \sigma^1(g)(x_{p+1} \cdots x_{p+q+1})$$
$$+ (-1)^p \sigma^1(f)(x_0 x_1 \cdots x_p) \sigma^1(g)[d(x_p \cdots x_{p+q+1})]$$
$$= f[\sigma_1 d(x_0 x_1 \cdots x_{p+1})] g[\sigma_1(x_{p+1} \cdots x_{p+q+1})]$$
$$+ (-1)^p f[\sigma_1(x_0 x_1 \cdots x_p)] g[\sigma_1 d(x_p \cdots x_{p+q+1})]$$
$$= f[d\sigma_1(x_0 x_1 \cdots x_{p+1})] g[\sigma_1(x_{p+1} \cdots x_{p+q+1})]$$
$$+ (-1)^p f[\sigma_1(x_0 x_1 \cdots x_p)] g[d\sigma_1(x_p \cdots x_{p+q+1})]$$

(Theorem 1-2)

$$= \delta f[\sigma_1(x_0 x_1 \cdots x_{p+1})] g[\sigma_1(x_{p+1} \cdots x_{p+q+1})]$$
$$+ (-1)^p f[\sigma_1(x_0 x_1 \cdots x_p)] \delta g[\sigma_1(x_p \cdots x_{p+q+1})]$$

(Definition of δ)

$$= \delta f[\sigma(x_0 x_1 \cdots x_{p+q})] g[\sigma(x_{p+1} \cdots x_{p+q+1})]$$
$$+ (-1)^p f[\sigma(x_0 x_1 \cdots x_p)] \delta g[\sigma(x_p \cdots x_{p+q+1})]$$

(Definition 1-9)

$$= (\delta f \cup g)(\sigma) + (-1)^p (f \cup \delta g)(\sigma)$$

In the last step (4-6) was used, replacing F_p and L_q by $(x_0 x_1 \cdots x_{p+1})$ and $(x_{p+1} \cdots x_{p+q+1})$ in the first term and by $(x_0 x_1 \cdots x_p)$ and $(x_p x_{p+1} \cdots x_{p+q+1})$ in the second. It follows then that

$$\delta(f \cup g) = \delta f \cup g + (-1)^p f \cup \delta g$$

as was to be shown. ∎

4-3. DEFINITION OF THE SINGULAR COHOMOLOGY RING

The cup product so far defined on singular cochains will now be shown to induce a product of cohomology classes. The discussion will be similar to the discussion of ordered cohomology in Section 4-1.

Lemma 4-4. (1) *If f and g are singular cocycles in E, so is $f \cup g$.*

(2) *If f is a cocycle and g a coboundary on E then $f \cup g$ is a coboundary. If f is a coboundary and g is a cocycle then $f \cup g$ is a coboundary.*

Proof. The argument is as in Section 4-1. ∎

Again as in Section 4-1, if f, f', g, g' are cocycles on E such that f and f' are cohomologous and g and g' are cohomologous, it follows that $f \cup g$ is cohomologous to $f' \cup g'$.

Definition 4-2. Let $\bar{f} \in H^p(E)$, $\bar{g} \in H^q(e)$ and let f and g be represented by cocycles f and g respectively. Define $\bar{f} \cup \bar{g}$ to be the cohomology class in $H^{p+q}(E)$ represented by $f \cup g$.

The remark just made shows that $\bar{f} \cup \bar{g}$ depends only on the cohomology classes $\bar{f}$ and $\bar{g}$ and not on the choice of representative cocycles.

Theorem 4-5. *Let $H^*(E)$ be the direct sum $\sum H^p(E)$. Then the $\cup$ product of Definition 4-2 makes $H^*(E)$ into a ring. The identity element is represented by the cocycle taking the value* 1 *on each zero-dimensional singular simplex.*

Proof. Since the product on cohomology classes is defined by using representative cocycles, this follows at once from Theorem 4-1. ∎

Definition 4-3. The ring $H^*(E)$ so constructed is called *the singular cohomology ring of E.*

Theorem 4-6. *Let E and F be topological spaces and $f: E \to F$ a continuous map. Let $f^*: H^p(E) \to H^p(E)$, for each p, be the induced homomorphism on cohomology groups. Then f^* is a ring homomorphism of $H^*(F)$ into $H^*(E)$.*

Proof. This follows at once from Theorem 4-2. ∎

4-4. SOME EXAMPLES OF COMPUTATIONS

If E is a simplicial complex then, in principle at least, the cohomology groups can be computed as described in Section 3-8, and again in principle, it should be possible to compute the cohomology ring since only a finite number of generating cocycles and their products need be considered. In general, however, the computation is hopelessly complicated, and special devices usually have to be used to make it at all manageable. One of the difficulties is that, while the calculation of cohomology groups may be simplified by using cell complexes, the definition itself of the $\cup$ product is closely tied to the use of simplexes. There is a situation, however, that sometimes provides the best of both worlds, namely, that in which the cells of a cell decomposition are actually simplexes (without forming a simplicial decomposition).

Suppose then that E is a simplicial complex with a cell decomposition that satisfies the following conditions.

(1) Let $X - Y$ be a p cell of E. Here the notation of Definition 2-18 is used; X is a subcomplex of E and Y a subcomplex consisting of cells of dimension less than p. Then corresponding to $X - Y$ there is a continuous map $\sigma: (\Delta_p, \dot{\Delta}_p) \to (X, Y)$ that induces isomorphisms of the homology groups. (In practice, σ usually induces a homeomorphism of $\Delta_p - \dot{\Delta}_p$ onto $X - Y$.)

Note that in this condition σ is in fact a singular simplex on E and is a relative cycle of X modulo Y representing a generator of $H_p(X, Y)$. Now let c be an oriented cell associated with $X - Y$ and write $i(c) = \sigma$. Here c and σ are generators of the cell chain group $\mathscr{C}_p(E)$ and the singular chain group $C_p(E)$, respectively, so i becomes a homomorphism

$$i: \overline{\mathscr{C}}_p(E) \to C_p(E)$$

This homomorphism has already been studied in Exercise 3-3. It is a chain homomorphism that identifies the cellular chain complex $\overline{\mathscr{C}}(E)$ with a subcomplex of the singular chain complex $C(E)$, and it induces isomorphisms of the corresponding homology and cohomology groups.

The second condition can now be stated:

(2) Let $C = X - Y$ be, as before, a $(p + q)$ cell of E, c an associated oriented cell, and $\sigma = i(c)$ as in (1). Let F_p and L_q be the linear maps taking the standard simplexes Δ_p and Δ_q onto the faces $[x_0 x_1 \cdots x_p]$ and $[x_p x_{p+1} \cdots x_{p+q}]$ of Δ_{p+q}. Then there are cells C_1 and C_2 contained in Y (we can think of them as faces of C) which carry oriented cells c_1 and c_2 such that

$$i(c_1) = \sigma_1 = \sigma F_p \qquad i(c_2) = \sigma_2 = \sigma L_q$$

As pointed out, i induces an isomorphism i^* of the cohomology groups. The next step is to define a product on $\overline{\mathscr{C}}(E)$ inducing a product on cohomology classes, thereby making i^* into a ring isomorphism. Let f and g be cochains on $\overline{\mathscr{C}}(E)$ of dimensions p and q, respectively, and let c be a $(p + q)$-dimensional oriented cell on E, a generator of $\overline{\mathscr{C}}_{p+q}(E)$. Let c_1 and c_2 be the oriented cells appearing in condition (2) that correspond to the cell c and to the expression of $p + q$ as the sum of p and q. Define $f \cup g$ by the formula

$$(f \cup g)(c) = f(c_1)g(c_2) \tag{4-9}$$

Suppose, on the other hand, that f and g are cochains on $C(E)$ and let $i': C^p(E) \to \overline{\mathscr{C}}^p(E)$, for each p, be the dual of i. Then, for an oriented $(p + q)$ cell c of E,

$$\begin{aligned} i'(f \cup g)(c) &= (f \cup g)i(c) = (f \cup g)(\sigma) && \text{(in notation of Condition (1))} \\ &= f(\sigma F_p)g(\sigma L_q) && \text{(Definition 4-1)} \\ &= fi(c_1)gi(c_2) && \text{(Condition (2))} \\ &= i'f(c_1)i'g(c_2) \\ &= (i'f \cup i'g)(c) && \text{[by Eq. (4-9)]} \end{aligned}$$

Hence

$$i'(f \cup g) = i'f \cup i'g \tag{4-10}$$

for any cochains f and g on $C(E)$.

This equation shows that i' is a ring homomorphism, provided that it is determined that the product defined by (4-9) makes $\sum \overline{\mathscr{C}}^p(E)$ into a ring. To check this, note first that i' is onto; that is, if f is a p cochain on $\overline{\mathscr{C}}(E)$ define f_1 as a p cochain on $C(E)$ by writing $f_1(\sigma) = f(c)$ if $\sigma = i(c)$ for some cell of E and make $f_1(\sigma) = 0$ if σ is a singular simplex not in the image of i. Then $f = i'f_1$, meaning that i' is onto. Now the associative law can be verified for the product (4-9). Let f, g, and h be cochains on $\overline{\mathscr{C}}(E)$. Find cochains f_1, g_1, h_1, on $C(E)$ such that $f = i'f_1$, $g = i'g_1$, $h = i'h_1$. Then

$$\begin{aligned} f \cup (g \cup h) &= i'f_1 \cup (i'g_1 \cup i'h_1) \\ &= i'[f_1 \cup (g_1 \cup h_1)) && \text{[two applications of (4-10)]} \\ &= i'[(f_1 \cup g_1) \cup h_1] && \text{(Theorem 4-1)} \\ &= i'f_1 \cup i'g_1 \cup i'h_1 && \text{(by 4-10)} \\ &= (f \cup g) \cup h \end{aligned}$$

Similarly, the distributive laws can be checked. Thus the product (4-9) makes $\sum \overline{\mathscr{C}}^p(E)$ into a ring, and i' is a ring homomorphism.

Again let f and g be cochains on $\overline{\mathscr{C}}(E)$ of dimensions p and q, and suppose that $f = i'f_1$, $g = i'g_1$. Then

$$\begin{aligned} \delta(f \cup g) &= \delta(i'f_1 \cup i'g_1) \\ &= \delta i'(f_1 \cup g_1) && \text{[by (4-10)]} \\ &= i'\delta(f_1 \cup g_1) && \text{(i' is a cochain homomorphism)} \\ &= i'[\delta f_1 \cup g_1 + (-1)^p f_1 \cup \delta g_1] \\ &= \delta f \cup g + (-1)^p f \cup \delta g \end{aligned}$$

This coboundary formula is similar to the formula for the coboundary of the cup product of singular cochains (Theorem 4-3), so conclusions similar to those of Theorem 4-5 can be drawn; that is, the $\cup$ product on $\sum \overline{\mathscr{C}}^p(E)$ that is defined by (4-9) induces a product of cohomology classes (in the cohomology of the cochain complex $\bar{\mathscr{C}}^*(E)$). This makes the direct sum $\sum \bar{\mathscr{H}}^p(E) = \mathscr{H}^*(E)$ into a ring. Then, by (4-10), the induced homomorphism $i^*: H^*(E) \to \mathscr{H}^*(E)$ becomes a ring homomorphism and, in fact, a ring isomorphism.

All this means that the singular cohomology ring of E can be computed by working only with the cellular cochains on E, defining the product by (4-9). In using this result it is sometimes convenient simply to identify an oriented cell c with the corresponding singular simplex $\sigma = i(c)$. The cochain groups $\bar{\mathscr{C}}^p(E)$ are then generated by the corresponding cosimplex, the cochains taking the value 1 on a simplex σ and 0 on all others, and the $\cup$ product being defined by (4-9). Of course, it will not be necessary to calculate all the $\cup$ products, but only those of cocycles representing generators of the cohomology groups.

Examples

4-1. Let E be a simplicial complex whose vertexes are given in a standard order. Then E can be thought of as a cell complex with the simplexes as cells. If $[x_0 x_1 \cdots x_p]$ is a p simplex of E with the vertexes in the standard order, c an associated oriented cell (here an oriented simplex) define $i(c)$ as $(x_0 x_1 \cdots x_p)$. Then (1) and (2) are automatically satisfied. The formula (4-9) for the product thus coincides with the formula for the cup product in oriented simplicial cohomology (cf. Section 4-1).

4-2. There are occasional situations where the structure of the cohomology ring can be seen without any computation at all. For example, take the sphere S^n. Here the only nonzero cohomology groups are $H^0(S^n)$ and $H^n(S^n)$, which have generators e and γ, respectively, and e acts as identity of the cohomology ring. Thus $e \cup \gamma = \gamma \cup e = \gamma$, and $e \cup e = e$. The only other possible product would be $\gamma \cup \gamma$. This, however, would be of dimension $2n$, so it is automatically zero.

4-3. Let E be the torus. This can be obtained from a square (Fig. 13) by identifying opposite sides. The images of the open triangles, segments, and

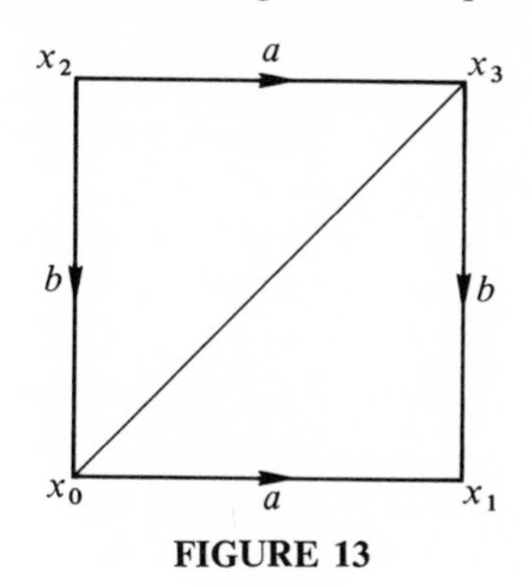

FIGURE 13

vertexes in this figure give a cell decomposition of the torus. It is easy to check the conditions of Definition 2-18.

On the other hand, consider the singular simplexes

$$(x_0 x_1 x_3), (x_0 x_2 x_3), (x_0 x_1), (x_0 x_2), (x_0 x_3), (x_0) \tag{4-11}$$

on the square. Their images on the torus are denoted by the same symbols and are in one-to-one correspondence with the cells of the torus. Each of them can be taken as the image under i of the corresponding oriented cell in the notation of the conditions (1) and (2), and it is easy to see that these conditions are satisfied. In accordance with the remark made at the end of the preceding discussion, the simplexes (4-11) on the torus are identified with the corresponding cells. The symbols

$$(x_0 x_1 x_3)', (x_0 x_2 x_3)', (x_0 x_1)', (x_0 x_2)', (x_0 x_3)', (x_0)' \tag{4-12}$$

denote the corresponding cosimplexes. Thus, the cohomology ring of the torus can be computed by finding the products of these cosimplexes, using the formula (4-9).

There are some obvious remarks that can be made about these products. In the first place, it is easy to see that $(x_0)'$ acts as the identity. It is also a cocycle on $\mathscr{C}(E)$, so its cohomology class is the identity of the ring $\mathscr{H}^*(E)$. Also it is known (Exercise 3-7) that $\mathscr{H}^p(E) = 0$ for $p > 2$ and, in fact, the only nonzero products are products of 1 cochains. Consider then a product $(x_i x_j)' \cup (x_h x_k)'$. When its operation is checked on $(x_0 x_1 x_3)$ and $(x_0 x_2 x_3)$ it can easily be seen that the result will be nonzero if and only if either $i = 0$, $j = h = 1$, $k = 3$, or $i = 0$, $j = h = 2$, $k = 3$. In these cases

$$\begin{aligned} (x_0 x_1)' \cup (x_1 x_3)' &= (x_0 x_1 x_3)' = (x_0 x_1)' \cup (x_0 x_2)' \\ (x_0 x_2)' \cup (x_2 x_3)' &= (x_0 x_2 x_3) \ = (x_0 x_2)' \cup (x_0 x_1)' \end{aligned} \tag{4-13}$$

The last entry in each line takes into account the identification of opposite sides of the square. All other products are zero.

Now the one-dimensional coboundary group of $\mathscr{C}(E)$ is zero, so $\mathscr{H}^1(E)$ is isomorphic to the one-dimensional cocycle group. It is easy to check the following formulas (remembering the identifications $(x_0 x_1) = (x_2 x_3)$, etc.):

$$\begin{aligned} \delta(x_0 x_1)' &= (x_0 x_1 x_3)' + (x_0 x_2 x_3)' \\ \delta(x_0 x_2)' &= (x_0 x_1 x_3)' + (x_0 x_2 x_3)' \\ \delta(x_0 x_3)' &= -(x_0 x_1 x_3)' - (x_0 x_2 x_3)' \end{aligned} \tag{4-14}$$

It follows that

$$\alpha = (x_0 x_1)' + (x_0 x_3)'$$
$$\beta = (x_0 x_2)' + (x_0 x_3)'$$

are cocycles and that every 1 cocycle of $\mathscr{C}(E)$ is a linear combination of these. The formulas (4-14) imply that

$$\alpha \cup \alpha = 0$$
$$\beta \cup \beta = 0$$
$$\alpha \cup \beta = (x_0 x_1 x_3)'$$
$$\beta \cup \alpha = (x_0 x_2 x_3)'$$

Finally, $(x_0 x_1 x_3)'$ and $(x_0 x_2 x_3)'$ are both cocycles, since there are no 3 cells, and their sum [cf. (4-14)] is a coboundary. Hence $\mathscr{H}^*(E)$ is generated by the elements e [class of $(x_0)'$, the identity element], $\bar{\alpha}$, $\bar{\beta}$, the classes of α, β, and $\bar{\gamma}$, the cohomology class of $(x_0 x_1 x_3)'$ and the product formulas become

$$\bar{\alpha} \cup \bar{\alpha} = \bar{\beta} \cup \bar{\beta} = 0$$
$$\bar{\alpha} \cup \bar{\beta} = -\bar{\beta} \cup \bar{\alpha} = \bar{\gamma}$$

This completes the computation of the ring $\mathscr{H}^*(E) = H^*(E)$.

4-4. Consider the projective space P^n, obtained by identifying antipodal pairs of points on the n sphere S^n, simplicially subdivided as described in Example 2-9. Consider too the cell decomposition of P^n obtained by taking as cells the images of the open simplexes of S^n; the conditions of the definition of a cell complex can easily be verified. Next, suppose c is an oriented cell of P^n carried by the image of the simplex $[x_i x_j \cdots x_k]$. Then define $\sigma = i(c)$ to be the image in P^n of the singular simplex $(x_i x_j \cdots x_k)$ on S. σ also is denoted, without danger of confusion, by $(x_i x_j \cdots x_k)$. If the orientation of c is chosen correctly, in fact if it is defined by the order of appearance of the vertexes $x_i, x_j, \ldots, x_k$, then the map i satisfies condition (1) above. If $(x_i x_j \cdots x_h \cdots x_k) = i(c)$ is of dimension $p + q$, and $(x_i x_j \cdots x_h) = i(c_1)$ and $(x_h \cdots x_k) = i(c_2)$ are of dimensions p and q, respectively, then c_1 and c_2 correspond to c as in condition (2). To simplify the notation, if $i(c) = (x_i x_j \cdots x_k)$ then c is written as $(x_i x_j \cdots x_k)$. So $\mathscr{H}^*(P^n)$ can be computed from the chain complex $\mathscr{C}(P^n)$ generated by the $(x_i x_j \cdots x_k)$. $\mathscr{C}^*(P^n)$ is then generated by the cocells $(x_i x_j \cdots x_k)'$, and the product formula (4-9) becomes

$$(x_i x_j \cdots x_h)' \cup (x_h x_k \cdots x_l)' = (x_i x_j \cdots x_l)'$$

all other products being zero.

The real problem now is to find the products of cocycles representing the generators of the cohomology groups $\mathscr{H}^p(P)$ of the chain complex $\mathscr{C}^*(P)$. The calculation can be simplified considerably by using previous results on the cohomology groups of P^n (cf. Exercise 2-18).

In Section 2-9 the cohomology groups of P^n were computed by using a different cell decomposition from that of the present example. Denote by $\bar{\bar{\mathscr{C}}}(P^n)$ the cellular chain complex corresponding to the decomposition of Example 2-9. Each oriented cell of $\bar{\bar{\mathscr{C}}}(P^n)$ can be thought of in an obvious way as a sum of oriented cells of $\bar{\mathscr{C}}(P^n)$, so that there is a sequence of inclusion maps

$$\bar{\bar{\mathscr{C}}}(P^n) \to \bar{\mathscr{C}}(P^n) \to \mathscr{C}(P^n)$$

where the last complex is that of oriented simplicial chains. These maps have algebraic homotopy inverses. It was found in Exercise 3-8 that, for each even dimension p, $\bar{\bar{\mathscr{H}}}^p(P^n)$, the cohomology computed from $\bar{\bar{\mathscr{C}}}(P^n)$, has a generator represented by a cocycle taking the value 1 on the single oriented p cell c_p on $\bar{\bar{\mathscr{C}}}^p(P^n)$. To find a cocycle that represents a generator of $\bar{\mathscr{H}}^p(P)$, the cohomology computed from $\bar{\mathscr{C}}(P^n)$, a cocycle on $\bar{\mathscr{C}}(P^n)$ must be found, which, when restricted to $\bar{\bar{\mathscr{C}}}(P^n)$, takes the value 1 on c_p. Now c_p is a sum of oriented cells of $\bar{\mathscr{C}}(P^n)$, so the required cocycle can be constructed if some more or less natural way can be found of picking out one of these cells, then defining a function which has the value 1 on that cell and zero on the rest. After a bit of trial and error it appears that the simplest procedure is to define for each p (odd or even) a cochain c^p of $\bar{\mathscr{C}}(P^n)$ which takes the value 1 on an oriented cell $(x_i x_j \cdots x_k)$ if and only if the indices are alternately 1 and -1. It is not hard to see that c_p, expressed as a sum of cells of $\bar{\mathscr{C}}(P^n)$, contains exactly one cell of this kind and c^p takes the value 1 on c_p as required. The next thing is to check that c^p is a cocycle for odd p and also to compute the products of the c^p for various p. Actually, these two operations can be carried out simultaneously.

Consider the value of $c^p \cup c^q$ on the $(p+q)$ cell $(x_i x_j \cdots x_h \cdots x_k)$ where there are $p+1$ vertexes up to x_h. The product formula (4-9) gives

$$(c^p \cup c^q)(x_i x_j \cdots x_h \cdots x_k) = c^p(x_i x_j \cdots x_h) c^q(x_h \cdots x_k)$$

where each factor on the right is 1 if and only if the indices of the x are alternately 1 and -1; otherwise the value is zero. It follows at once that

$$c^p \cup c^q = c^{p+q} \tag{4-15}$$

All the c^p can be obtained by taking powers of c^1, and the coboundary of c^p can be computed by finding δc^1 and then using repeatedly the coboundary formula.

Consider the value of δc^1 on $(x_i^\varepsilon x_j^\eta x_k)$ where ε and η are ± 1 (remember that, by virtue of the identification of antipodal points, it is only necessary to consider cells in whose symbols the last x has the index 1). Then

$$\begin{aligned}\delta c^1(x_i^\varepsilon x_j^\eta x_k) &= c^1(d(x_i^\varepsilon x_j^\eta x_k)) \\ &= c^1(x_j^\eta x_k) - c^1(x_i^\varepsilon x_k) + c^1(x_i^\varepsilon x_j^\eta)\end{aligned}$$

Now calculate the right-hand side for all possible pairs of values of ε and η:

$$\begin{aligned}
\varepsilon &= 1, & \eta &= 1: & \delta c^1(x_i x_j x_k) &= 0 \\
\varepsilon &= 1, & \eta &= -1: & \delta c^1(x_i x_j^{-1} x_k) &= 2 \\
\varepsilon &= -1, & \eta &= 1: & \delta c^1(x_i^{-1} x_j x_k) &= 0 \\
\varepsilon &= -1, & \eta &= -1: & \delta c^1(x_i^{-1} x_j^{-1} x_k) &= 0
\end{aligned}$$

Thus δc^1 takes the values zero on $(x_i^\varepsilon x_j^\eta x_k)$ unless the indices are alternately 1 and -1, in which case it takes the value 2. Hence

$$\delta c^1 = 2c^2 \tag{4-16}$$

Using the coboundary formula

$$\begin{aligned}
\delta c^2 &= \delta(c^1 \cup c^1) \qquad \text{(by (4-15))} \\
&= \delta c^1 \cup c^1 - c^1 \cup \delta c^1 \\
&= 2c^2 \cup c^1 - 2c^1 \cup c^2 \qquad \text{(by (4-16))} \\
&= 2(c^3 - c^3) \qquad \text{(by (4-15))} \\
&= 0
\end{aligned} \tag{4-17}$$

Again using the coboundary formula for a product $c^{p+2} = c^p \cup c^2$.

$$\begin{aligned}
\delta c^{p+2} &= \delta(c^p \cup c^2) \\
&= \delta c^p \cup c^2 + (-1)^p c^p \cup \delta c^2 \\
&= \delta c^p \cup c^2
\end{aligned}$$

Note. Since c^p takes the value 1 on c_p it can be identified with the dual of c_p in constructing the cochain complex $\bar{\bar{\mathscr{C}}}^*(P^n)$. If products in this complex are defined by thinking of cochains in $\bar{\bar{\mathscr{C}}}^*(P^n)$ as restrictions of cochains in $\bar{\mathscr{C}}^*(P^n)$, then it makes sense to use the formulas $c^p \cup c^q = c^{p+q}$ as defining the products in $\bar{\bar{\mathscr{C}}}^*(P^n)$.

By induction, starting off with (4-16) and (4-17),

$$\delta c^p = 2c^{p+1} \qquad \text{if } p \text{ is odd}$$

$$\delta c^p = 0 \qquad \text{if } p \text{ is even}$$

Of course, in any case

$$\delta c^n = 0$$

In particular, this means that $\overline{\mathscr{H}}^2(p^n)$ has a generator ξ, represented by c^2, and $\overline{\mathscr{H}}^{2q}(P^n)$ has a generator represented by c^{2q}. By the application of (4-15), this generator is the qth power, under the $\cup$ product, of ξ.

To sum up, now in terms of singular cohomology, all the odd dimensional cohomology groups of P^n are zero, while $H^{2q}(P^n)$ has a generator ξ^q, the qth power of a generator ξ of $H^2(P^n)$. Also, the relation $2\xi^q = 0$ hold for each q.

4-5. Sometimes, when a space is built up from other spaces, it is possible to express the cohomology ring of the new space in terms of those of the old ones. An example of this situation is provided by one-point unions. Let E and F be two spaces, x a point of E and y a point of F. Form the union of E and F, the pair of points x and y being identified. The space so obtained is called the one-point union of E and F and is denoted by $E \vee F$. Note that, in general, $E \vee F$ depends on the pair of points which has been identified. In the examples that appear here this is not the case, so the symbol $E \vee F$ is unambiguous. Also note that there is another way of constructing $E \vee F$: in the product $E \times F$, $E \vee F$ is the subspace $(E \times \{y\}) \cup (\{x\} \times F)$.

Suppose that E and F are cell complexes, both having the properties (1) and (2) and suppose that x and y are vertexes of E and F, respectively; that is, they are zero cells. Use the symbol $\overline{\mathscr{C}}$, as usual, to denote the oriented cellular chain complex. Then it is easy to see that, for each p, $\overline{\mathscr{C}}^p(E \vee F)$ is the direct sum $\overline{\mathscr{C}}^p(E) \oplus \overline{\mathscr{C}}^p(F)$ and that this direct sum decomposition induces direct sum decompositions of the cocycle and coboundary groups. Hence $\overline{\mathscr{H}}^p(E \vee F) = \overline{\mathscr{H}}^p(E) \oplus \overline{\mathscr{H}}^p(F)$. On the other hand, let c_1 and c_2 be cocells on E and F, respectively, of dimensions p and q, both dimensions being greater than 0. c_1 becomes a cocell on $E \vee F$ when it is assigned the value 0 on any cell of F. Similarly, c_2 becomes a cocell on $E \vee F$. Since there is no $(p+q)$ cell having a p face in E and a q face in F (using the terminology of condition (1)) it follows that $c_1 \cup c_2 = 0$. Of course, the product of cochains on E (or F) is the same whether they are regarded as on E (or F) or on $E \vee F$. So, in terms of singular cohomology (which is isomorphic to the cellular cohomology), it follows that the ring $H^*(E \vee F)$ is the direct sum of the rings $H^*(E)$ and $H^*(F)$, the product of any elements of positive dimensions, one from $H^*(E)$ and the other from $H^*(F)$, being zero.

4-5. TENSOR PRODUCTS AND CUP PRODUCTS

The relations between three kinds of products, namely, tensor, cup, and topological products, are the concern of most of the remainder of this chapter. First, the notion of tensor product of chain complexes will be described and related, in a purely algebraic way, to the cup product. The relations obtained will suggest a geometric way of describing the cup product in terms of topological products of spaces.

The cup product appears first as a function of two variables, cochains on a space E, bilinear in both variables over the coefficient ring. This means that the cup product acts as a homomorphism

$$\cup : C^p(E) \otimes C^q(E) \to C^{p+q}(E) \tag{4-18}$$

On the other hand, $\cup$ induces a homomorphism

$$\cup : H^*(E) \otimes H^*(E) \to H^*(E) \tag{4-19}$$

so it is natural to try to make the connection between (4-18) and (4-19) similar to the usual induced homomorphism situation, namely, to make the collection of $C^p(E) \otimes C^q(E)$ into a cochain complex in such a way that (4-18) is a cochain homomorphism and so that (4-19) is the induced homomorphism on cohomology groups. If $\cup$ is to be a cochain homomorphism it must preserve dimension, so all the $C^p(E) \otimes C^q(E)$ with $p + q$ fixed equal to n, for instance, must be taken together as the n-dimensional cochain group. To make $\cup$ commute with the coboundary operator, the appropriate definition for $\delta(\alpha \otimes \beta)$ should be $\delta\alpha \otimes \beta + (-1)^p \alpha \otimes \delta\beta$, where $\dim \alpha = p$. This motivates the following sequence of definitions.

4-6. TENSOR PRODUCTS OF CHAIN AND COCHAIN COMPLEXES

Let K and L be chain complexes. Their p-dimensional groups are denoted by K_p and L_p, respectively, and their boundary operators are both denoted by d. Define T_p by

$$T_p = \sum_{i+j=p} K_i \otimes L_j \tag{4-20}$$

where the sum on the right is a direct sum and the tensor product is over Z. If $x \in K_p$, $y \in L_q$ define

$$d(x \otimes y) = dx \otimes y + (-1)^p x \otimes dy \tag{4-21}$$

This operator d is then extended by linearity to a homomorphism of T_p into T_{p-1}. Note that the first d on the right is the boundary operator on K while the second is that on L.

Lemma 4-7. *The collection of T_p along with the boundary operator defined by (4-21) forms a chain complex.*

Proof. It has to be shown that d as defined by (4-21) satisfies the equation $d^2 = 0$. It is sufficient to show that $d^2(x \otimes y) = 0$ for any $x \in K_p$ and $y \in L_q$.

$$\begin{aligned} d^2(x \otimes y) &= d[dx \otimes y + (-1)^p x \otimes dy] \\ &= d^2x \otimes y + (-1)^{p-1}\, dx \otimes dy + (-1)^p\, dx \otimes dy + (-1)^{2p} x \otimes d^2 y \end{aligned}$$

where (4-21) is applied repeatedly. Here the first and last terms are zero since $d^2x = 0$ in K and $d^2y = 0$ in L, and the terms in the middle cancel. This completes the proof. ∎

Definition 4-4. The chain complex T whose p-dimensional group is T_p, which is defined in (4-20) and whose boundary operator is defined by (4-21), is called the *tensor product of K and L* and is denotd by $K \otimes L$.

The notion of tensor product of chain homomorphisms is derived directly from the corresponding notions for group or ring homomorphisms. Let K, L, M, and N be chain complexes. Write $T = K \otimes L$ and $U = M \otimes N$. Let

$$f: K \to L$$

$$g: L \to N$$

be chain homomorphisms. Then for $x \in K_p$ and $y \in L_q$ define

$$(f \otimes g)(x \otimes y) = f(x) \otimes g(y) \tag{4-22}$$

Thus $f \otimes g$ becomes a homomorphism of T into U for each r.

Lemma 4-8. *$f \otimes g$ is a chain homomorphism.*

Proof. It has to be checked that $f \otimes g$ commutes with the boundary operator. Of course it is sufficient to check this by looking at the operation on any element $x \otimes y$ with $x \in K_p$ and $y \in L_q$. Then

$$\begin{aligned} d(f \otimes g)(x \otimes y) &= d[f(x) \otimes g(y)] \\ &= df(x) \otimes g(y) + (-1)^p f(x) \otimes dg(y) \\ &= f(dx) \otimes g(y) + (-1)^p f(x) \otimes gd(y) \\ &\qquad \text{(since } f \text{ and } g \text{ are chain homomorphisms)} \\ &= (f \otimes g)(dx \otimes y) + (-1)^p (f \otimes g)(x \otimes dy) \\ &= (f \otimes g)d(x \otimes y) \end{aligned}$$

This is the required result. ∎

Definition 4-5. The chain homomorphism $f \otimes g$ defined by (4-22) is called the *tensor product of the chain homomorphisms f and g.*

Exercises. 4-3. Let K, L, M, X, Y, Z be chain complexes and let

$$K \xrightarrow{f} L \xrightarrow{g} M$$
$$X \xrightarrow{i} Y \xrightarrow{j} Z$$

be sequences of chain homomorphisms. Prove that

$$gf \otimes ji = (g \otimes j)(f \otimes i)$$

4-4. Let K, L, X, Y be chain complexes and let $f: K \to L$ and $g: K \to L$ and $h: X \to Y$ be chain homomorphisms. Suppose that f and g are algebraically homotopic. Prove that $f \otimes h$ and $g \otimes h$ are algebraically homotopic.

(The given algebraic homotopy means that there is a homomorphism $H: K_p \to L_{p+1}$ for each p such that $f - g = dH + Hd$. Define $\bar{H}$ by setting $\bar{H}(x \otimes y) = Hx \otimes hy$ where $x \in K_p$ and $y \in L_q$ and show that $\bar{H}$ defines the algebraic homotopy of $f \otimes h$ and $g \otimes h$.)

All that has just been said will work equally well for cochain complexes. Thus if K and L are cochain complexes, both coboundary operators being denoted by δ, $K \otimes L$ is defined as the cochain complex whose n-dimensional group is

$$T^n = \sum_{p+q=n} K^p \otimes L^q$$

with coboundary operator defined by

$$\delta(x \otimes y) = \delta x \otimes y + (-1)^p x \otimes \delta y$$

where $\dim x = p$. The proof that this is a cochain complex, that is, that $\delta^2 = 0$, is the same as that of Lemma 4-7. The tensor product of cochain homomorphisms is defined by the formula (4-22) and is proved to be a cochain homomorphism as in Lemma 4-8. The results of Exercises 4-3 and 4-4 hold similarly for the compositions of cochain homomorphisms and for cochain homotopy.

Special attention must be given to the situation in which cochains are obtained as duals, with respect to some given coefficient group $\mathscr{G}$, of chain complexes. So let K and L be chain complexes and let K' and L' be their duals with respect to $\mathscr{G}$. The p-dimensional groups of K and L are written as K_p and L_p, respectively, and those of K' and L' as K^p and L^p. On the one hand, the cochain complex $K' \otimes L'$ can be constructed, on the other, the dual (with respect to $\mathscr{G}$) of chain complex $K \otimes L = T$. The question is: what relation holds between these two cochain complexes? An answer is given here for the case in which the group $\mathscr{G}$ is a ring.

Take elements $\alpha \in K^p$, $\beta \in L^q$, $x \in K_p$, $y \in L_q$. For fixed α and β, the product $\alpha(x)\beta(y)$ is bilinear in x and y with values in $\mathcal{G}$. It follows that a homomorphism ϕ can be defined of $K_p \otimes L_q$ (the tensor products here are all over the integers) into $\mathcal{G}$ by the formula

$$\phi(x \otimes y) = \alpha(x)\beta(y)$$

Thus ϕ is an element of the dual of $K_p \otimes L_q$ and so can be extended (by making it act as zero on all other $K_i \otimes L_j$) to an element of the dual T' of $K \otimes L$. On the other hand, ϕ depends on α and β; in fact, it is bilinear in α and β, so it can be written as $\phi = \pi(\alpha \otimes \beta)$ where π is a homomorphism of $K^p \otimes L^q$ into T'. π can be defined in this way for all p and q and so defines a homomorphism

$$\pi\colon K' \otimes L' \to T' = (K \otimes L)'$$

π is given explicitly by the formula

$$\pi(\alpha \otimes \beta)(x \otimes y) = \alpha(x)\beta(y) \tag{4-23}$$

where $\dim \alpha = \dim x$ and $\dim \beta = \dim y$, while $\pi(\alpha \otimes \beta)$ has the value 0 on any $x \otimes y$ that does not satisfy these conditions on dimensions.

Now it will be proved that π is a cochain homomorphism. To show this it is sufficient to verify that π and δ commute when they act on an element of $K' \otimes L'$ of the form $\alpha \otimes \beta$ with $\alpha \in K^p$, $\beta \in L^q$. Then to show that

$$\pi\delta(\alpha \otimes \beta) = \delta\pi(\alpha \otimes \beta) \tag{4-24}$$

it must be shown that both sides, which are elements of T', have the same value on any element of T, in particular on any generating element $x \otimes y$ with $x \in K_r$ and $y \in L_s$.

$$[\pi\delta(\alpha \otimes \beta)](x \otimes y) = \pi[\delta\alpha \otimes \beta + (-1)^p\alpha \otimes \delta\beta](x \otimes y)$$

Here the right-hand side has nonzero value if and only if either $x \in K_{p+1}$, $y \in L_q$, when it has the value $\delta\alpha(x)\beta(y) = \alpha(dx)\beta(y)$, or $x \in K_p$, $y \in L_{q+1}$, when it has the value $(-1)^p\alpha(x)\delta\beta(y) = (-1)^p\alpha(x)\beta(dy)$.

On the other hand,

$$\delta\pi(\alpha \otimes \beta)(x \otimes y) = \pi(\alpha \otimes \beta)d(x \otimes y) = \pi(\alpha \otimes \beta)[dx \otimes y + (-1)^r x \otimes dy]$$

Again the right-hand side has a nonzero value if and only if either $x \in K_{p+1}$, $y \in L_q$, when it is equal to $\alpha(dx)\beta(y)$, or $x \in K_p$, $y \in L_{q+1}$, when it is equal to $(-1)^p\alpha(x)\beta(dy)$. These are the same values as those obtained for $\pi\delta(\alpha \otimes \beta)(x \otimes y)$. Hence (4-24) holds as required.

The following lemma sums up the discussion.

Lemma 4-9. *The formula* (4-23) *defines a cochain homomorphism*

$$\pi: K' \otimes L' \to T' = (K \otimes L)'$$

Exercises. 4-5. In general, there is no reason to believe that π is onto or is an isomorphism. There is, however, a case in which both of these conditions are satisfied: show that if K and L are complexes consisting of finitely generated free Abelian groups then π is an isomorphism onto.

4-6. Let K, L, X, Y be chain complexes with duals K', L', X', Y'. Let $f: K \to X$ and $g: L \to Y$ be chain homomorphisms with duals f' and g'. Show that the following diagram is commutative.

$$\begin{array}{ccc} K' \otimes L' & \xrightarrow{\pi} & (K \otimes L)' \\ {\scriptstyle f' \otimes g'} \uparrow & & \uparrow {\scriptstyle (f' \otimes g')} \\ X' \otimes Y' & \xrightarrow{\pi} & (X \otimes Y)' \end{array}$$

4-7. COHOMOLOGY OF A TENSOR PRODUCT

In Section 4-2 $\cup$ appears as a homomorphism of $C^*(E) \otimes C^*(E)$ into $C^*(E)$. Clearly the definition (Section 4-6) of $C^*(E) \otimes C^*(E)$ as a cochain complex automatically makes $\cup$ into a cochain homomorphism. It follows that there is an induced homomorphism

$$\cup^*: H^*[C^*(E) \otimes C^*(E)] \to H^*(E) \tag{4-25}$$

where the cohomology on the left is computed for the cochain complex $C^*(E) \otimes C^*(E)$. On the other hand, Section 4-3 exhibits $\cup$ as a homomorphism

$$\cup: H^*(E) \otimes H^*(E) \to H^*(E) \tag{4-26}$$

The question posed in Section 4-5 can now be reformulated: what is the connection between $\cup$ in (4-26) and $\cup^*$ in (4-25)? To answer this the left-hand sides of (4-25) and (4-26) must be compared.

More generally, let K and L be cochain complexes and let $K \otimes L$ be their tensor product. Consider the comparison of $H^*(K) \otimes H^*(L)$ with $H^*(K \otimes L)$. There is a natural way of mapping the first of these into the second. Let $\bar{\alpha} \in H^p(K)$, $\bar{\beta} \in H^q(L)$ and let α and β be representative cocycles of $\bar{\alpha}$ and $\bar{\beta}$, respectively. Then, by the coboundary formula, it is clear that $\alpha \otimes \beta$ is a cocycle in the complex $K \otimes L$. On the other hand, suppose that γ is a coboundary in L, $\gamma = \delta\theta$, for example. Then

$$\begin{aligned}\delta(\alpha \otimes \beta) &= \delta\alpha \otimes \theta + (-1)^p \alpha \otimes \delta\theta \\ &= (-1)^p \alpha \otimes \gamma\end{aligned}$$

Thus $\alpha \otimes \gamma$ is a coboundary. It follows that the cohomology class of $\alpha \otimes \beta$ in $K \otimes L$ is the same as that of $\alpha \otimes (\beta + \gamma)$ and so depends only on the cohomology class of β and not on the representative cocycle. Similarly, the cohomology class of $\alpha \otimes \beta$ depends only on $\bar{\alpha}$ and not on the choice of representative cocycle. It follows that a map

$$H^p(K) \otimes H^q(L) \to H^{p+q}(K \otimes L)$$

can be defined by mapping $\bar{\alpha} \otimes \bar{\beta}$ on the cohomology class in $K \otimes L$ of $\alpha \otimes \beta$, where α and β are representative cocycles of $\bar{\alpha}$ and $\bar{\beta}$. Applied to all dimensions p and q this gives a homomorphism

$$\Phi\colon H^*(K) \otimes H^*(L) \to H^*(K \otimes L)$$

In particular, taking $K = L = C^*(E)$ for a topological space E yields a homomorphism

$$\Phi\colon H^*(E) \otimes H^*(E) \to H^*[C^*(E) \otimes C^*(E)]$$

The main result of this section can now be stated.

Theorem 4-10. *The following diagram is commutative.*

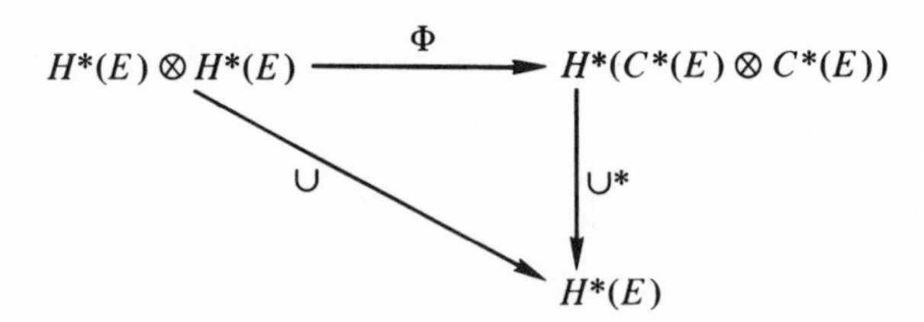

where $\cup$ *is as in* (4-26) *and* $\cup^*$ *is as in* (4-25).

Proof. Take $\bar{\alpha} \in H^p(E)$, $\bar{\beta} \in H^q(E)$. $\cup(\bar{\alpha} \otimes \bar{\beta})$ or, in product notation, $\bar{\alpha} \cup \bar{\beta}$ is defined as the cohomology class in E of $\alpha \cup \beta$, where α, β are representative cocycles of $\bar{\alpha}$, $\bar{\beta}$, respectively (cf. Section 4-3). On the other hand, $\Phi(\bar{\alpha} \otimes \bar{\beta})$ is the cohomology class $\overline{\alpha \otimes \beta}$ in the cochain complex $C^*(E) \otimes C^*(E)$ of $\alpha \otimes \beta$, and $\cup^*(\overline{\alpha \otimes \beta})$ is the cohomology class in $C^*(E)$ of $\cup(\alpha \otimes \beta)$ (i.e., $\cup$ operates on a cocycle representing $\overline{\alpha \otimes \beta}$) which is $\bar{\alpha} \cup \bar{\beta}$. It follows that $\cup^*\Phi = \cup$, as required. ∎

Exercises. 4-7. The algebraic properties of Φ depend on the algebraic nature of the complexes K and L. Prove, for example, that if the cochain groups forming K and L are all free Abelian then Φ identifies $H^*(K) \otimes H^*(L)$ with a subcomplex of $H^*(K \otimes L)$ and, in fact, this subcomplex is a direct summand.

4-8. If $H^*(K)$ or $H^*(L)$ is torsion free, prove that Φ is an isomorphism.

4-9. In a similar way set up a homomorphism Φ for chain complexes.

4-10. Let K, L, X, Y, be cochain complexes and let $f\colon K \to X$, $g\colon L \to Y$ be cochain homomorphisms. Prove that the following diagram is commutative.

$$\begin{array}{ccc} H^*(K) \otimes H^*(L) & \xrightarrow{\Phi} & H^*(K \otimes L) \\ {\scriptstyle f^* \otimes g^*}\downarrow & & \downarrow{\scriptstyle (f \otimes g)^*} \\ H^*(X) \otimes H^*(Y) & \xrightarrow{\Phi} & H^*(X \otimes Y) \end{array}$$

4-8. *TENSOR PRODUCTS AND TOPOLOGICAL PRODUCTS*

The discussion of Section 4-5 related the $\cup$ product to a study of the tensor product $C^*(E) \otimes C^*(E)$ of the singular cochain complex of E with itself, and this in turn was linked, by the map π of Section 4-6, with the cochain complex $[C(E) \otimes C(E)]^*$. Now some rather informal geometric remarks can be made about the chain complex $C(E) \otimes C(E)$ or, more generally, $C(E) \otimes C(F)$, where E and F are any two topological spaces. If σ and τ are singular simplexes in E and F, respectively, they can be thought of as simplexes embedded in their respective spaces. Then $\sigma \otimes \tau$ can be regarded as representing algebraically the product $\sigma \times \tau$ of these simplexes in E and F. Geometrically, $\sigma \times \tau$ is a cell in $E \times F$. The boundary of the cell $\sigma \times \tau$ should be made up of the product of σ with the boundary of τ along with the product of τ with the boundary of σ. This corresponds to the algebraic statement that, by definition, $d(\sigma \otimes \tau)$ is a linear combination of $d\sigma \otimes \tau$ and $\sigma \otimes d\tau$. Thus, it appears that a study of the homology and cohomology of $C(E) \otimes C(F)$ should be pictured geometrically as a study of the homology and cohomology of $E \times F$, forming chains not with simplexes but with cells that are products of simplexes. This in turn suggests a comparison of the complexes $C(E) \otimes C(F)$ and $C(E \times F)$. In fact, it will now be shown that these two complexes can be identified up to algebraic homotopy. It follows that the map π of Section 4-6 (Eq. 4-23) taking $C^*(E) \otimes C^*(E)$ into $(C(E) \otimes C(E))^*$ can be replaced by a map

$$C^*(E) \otimes C^*(E) \to C^*(E \times E) \tag{4-27}$$

For the construction of the $\cup$ product a map of $C^*(E) \otimes C(E)$ into $C^*(E)$ is needed, so the question arises whether this can be achieved by composing (4-27) with some map

$$C^*(E \times E) \to C^*(E) \tag{4-28}$$

The answer happens to be yes, at least up to algebraic homotopy. The required map (4-28) has a very simple geometric meaning: namely it is induced by the diagonal map $\Delta: E \to E \times E$, defined by $\Delta(x) = (x, x)$.

The program sketched here will be carried out in the next few sections.

4-9. SINGULAR CHAINS ON A PRODUCT

The main part of the program outlined in the last section is the comparison of the complex $C(E) \otimes C(F)$ with $C(E \times F)$ where E and F are two given topological spaces. There is a natural identification between the two complexes in dimension zero, for a zero-dimensional singular simplex on a space can be thought of as a point of the space. Thus $C_0(E) \otimes C_0(F)$ is generated by elements $x \otimes y$, where $x \in E$ and $y \in F$, while $C_0(E \times F)$ is generated by elements (x, y). In addition, a map

$$g_0: C_0(E \times F) \to C_0(E) \otimes C_0(F)$$

can be defined by setting

$$g_0(x, y) = x \otimes y$$

Clearly, this is an isomorphism. The idea now is to extend g_0 to a chain homomorphism

$$g: C(E \times F) \to C(E) \otimes C(F)$$

which will have an algebraic homotopy inverse. The construction is motivated by the following considerations.

$$g(x, y) = g_0(x, y)$$

is already defined on any zero-dimensional singular simplex (x, y) of $E \times F$. Think of the singular one simplex σ on $E \times F$ as represented by a segment joining (x, y) and (u, v) (cf. Fig. 14). The map g is defined on $d\sigma$:

$$\begin{aligned} g(d\sigma) &= g(u, v) - g(x, y) \\ &= u \otimes v - x \otimes y \\ &= (u - x) \otimes v + x \otimes (v - y) \\ &= d(xu) \otimes v + x \otimes d(yv) \end{aligned} \tag{4-29}$$

where (xu) and (yv) denote simplexes on E and F, respectively, which are obtained by projecting σ into these spaces. Also, by the formula (4-21)

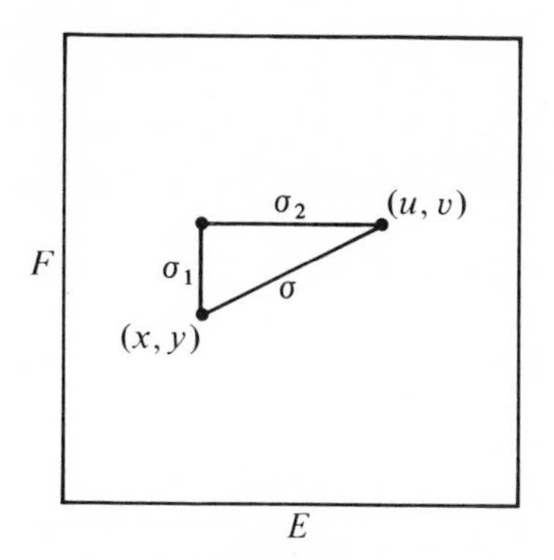

FIGURE 14

for the boundary operator in $C(E) \otimes C(F)$, the right-hand side of (4-29) can be written as

$$d[x \otimes (yv)] + d[(xu) \otimes v]$$

Thus, in order that the commutativity rule $d(g\sigma) = g(d\sigma)$ holds, $g(\sigma)$ should be defined by

$$g(\sigma) = x \otimes (yv) + (xu) \otimes v$$

Think of this geometrically as representing the replacement of the segment σ by the sum of the segments σ_1 and σ_2 in Fig. 14; these segments are the geometric equivalents of the two terms on the right of the last equation. This suggests the following definition, in general.

Let Δ_p and Δ_q be the standard Euclidean simplexes of dimensions p and q and let

$$F_q \colon \Delta_q \to \Delta_p$$

$$L_q \colon \Delta_q \to \Delta_p$$

be the linear maps carrying Δ_q onto the faces of Δ_p spanned by the first and the last $q + 1$ vertexes, respectively (cf. Section 4-2). For any topological spaces E and F denote by

$$p_E \colon E \times F \to E$$

$$p_F \colon E \times F \to F$$

the two projection maps. Finally, if σ is a singular p simplex on $E \times F$, write

$$g(\sigma) = \sum_{i=0}^{p} (p_E \sigma F_i) \otimes (p_F \sigma L_{p-i}) \tag{4-30}$$

Note that this defines $g(\sigma)$ as a p-dimensional element of $C(E) \otimes C(F)$. Also, the formula (4-30) reduces to the motivating formula when $p = 1$. That g is a chain homomorphism will now be shown. It is more convenient, however, to show first that g commutes with induced homomorphisms.

Lemma 4-11. *Let ϕ: $E \to E'$ and ψ: $F \to F'$ be continuous maps and denote by $\phi \times \psi: E \times F \to E' \times F'$ the continuous map defined by setting $(\phi \times \psi)(x, y) = [\phi(x), \psi(y)]$ for any $(x, y) \in E \times F$. Then the diagram*

$$\begin{array}{ccc} C(E \times F) & \xrightarrow{g} & C(E) \otimes C(F) \\ {\scriptstyle (\phi \times \psi)_1} \downarrow & & \downarrow {\scriptstyle \phi_1 \otimes \psi_1} \\ C(E' \times F') & \xrightarrow{g} & C(E') \otimes C(F') \end{array} \tag{4-31}$$

is commutative, where the subscript 1 *denotes the induced homomorphism on singular chain complexes, $\phi_1 \otimes \psi_1$ is a tensor product of chain homomorphisms as in Definition* 4-5 *and the horizontal maps are defined by* (4-30).

Proof. It is sufficient to check the commutativity by operating on a singular p simplex σ on $E \times F$. $g(\sigma)$ is given by (4-30), so

$$(\phi_1 \otimes \psi_1)g(\sigma) = (\phi_1 \otimes \psi_1)\sum_{i=0}^{p} (p_E \sigma F_i) \otimes (p_F \sigma L_{p-i})$$

$$= \sum_{i=0}^{p} \phi_1(p_E \sigma F_i) \otimes \psi_1(p_F \sigma L_{p-i})$$

(definition of tensor product of homomorphisms)

$$= \sum (\phi p_E \sigma F_i) \otimes (\psi p_F \sigma L_{p-i}) \qquad \text{(Definition 1-9)} \tag{4-32}$$

On the other hand, writing $\sigma' = (\phi \times \psi)_1(\sigma)$,

$$g(\sigma') = \sum (p_{E'} \sigma' F_i) \otimes (p_{F'} \sigma' L_{p-i}) \tag{4-33}$$

where $p_{E'}$ and $p_{F'}$ are the projections of $E' \times F'$ on E' and F'. But σ' is the composition of σ with $\phi \times \psi$. Thus, for any x

$$\sigma'(x) = (\phi \times \psi)\sigma(x) = (\phi \times \psi)\{p_E[\sigma(x)], p_F[\sigma(x)]\}$$
$$= [\phi p_E \sigma(x), \psi p_F \sigma(x)]$$

so

$$p_{E'} \sigma' = \phi p_E \sigma, p_{F'} \sigma' = \psi p_F \sigma$$

Hence (4-32) is the same as (4-33), as required. ∎

Lemma 4-12. *The map g defined by* (4-30) *is a chain homomorphism.*

Proof. g must be shown to commute with d. As usual, it is sufficient to check the operation on a singular p simplex σ on $E \times F$. First note that F_i, in (4-30), is a singular i simplex on Δ_p and, in fact, is equal to $(x_0 x_1 \cdots x_i)$, in the notation used for singular simplexes that are linear maps. Also, $p_E \sigma$ is a continuous map of Δ_p into E, with induced homomorphism $(p_E \sigma)_1$ on the chain groups, so that $p_E \sigma F_i$ can be written as $(p_E \sigma)_1(x_0 x_1 \cdots x_i)$. Similarly, $p_F \sigma L_{p-i} = (p_F \sigma)_1(x_i x_{i+1} \cdots x_p)$. Then (4-31) can be written

$$g(\sigma) = (p_E \sigma)_1 \otimes (p_F \sigma)_1 \sum (x_0 x_1 \cdots x_i) \otimes (x_i x_{i+1} \cdots x_p)$$

using the tensor product of chain homomorphisms as defined by Definition 4-5. Since the tensor product of chain homomorphisms is a chain homomorphism (Lemma 4-8), it follows that

$$dg(\sigma) = (p_E \sigma)_1 \otimes (p_F \sigma)_1 \, d \sum (x_0 x_1 \cdots x_i) \otimes (x_i x_{i+1} \cdots x_p) \tag{4-34}$$

On the other hand, the continuous map $\sigma\colon \Delta_p \to E \times F$ can be expressed as the composition

$$\Delta_p \to \Delta_p \times \Delta_p \to E \times F$$

where the first map is the diagonal Δ, defined by $\Delta(x) = (x, x)$ and the second map is $(p_E \sigma) \times (p_F \sigma)$ (in the notation of the last lemma). Therefore, $\sigma = [(p_E \sigma) \times (p_F \sigma)]\, \Delta$, hence

$$\begin{aligned} g(d\sigma) &= g\, d(p_E \sigma \times p_F \sigma)_1 \,\Delta \\ &= g(p_E \sigma \times p_F \sigma)_1 \, d\,\Delta \qquad \text{(property of chain homomorphisms)} \\ &= (p_E \sigma)_1 \otimes (p_F \sigma)_1 g \, d\,\Delta \qquad \text{(Lemma 4-11)} \end{aligned} \tag{4-35}$$

So, by using (4-34) and (4-35), the proof of the lemma will be completed if it can be shown that

$$d \sum (x_0 x_1 \cdots x_i) \otimes (x_i x_{i+1} \cdots x_p) = g\, d\,\Delta \tag{4-36}$$

So, starting with the left-hand side,

$$\begin{aligned} & d \sum (x_0 x_1 \cdots x_i) \otimes (x_i x_{i+1} \cdots x_p) \\ &\quad = \sum [d(x_0 x_1 \cdots x_i) \otimes (x_i \cdots x_p) + (-1)^i (x_0 x_1 \cdots x_i) \otimes d(x_i \cdots x_p)] \\ &\quad = \sum_{i=0}^{p} \sum_{j=0}^{p} (-1)^j (x_0 x_1 \cdots \hat{x}_j \cdots x_i) \otimes (x_i \cdots x_p) \\ &\qquad + \sum_{i=0}^{p} \sum_{j=i}^{p} (-1)^j (x_0 \cdots x_i) \otimes (x_i \cdots \hat{x}_j \cdots x_p) \end{aligned}$$

$$
\begin{aligned}
&= \sum_{j<i} (-1)^j (x_0 x_1 \cdots \hat{x}_j \cdots x_i) \otimes (x_i \cdots x_p) \\
&\quad + \sum_{i=1}^{p} (-1)^i (x_0 \cdots x_{i-1}) \otimes (x_i \cdots x_p) \\
&\quad + \sum_{j>i} (-1)^j (x_0 x_1 \cdots x_i) \otimes (x_i \cdots \hat{x}_j \cdots x_p) \\
&\quad + \sum_{i=0}^{p-1} (-1)^i (x_0 x_1 \cdots x_i) \otimes (x_{i+1} \cdots x_p) \\
&= \sum_{j<i} (-1)^j (x_0 \cdots \hat{x}_j \cdots x_i) \otimes (x_i \cdots x_p) \\
&\quad + \sum_{j>i} (-1)^j (x_0 \cdots x_i) \otimes (x_i \cdots \hat{x}_j \cdots x_p) \qquad (4\text{-}37)
\end{aligned}
$$

since the second and fourth sums cancel. Now consider the right-hand side of (4-36).

$$d\,\Delta = \Delta_1 \sum (-1)^j (x_0 x_1 \cdots \hat{x}_j \cdots x_p) \qquad \text{(Definition 1-12)}$$

Here $\Delta_1(x_0 x_1 \cdots \hat{x}_j \cdots x_p)$ is the linear map $(x_0 x_1 \cdots \hat{x}_j \cdots x_p)$ composed with the diagonal map Δ. On the other hand, Δ composed with either projection map is the identity. It follows that (4-31) becomes here

$$
\begin{aligned}
g\,\Delta_1(x_0 x_1 \cdots \hat{x}_j \cdots x_p) &= g\,\Delta(x_0 x_1 \cdots \hat{x}_j \cdots x_p) \\
&= \sum_{i<j} (x_0 x_1 \cdots x_i) \otimes (x_i \cdots \hat{x}_j \cdots x_p) \\
&\quad + \sum_{i>j} (x_0 \cdots \hat{x}_j \cdots x_i) \otimes (x_i \cdots x_p)
\end{aligned}
$$

and so

$$
\begin{aligned}
g\,d\,\Delta &= \sum (-1)^j g\,\Delta_1(x_0 x_1 \cdots \hat{x}_j \cdots x_p) \\
&= \sum_{i<j} (-1)^j (x_0 \cdots x_i) \otimes (x_i \cdots \hat{x}_j \cdots x_p) \\
&\quad + \sum_{i>j} (-1)^j (x_0 \cdots \hat{x}_j \cdots x_i) \otimes (x_i \cdots x_p) \qquad (4\text{-}38)
\end{aligned}
$$

From a comparison of (4-37) and (4-38) it follows that the two sides of (4-36) are equal, as required, and the proof of the lemma is complete. ∎

4-10. SOME ACYCLICITY ARGUMENTS

The principal aim of this section is to show that the chain homomorphism g of the last section has an algebraic homotopy inverse. Here it is not so convenient to set up an explicit formula, and after all, such a formula is not really necessary. The argument used here is closely analogous to the proof of algebraic homotopy of maps with a common acyclic carrier function (cf. Definition 3-18). This kind of argument will be used first to construct a chain homomorphism $f: C(E) \otimes C(F) \to C(E \times F)$ and again to show that both fg and gf are algebraically homotopic to the appropriate identities.

Ignoring the actual formula for g, note that its essential properties are given by Lemmas 4-11 and 4-12; that is, for every pair of spaces E and F there is a chain homomorphism g with the property that for any continuous maps $\phi: E \to E'$ and $\psi: F \to F'$ the following diagram is commutative.

$$\begin{array}{ccc} C(E \times F) & \xrightarrow{g} & C(E) \otimes C(F) \\ {\scriptstyle (\phi \times \psi_1)} \downarrow & & \downarrow {\scriptstyle \phi_1 \otimes \psi_1} \\ C(E' \times F') & \xrightarrow{g} & C(E') \otimes C(F') \end{array}$$

Also, for any two spaces, g is given in dimension zero by

$$g(x, y) = x \otimes y$$

The next step is to prove inductively (on dimension) the existence of a similar family of chain homomorphisms going in the opposite direction and having a similar commutativity property. Begin by defining the zero-dimensional map for any two spaces E and F by

$$f(x \otimes y) = (x, y) \tag{4-39}$$

where, as before, a zero-dimensional singular simplex on a space is identified with a point of the space. The desired result can now be stated as follows.

Lemma 4-13. *For every pair of spaces E and F there is a chain homomorphism*

$$f: C(E) \otimes C(F) \to C(E \times F)$$

given in dimension zero by (4-39), *such that, for every pair of continuous maps* $\phi: E \to E'$ *and* $\psi: F \to F'$ *the following diagram is commutative.*

$$\begin{array}{ccc} C(E) \otimes C(F) & \xrightarrow{f} & C(E \times F) \\ {\scriptstyle \phi_1 \otimes \psi_1} \downarrow & & \downarrow {\scriptstyle (\phi \times \psi)_1} \\ C(E') \otimes C(F') & \xrightarrow{f} & C(E' \times F') \end{array} \tag{4-40}$$

Proof. Suppose that f has already been constructed for every pair of spaces E and F and for chains of dimension up to $p - 1$. Let Δ_h and Δ_k be the standard h- and k-dimensional Euclidean simplexes with $h + k = p$, and let λ and μ be the identity maps of Δ_h and Δ_k, respectively, on themselves (i.e., $\lambda = (x_0 x_1 \cdots x_h)$ and $\mu = (x_0 x_1 \cdots x_k)$ in the usual notation). Then $d(\lambda \otimes \mu)$ is an element of dimension $p - 1$ of $C(\Delta_h) \otimes C(\Delta_k)$, so by the induction hypothesis, $fd(\lambda \times \mu)$ is defined as an element of $C(\Delta_h \times \Delta_k)$. On the other hand, f is a chain homomorphism so

$$df\,d(\lambda \otimes \mu) = fd^2(\lambda \otimes \mu) = 0$$

Thus $fd(\lambda \otimes \mu)$ is a cycle on $\Delta_h \times \Delta_k$. But $\Delta_h \times \Delta_k$ is acyclic; that is, its homology groups in positive dimensions are all zero. It follows that the cycle $fd(\lambda \otimes \mu)$ can be written as dv, for some element $v \in C_p(\Delta_h \times \Delta_k)$. Now define $f(\lambda \otimes \mu)$ by the formula

$$f(\lambda \otimes \mu) = v \tag{4-41}$$

Note that automatically

$$df(\lambda \otimes \mu) = dv = fd(\lambda \otimes \mu) \tag{4-42}$$

Now take any two spaces E and F and let σ and τ be singular simplexes on E and F, respectively, of dimensions h and k, again with $h + k = p$. Remember that the p-dimensional group of $C(E) \otimes C(F)$ is generated by elements of type $\sigma \times \tau$ so, to extend f to dimension p, it need only be defined on such elements. Set

$$f(\sigma \otimes \tau) = (\sigma \times \tau)_1 v \tag{4-43}$$

where v is as in (4-41), and in the right-hand side the symbol $(\sigma \times \tau)_1$ is as in (4-40), σ and τ being continuous maps of Δ_h and Δ_k into E and F, respectively. As defined by (4-43), f is a chain homomorphism. For

$$\begin{aligned}
df(\sigma \otimes \tau) &= d(\sigma \times \tau)_1 v \\
&= (\sigma \times \tau)_1\, dv \qquad \text{(Theorem 1-2)} \\
&= (\sigma \times \tau)_1 f\,d(\lambda \otimes \mu) \qquad \text{(by (4-42))} \\
&= f(\sigma_1 \otimes \tau_1)\, d(\lambda \otimes \mu) \\
&= f\,d(\sigma_1 \otimes \tau_1)(\lambda \otimes \mu) \qquad \text{(since } \sigma_1 \otimes \tau_1 \text{ is} \\
&= f\,d[\sigma_1(\lambda) \otimes \tau_1(\mu)] \qquad \text{a chain homomorphism)}
\end{aligned}$$

Here $\sigma_1(\lambda) = \sigma\lambda = \sigma$ and $\tau_1(\mu) = \tau\mu = \tau$, so

$$df(\sigma \otimes \tau) = f\,d(\sigma \otimes \tau)$$

as required.

Next, let $\phi: E \to E'$ and $\psi: F \to F'$ be continuous maps, and consider the commutativity of the diagram (4-40). It is sufficient to check this by operating on $\sigma \otimes \tau$, where σ and τ are singular simplexes on E and F, respectively of dimensions h and k, with $h + k = p$. Then

$$\begin{aligned}(\phi \times \psi)_1 f(\sigma \otimes \tau) &= (\phi \times \psi)_1 (\sigma \times \tau)_1 \nu \qquad \text{(by (4-43))} \\ &= (\phi\sigma \times \psi\tau)_1 \nu \\ &= f(\phi\sigma \otimes \psi\tau) \qquad \text{((4-43) applied to } \phi\sigma \text{ and } \psi\tau) \\ &= f[\phi_1(\sigma) \otimes \psi_1(\tau)] \\ &= f(\phi_1 \otimes \psi_1)(\sigma \otimes \tau)\end{aligned}$$

This verifies the commutativity of (4-40) for dimension p, as required. ∎

The next step is to show that f and g are algebraic homotopy inverses of each other. The construction of the algebraic homotopies will be done inductively, very much on the same pattern as the construction of Section 3-7 for maps with a common acyclic carrier.

Lemma 4-14. *fg: $C(E \times F) \to C(E \times F)$ is algebraically homotopic to the identity for each pair of spaces E and F.*

Proof. Write $fg = h$ and denote the identity by i. Then it must be shown that there is a homomorphism $H: C_p(E \times F) \to C_{p+1}(E \times F)$, for each dimension p, such that

$$h - i = Hd + dH \tag{4-44}$$

In fact, the homomorphisms H will be constructed simultaneously for all pairs of spaces and will satisfy a commutativity condition with respect to induced homomorphisms. So assume that, for each pair of spaces E and F, the homomorphisms H have been constructed up to dimension $p - 1$, satisfying (4-44), and suppose that, for any continuous maps $\phi: E \to E'$ and $\psi: F \to F'$, the following diagram, with $j \leqq p - 1$, is commutative.

$$\begin{array}{ccc} C_j(E \times F) & \xrightarrow{H} & C_{j+1}(E \times F) \\ {\scriptstyle (\phi\times\psi)_1}\downarrow & & \downarrow{\scriptstyle (\phi\times\psi)_1} \\ C_j(E' \times F') & \xrightarrow{H} & C_{j+1}(E' \times F') \end{array} \tag{4-45}$$

This is certainly true for $j = 0$, for on dimension 0, $h = i$ and H can be taken as 0.

First consider the diagonal map

$$\Delta: \Delta_p \to \Delta_p \times \Delta_p$$

defined by $\Delta(x) = (x, x)$, where Δ_p is the standard Euclidean p simplex. Δ is a singular p simplex on $\Delta_p \times \Delta_p$ and $d\,\Delta$ is a $(p-1)$ chain. Thus, by the induction hypothesis, $Hd\Delta$ is defined and (4-44) becomes

$$hd\Delta - d\Delta = Hdd\Delta + dHd\Delta$$

That is,

$$hd\Delta - d\Delta = dHd\Delta \tag{4-46}$$

$H\Delta$ needs to be defined so that (4-44), operating on Δ, is satisfied, that is, so that

$$h\Delta - \Delta = Hd\Delta + dH\Delta \tag{4-47}$$

Now $h\Delta - \Delta - Hd\Delta$ is a p chain on $\Delta_p \times \Delta_p$, and in fact,

$$\begin{aligned} d(h\Delta - \Delta - Hd\Delta) &= dh\Delta - d\Delta - dHd\Delta \\ &= hd\Delta - d\Delta - dHd\Delta \qquad (h \text{ is a chain homomorphism}) \\ &= 0 \qquad [\text{by } (4\text{-}46)] \end{aligned}$$

Hence $h\Delta - \Delta - Hd\Delta$ is a cycle on $\Delta_p \times \Delta_p$. But the homology groups of positive dimension of $\Delta_p \times \Delta_p$ are zero, so there is a $(p+1)$ chain γ on $\Delta_p \times \Delta_p$ such that

$$h\Delta - \Delta - Hd\Delta = d\gamma$$

Now define

$$H\Delta = \gamma$$

and the required equation (4-47) is automatically satisfied.

Next H must be defined on a singular p simplex σ on $E \times F$, for any two spaces E and F. σ is a continuous map of Δ_p into $E \times F$, and if p_E and p_F are the projections of $E \times F$ onto the two factors E and F, σ can be expressed as the composition

$$\Delta_p \xrightarrow{\Delta} \Delta_p \times \Delta_p \xrightarrow{p_E\sigma \times p_F\sigma} E \times F$$

In other words,

$$\sigma = (p_E \sigma \times p_F \sigma)_1 \Delta$$

Define $H\sigma$ by the equation

$$H\sigma = (p_E \sigma \times p_F \sigma)_1 H\Delta = (p_E \sigma \times p_F \sigma)_1 \gamma \tag{4-48}$$

It must be shown that H so defined satisfies the commutativity relation (the diagram (4-45)) and Eq. (4-44). So let $\phi: E \to E'$ and $\psi: F \to F'$ be two continuous maps. The commutativity of (4-45) for dimension p will be checked by operating on a singular p simplex σ on $E \times F$.

$$\begin{aligned}(\phi \times \psi)_1 H\sigma &= (\phi \times \psi)_1 (p_E \sigma \times p_F \sigma)_1 \gamma \qquad \text{(by (4-48))} \\ &= (\phi p_E \sigma \times \psi p_F \sigma)_1 \gamma\end{aligned}$$

Here $\phi p_E \sigma = p_{E'}(\phi \times \psi)\sigma = p_{E'} \sigma'$, where $\sigma' = (\phi \times \psi)\sigma = (\phi \times \psi)_1 \sigma$. Similarly $\psi p_F \sigma = p_{F'} \sigma'$. Hence

$$\begin{aligned}(\phi \times \psi)_1 H\sigma &= (p_{E'} \sigma' \times p_{F'} \sigma')_1 \qquad \text{[from (4-48)]} \\ &= H\sigma' \qquad \text{[(4-48) applied to } \sigma'] \\ &= H(\phi \times \psi)_1 \sigma\end{aligned}$$

Thus, the commutativity of (4-45) is proved for dimension p.

Finally, it must be shown that, for any singular p simplex σ on $E \times F$

$$h\sigma - \sigma = Hd\sigma + dH\sigma$$

Start by computing $dH\sigma$.

$$\begin{aligned}dH\sigma &= d(p_E \sigma \times p_F \sigma)_1 \gamma \qquad \text{(by (4-48))} \\ &= (p_E \sigma \times p_F \sigma)_1 d\gamma \\ &= (p_E \sigma \times p_F \sigma) h\Delta - (p_E \sigma \times p_F \sigma)_1 \Delta \\ &\quad - (p_E \sigma \times p_F \sigma)_1 H d\Delta\end{aligned} \tag{4-49}$$

Here

$$(p_E \sigma \times p_F \sigma)_1 = (p_E \sigma \times p_F \sigma)\Delta = \sigma \tag{4-50}$$

h is the composition fg. The commutativity properties of f and g with induced homomorphisms (that is, the commutativity of the diagrams of

Lemmas 4-11 and 4-13) imply that

$$\begin{aligned}(p_E\sigma \times p_F\sigma)_1 h\Delta &= h(p_E\sigma \times p_F\sigma)\Delta \\ &= h\sigma \qquad \text{[by (4.0)]}\end{aligned}$$

Lastly $d\Delta$ is of dimension $p-1$, so the commutativity of H with induced homomorphisms is already assumed on $d\Delta$. Thus

$$\begin{aligned}(p_E\sigma \times p_F\sigma)_1 Hd\Delta &= H(p_E\sigma \times p_F\sigma)_1 d\Delta \\ &= Hd(p_E\sigma \times p_F\sigma)_1\Delta \\ &= Hd\sigma\end{aligned}$$

When all these computations are put together, (4-49) becomes

$$dH\sigma = h\sigma - \sigma - Hd\sigma$$

as was to be shown. ∎

Before considering the composition gf it is worthwhile to pause for a moment to see why the proof of the last lemma works. A family $\mathscr{K}$ of complexes is given, namely, complexes of the type $C(E \times F)$, along with a certain family $\mathscr{H}$ of chain homomorphisms, namely, those of the type $(\phi \times \psi)_1$ where $\phi: E \to E'$ and $\psi: F \to F'$ are continuous maps. The family of maps is closed under composition. Moreover, a subfamily $\mathscr{K}_0$ of $\mathscr{K}$ can be picked out, namely, complexes of the form $C(\Delta_p \times \Delta_p)$, which have the property that they are acyclic. Each chain complex of the family $\mathscr{K}$ is generated by images of members of the family $\mathscr{K}_0$ under certain maps of the family $\mathscr{H}$, in this case, maps of the type $(p_E\sigma \times p_F\sigma)_1$. For each complex of $\mathscr{K}$, a chain homomorphism into itself is given; here $h: C(E \times F) \to C(E \times F)$. These chain homomorphisms commute with the members of the family $\mathscr{H}$, and for each member of $\mathscr{K}$, the appropriate h is the identity on the zero-dimensional group. The argument of Lemma 4-14 then shows that the h are all algebraically homotopic to the identity.

Exercises. 4-11. Prove the statement made in the last sentence.

4-12. Use a similar argument to show that if, for each member K of $\mathscr{K}$, we have two chain homomorphisms h and k of K onto itself, commuting with the maps of $\mathscr{H}$, and if $h = k$ on the zero-dimensional groups, then h and k are algebraically homotopic.

It happens in the next lemma that the same general situation holds as that which has just been described, but with a different family of chain complexes and chain homomorphisms.

Lemma 4-15. *In the notation of Lemmas* 4-13 *and* 4-14

$$gf: C(E) \otimes C(F) \to C(E) \times C(F)$$

is algebraically homotopic to the identity for each pair of spaces E and F.

Proof. It will be shown (without full details) how the argument follows the general pattern just described, so that Exercise 4-11 above fills in all the steps. It might be wise, on a first approach to this subject, to write out all the details of the proof of this lemma without using Exercise 4-11, but following the pattern of that exercise.

The family $\mathscr{K}$ consists here of complexes of the form $C(E) \otimes C(F)$ and the maps $\mathscr{H}$ are of the form $\phi_1 \times \psi_1$, where $\phi: E \to E'$ and $\psi: F \to F'$ are continuous maps. The complex $C(E) \otimes C(F)$ is generated by elements of the type $\sigma \otimes \tau$, where σ and τ are p- and q-dimensional singular simplexes on E and F, respectively. In other words, $C(E) \otimes (F)$ is generated by images of maps of the type

$$\sigma_1 \otimes \tau_1: C(\Delta_p) \otimes C(\Delta_q) \to C(E) \otimes C(F)$$

So the subfamily $\mathscr{K}_0$ can be taken to be the set of chain complexes of the form $C(\Delta_p) \otimes C(\Delta_q)$. Exercise 4-4 shows that such complexes are acyclic. (Apply the exercise along with homotopies shrinking Δ_p, Δ_q to points!)

Then, for each E and F, $h = gf$ is a chain homomorphism of $C(E) \otimes C(F)$ into itself, equal to the identity on dimension zero and the h commute with the homomorphisms of the type $\phi_1 \otimes \psi_1$ (this follows from the commuting properties of f and g). It follows (Exercise 4-11) that each h is algebraically homotopic to the identity, as was to be shown. ∎

The results of this section are summed up in the following theorem.

Theorem 4-16. *For each pair of spaces E and F the chain homomorphisms*

$$f: C(E) \otimes C(F) \to C(E \times F)$$

and

$$g: C(E \times F) \to C(E) \otimes C(F)$$

are algebraic homotopy inverse to each other.

Actually, more has been proved. Note that the essential property of g, although given by explicit formula, is that it extends the zero-dimensional map $g(x, y) = x \otimes y$ and makes diagrams of the type (4-31) commutative.

If any other family of chain homomorphisms g' could be constructed with properties, g' could be used instead of g in the proofs of Lemmas 4-14 and 4-15. Thus, $g'f$ and fg' would be algebraically homotopic to the identity. Then $g'fg$ would be, on the one hand, algebraically homotopic to g and, on the other hand, to g'. In other words, the properties of g of making (4-31) commute and extending $g(x, y) = x \otimes y$ on dimension 0 determine the chain homomorphisms g uniquely up to algebraic homotopy. A similar remark holds for f.

Exercises. 4-13. At this point it is convenient to obtain a classical result on the homology (or cohomology) groups of a product space. By the results of this section, the homology of $E \times F$ is the same as that of the chain complex $C(E) \otimes C(F)$. This homology is to be compared with $H_*(E) \otimes H_*(F) = \sum_r H_r(E) \otimes H_{p-r}(F)$ for each p (cf. Section 4-7 and, in particular, Exercise 4-8). Prove that, if either $H_p(E)$ or $H_p(F)$ is torsion free for each p, then $H_p(E \times F) \cong \sum H_r(E) \otimes H_{p-r}(F)$ for each dimension.

4-14. When E and F are triangulable, compute the Betti numbers $b_p(E \times F)$ of $E \times F$ in terms of those of E and F to obtain the formula

$$b_p(E \times F) = \sum b_r(E) b_{p-r}(F)$$

This is known as the Künneth formula.

4-11. THE DIAGONAL MAP

Returning now to the geometric consideration of the cup product, remember that a cochain homomorphism

$$\pi\colon C^*(E) \otimes C^*(E) \to [C(E) \otimes C(E)]^*$$

was constructed in Section 4-6. Combine this with the dual g' of the chain homomorphism g of Section 4-9, with $F = E$, to obtain the diagram

$$C^*(E) \otimes C^*(E) \xrightarrow{\pi} [C(E) \otimes C(E)]^* \xrightarrow{g'} C^*(E \times E)$$

As pointed out in Section 4-8, this leads back to $C^*(E)$ very naturally by composing g' with the homomorphism

$$\Delta^1\colon C^*(E \times E) \to C^*(E)$$

induced by the diagonal map $\Delta\colon E \to E \times E$, which is defined by $\Delta(x) = (x,x)$. Thus, there is a sequence of homomorphisms

$$C^*(E) \otimes C^*(E) \to [C(E) \otimes C(E)]^* \to C^*(E \times E) \xrightarrow{\Delta^1} C^*(E) \qquad (4\text{-}51)$$

The main result of this section can now be stated.

Theorem 4-17. *The composition of the maps in* (4-51) *is the cup product or, in symbols,*

$$\cup = \Delta^1 g' \pi \tag{4-52}$$

Proof. Take $\alpha \in C^q(E)$, $\beta \in C^r(E)$ with $q + r = p$ and let σ be a singular p simplex on E. Then, by definition,

$$(\alpha \cup \beta)(\sigma) = \alpha(\sigma F_q)\beta(\sigma L_r)$$

Consider the right-hand side of (4-52). $\Delta^1 g'$ is the dual of $g\Delta_1$ and, for the singular simplex σ on E,

$$\begin{aligned} g\Delta_1\sigma &= g(\Delta\sigma) \\ &= \sum (p_1\Delta\sigma F_i) \otimes (p_2\Delta\sigma L_{p-i}) \end{aligned}$$

by (4-30), where p_1 and p_2 denote the projections of $E \times E$ on the first and second factors, respectively. But $p_1\Delta$ and $p_2\Delta$ are both the identity map of E on itself, so

$$g\Delta_1\sigma = \sum (\sigma F_i) \otimes (\sigma L_{p-i})$$

Now, applying the right-hand side of (4-52) to $\alpha \otimes \beta$ and operating on σ,

$$\begin{aligned} [\Delta^1 g'\pi(\alpha \otimes \beta)]\sigma &= \pi(\alpha \otimes \beta)(g\Delta_1\sigma) \\ &= \pi(\alpha \otimes \beta)[\sum (\sigma F_i) \otimes (\sigma L_{p-i})] \end{aligned}$$

Here, by the definition (4-23) of π, there is a nonzero contribution from the right-hand side only when $i = q$, $p - i = r$, and then the right-hand side becomes $\alpha(\sigma F_q)\beta(\sigma L_r)$; but this is $(\alpha \cup \beta)\sigma$ so

$$(\alpha \cup \beta)\sigma = [\Delta^1 g'\pi(\alpha \otimes \beta)]\sigma$$

as was to be shown. ∎

4-12. SUMMING UP

The whole discussion of Sections 4-5–4-11 began with the idea that the $\cup$ product structure of the cohomology ring should be induced in some way by a cochain homomorphism of $C^*(E) \otimes C^*(E)$ into $C^*(E)$. This led to the consideration, in a purely algebraic setting, of the notion of tensor product of chain and cochain complexes and, in particular, to the definition of the homomorphism

$$\pi\colon C^*(E) \otimes C^*(F) \to [C(E) \otimes C(F)]^*$$

Then, in the attempt to connect the chain complex $C(E) \otimes C(F)$ with something more directly geometrical, it was compared with $C(E \times F)$ and it turned out that chain homomorphisms

$$f: C(E) \otimes C(F) \to C(E \times F)$$
$$g: C(E \times F) \to C(E) \otimes C(F)$$

could be constructed which were algebraically homotopy inverse to each other. Moreover, the construction of f and g turned out to be very natural, in the sense that, if they obey all the desired commutativity conditions, then they are unique up to algebraic homotopy. The final, purely geometric step was the introduction of the diagonal map Δ and the proof of Theorem 4-17, showing that the cochain homomorphism $\Delta^1 g' \pi$ is exactly $\cup$.

The $\cup$ product on cohomology classes is constructed thus by composing the following sequence of homomorphisms:

$$H^*(E) \otimes H^*(E) \xrightarrow{\Phi} H^*[C^*(E) \otimes C^*(E)] \xrightarrow{\pi^*} H^*[C(E) \otimes C(E)]$$
$$\xrightarrow{g^*} H^*(E \times E) \xrightarrow{\Delta^*} H^*(E) \tag{4-53}$$

where the first is the homomorphism Φ of Section 4-7, $H^*[C^*(E) \otimes C^*(E)]$ denotes cohomology of the cochain complex $C^*(E) \otimes C^*(E)$, and π^*, g^*, and Δ^* are induced by π, g, and Δ.

4-13. THE x PRODUCT

In the sequence (4-52) of homomorphisms, the first three can be constructed for any two spaces E and F, the situation of (4-52) being obtained when $E = F$. Consider the general case. The composition of the homomorphisms

$$H^*(E) \otimes H^*(F) \xrightarrow{\Phi} H^*[C^*(E) \otimes C^*(F)] \xrightarrow{\pi^*} H^*[C(E) \otimes C(F)] \xrightarrow{g^*} H^*(E \times F)$$

will be denoted by **x**. It is convenient to denote the operation of **x** by the notation

$$\mathbf{x}(\alpha \otimes \beta) = \alpha \mathbf{x} \beta$$

where $\alpha \otimes \beta \in H^*(E) \otimes H^*(F)$.

Definition 4-6. The homomorphism **x**, so defined, is called *the* **x** *product on* $H^*(E)$ *and* $H^*(F)$. Note that it takes value in $H^*(E \times F)$.

The **x** product obeys a commutativity rule like that satisfied by the $\cup$ product.

Theorem 4-18. *Let $\phi: E \to E'$ and $\psi: F \to F'$ be continuous maps. Then the following diagram commutes.*

$$\begin{array}{ccc} H^*(E') \otimes H^*(F') & \xrightarrow{\mathbf{x}} & H^*(E' \times F') \\ \downarrow{\scriptstyle \phi^* \otimes \psi^*} & & \downarrow{\scriptstyle (\phi \times \psi)^*} \\ H^*(E) \otimes H^*(F) & \xrightarrow{\mathbf{x}} & H^*(E \times F) \end{array}$$

Proof. The fact that g^* commutes with induced homomorphisms follows from Lemma 4-11. That Φ and π^* commute with induced homomorphisms follows from Exercises 4-6 and 4-10. When these commutativity properties are put together, the stated result follows.

According to the last section, the $\cup$ product now becomes the composition of the homomorphisms

$$H^*(E) \otimes H^*(E) \xrightarrow{\mathbf{x}} H^*(E \times E) \xrightarrow{\Delta^*} H^*(E)$$

which can be stated formally as follows.

Theorem 4-19. $\cup = \Delta^* \mathbf{x}$.

This theorem expresses $\cup$ in terms of **x**. It is interesting to notice that there is an inverse relation expressing **x** in terms of $\cup$. This is obtained in the following way. The projections p_E and p_F, respectively, map $E \times F$ on E and F, so there are induced homomorphisms

$$p_E^*: H^*(E) \to H^*(E \times F)$$

$$p_F^*: H^*(F) \to H^*(E \times F)$$

Construct the sequence of homomorphisms

$$H^*(E) \times H^*(F) \xrightarrow{p_E^* \otimes p_F^*} H^*(E \times F) \otimes H^*(E \times F) \xrightarrow{\cup} H^*(E \times F) \tag{4-54}$$

It will be shown that the composition of these is **x**: the formal statement of this follows.

Theorem 4-20. *In* (4-53) *the composition* $\cup(p_E^* \otimes p_F^*)$ *is equal to* **x**.

Proof. In (4-53) express the homomorphism $\cup$ in terms of $\mathbf{x}$ by means of Theorem 4-19. The sequence (4-54) then becomes (replacing E by $E \times F$):

$$H^*(E) \otimes H^*(F) \xrightarrow{p_E^* \otimes p_F^*} H^*(E \times F) \times H^*(E \times F) \xrightarrow{\mathbf{x}} H^*(E \times F \times E \times F)$$

$$\xrightarrow{\Delta^*} H^*(E \times F) \qquad (4\text{-}55)$$

By its definition, $\mathbf{x}$ is a homomorphism that commutes with induced homomorphisms (Theorem 4-18), so the following commutative diagram can be set up.

$$\begin{array}{ccccc} H^*(E) \otimes H^*(F) & \xrightarrow{\quad\mathbf{x}\quad} & H^*(E \times F) & & \\ \downarrow{\scriptstyle p_E^* \otimes p_F^*} & & \downarrow{\scriptstyle (p_E \times p_F)^*} & & \\ H^*(E \times F) \otimes H^*(E \times F) & \xrightarrow{\mathbf{x}} & H^*(E \times F \times E \times F) & \xrightarrow{\Delta^*} & H^*(E \times F) \end{array} \qquad (4\text{-}56)$$

Note here the map $p_E \times p_F$ maps the element (x, y, u, v) of $E \times F \times E \times F$ on (x, v) of $(E \times F)$.

In the diagram (4-56) the composition of the left edge and the lower edge is the sequence (4-55), which must be shown to equal $\mathbf{x}$: $H^*(E) \otimes H^*(F) \to H^*(E \times F)$. By the commutativity of (4-56), it is sufficient to show that the composition $\Delta^*(p_E \times p_F)^*$ in (4-56) is the identity. This follows at once, however, from the fact that the composition of the maps

$$E \times F \xrightarrow{\Delta} E \times F \times E \times F \xrightarrow{p_E \times p_F} E \times F$$

is the identity. ∎

Note that, in this proof, the actual definition of $\mathbf{x}$ is not used; only its property of commuting with induced homomorphisms is used. This suggests setting up products of the same type as $\mathbf{x}$ and $\cup$, namely, homomorphisms from $H^*(E) \otimes H^*(F)$ to $H^*(E \times F)$ and from $H^*(E) \otimes H^*(E)$ to $H^*(E)$, respectively, which are bilinear and commute with induced homomorphisms, and which bear the same sort of relations to each other as $\mathbf{x}$ and $\cup$. It is of interest to consider such homomorphisms to have been induced by cochain homomorphisms, so the definitions are formulated in that way.

Definition 4-7. A $\cup$-type product is a cochain homomorphism for each space E

$$\cup_1 : C^*(E) \otimes C^*(E) \to C^*(E)$$

such that, for any continuous map $\phi: E \to E'$, the following diagram is homotopy commutative:

$$\begin{array}{ccc} C^*(E) \otimes C^*(E) & \xrightarrow{\cup_1} & C^*(E) \\ {\scriptstyle \phi^1 \otimes \phi^1} \uparrow & & \uparrow {\scriptstyle \phi^1} \\ C^*(E') \otimes C^*(E') & \xrightarrow{\cup_1} & C^*(E') \end{array}$$

(Homotopy commutative means that $\cup_1(\phi^1 \otimes \phi^1)$ and $\phi^1 \cup_1$ are algebraically homotopic.)

A **x**-type product is a cochain homomorphism for each pair of spaces E and F

$$\mathbf{x}_1: C^*(E) \otimes C^*(F) \to C^*(E \times F)$$

such that for any continuous maps $\phi: E \to E'$ and $\psi: F — F'$ the diagram

$$\begin{array}{ccc} C^*(E) \otimes C^*(F) & \xrightarrow{\mathbf{x}_1} & C^*(E \times F) \\ {\scriptstyle \phi^1 \otimes \psi^1} \uparrow & & \uparrow {\scriptstyle (\phi \times \psi)^1} \\ C^*(E') \otimes C^*(F') & \xrightarrow{\mathbf{x}_1} & C^*(E' \times F') \end{array}$$

is homotopy commutative.

With a $\cup$-type product $\cup_1$ there is associated a **x**-type product; namely, for any spaces E and F, the composition $\mathbf{x}_1$ of the sequence of homomorphisms

$$C^*(E) \otimes C^*(F) \xrightarrow{p_E^1 \otimes p_F^1} C^*(E \times F) \otimes C^*(E \times F) \xrightarrow{\cup_1} C^*(E \times F) \tag{4-57a}$$

With a **x**-type product $\mathbf{x}_1$ there is associated a cup-type product $\cup_1$, namely, for any space E, the composition of the following sequence of homomorphisms:

$$C^*(E) \otimes C^*(E) \xrightarrow{\mathbf{x}_1} C^*(E \times E) \xrightarrow{\Delta^1} C^*(E) \tag{4-57b}$$

The following two theorems give the reciprocal nature of the relation between the two types of products.

Theorem 4-21. *Let* $\mathbf{x}_1$ *be a* **x**-*type product and associate with it a* $\cup$-*type product* $\cup_1$ *(by* (4-57b)). *With* $\cup_1$ *associate a* **x**-*type product* $\mathbf{x}_2$ *(by* (4-57a)). *Then* $\mathbf{x}_1$ *and* $\mathbf{x}_2$ *are algebraically homotopic for any spaces E and F.*

Proof. Essentially, this is the argument of Theorem 4-20. $\mathbf{x}_2$ is the composition of the sequence of homomorphisms

$$C^*(E) \otimes C^*(F) \xrightarrow{p_E{}^1 \otimes p_F{}^1} C^*(E \times F) \otimes C^*(E \times F) \xrightarrow{\mathbf{x}_1} C^*(E \times F \times E \times F) \xrightarrow{\Delta^1} C^*(E \times F)$$

Corresponding to the maps p_E and p_F, there is, by Definition 4-7, a homotopy commutative diagram

$$\begin{array}{ccc} C^*(E \times F) \otimes C^*(E \times F) & \xrightarrow{\mathbf{x}_1} & C^*(E \times F \times E \times F) \\ {\scriptstyle p_E{}^1 \otimes p_F{}^1} \uparrow & & \uparrow {\scriptstyle (p_E \times p_F)^1} \\ C^*(E) \otimes C^*(F) & \xrightarrow{\mathbf{x}_1} & C^*(E \times F) \end{array}$$

Thus, $\mathbf{x}_1(p_E{}^1 \otimes p_F{}^1)$ is algebraically homotopic to $(p_E \times p_F)^1\mathbf{x}_1$, so $\mathbf{x}_2 = \Delta^1\mathbf{x}_1(p_E{}^1 \otimes p_F{}^1)$ is algebraically homotopic to $\Delta^1(p_E \times p_F)^1\mathbf{x}_1 = \mathbf{x}_1$ (since $(p_E \times p_F)\Delta =$ the identity) and this completes the proof. ∎

Theorem 4-22. *Let* $\cup_1$ *be a* $\cup$*-type product and associate with it a* $\mathbf{x}$*-type product* $\mathbf{x}_1$ *(by* (4-57a)). *Associate with* $\mathbf{x}_1$ *(by* (4-57b)) *a* $\cup$*-type product* $\cup_2$. *Then* $\cup_1$ *and* $\cup_2$ *are algebraically homotopic for any space E.*

Proof. In the diagram

$$C^*(E) \otimes C^*(E) \xrightarrow{p^1 \otimes q^1} C^*(E \times E) \otimes C^*(E \times E) \xrightarrow{\cup_1} C^*(E \times E) \xrightarrow{\Delta^1} C^*(E)$$

where p and q are the projections of $E \times E$ on the first and second factors, respectively, the composition of the first two homomorphisms is $\mathbf{x}_1$, and the composition of all three is $\cup_2$. Then corresponding to the diagonal map $\Delta\colon E \to E \times E$ there is, by Definition 4-7, a homotopy commutative diagram

$$\begin{array}{ccc} C^*(E \times E) \otimes C^*(E \times E) & \xrightarrow{\cup_1} & C^*(E \times E) \\ {\scriptstyle \Delta^1 \otimes \Delta^1} \downarrow & & \downarrow {\scriptstyle \Delta^1} \\ C^*(E) \otimes C^*(E) & \xrightarrow{\cup_1} & C^*(E) \end{array}$$

Thus $\Delta^1\cup_1$ is algebraically homotopic to $\cup_1(\Delta^1 \otimes \Delta^1)$, so $\Delta^1 \cup_1(p^1 \otimes q^1)$ is algebraically homotopic to $\cup_1(\Delta^1 \otimes \Delta^1)(p^1 \otimes q^1)$. But $p\Delta$ and $q\Delta$ are both the identity, so $(\Delta^1 \otimes \Delta^1)(p^1 \otimes q^1)$ is the identity. Thus $\cup_2 = \Delta^1\cup_1(p^1 \otimes q^1)$ is algebraically homotopic to $\cup_1$, as required. ∎

4-14. ANTICOMMUTATIVITY OF THE ∪ PRODUCT

In the computation of the cohomology ring of the torus it was seen that the product of the one-dimensional cohomology classes $\bar{\alpha}$, $\bar{\beta}$ satisfied the equation

$$\bar{\alpha} \cup \bar{\beta} = -\bar{\beta} \cup \bar{\alpha}$$

It will now be shown that an anticommutativity law of this type holds in general.

In the first place, consider the effect of reversing the factors in a tensor product $C(E) \otimes C(E)$, where $C(E)$ is the singular chain complex on a space E. Clearly, if $x \otimes y$ is mapped on $y \otimes x$, an isomorphism is obtained, but to give this significance in terms of homology and cohomology a chain homomorphism is needed. The way to obtain this is to insert a sign factor $(-1)^{pq}$ where p and q are the dimensions of x and y.

Lemma 4-23. *Let* $\rho: C(E) \otimes C(E) \to C(E) \otimes C(E)$ *be defined by*

$$\rho(x \otimes y) = (-1)^{pq} y \otimes x$$

where $\dim x = p$ *and* $\dim y = q$. *Then* ρ *is a chain homomorphism and is an isomorphism onto.*

Proof. Clearly, ρ is an isomorphism onto. To check that it commutes with the boundary operator let x and y be of dimensions p and q, respectively.

$$\begin{aligned} d\rho(x \otimes y) &= (-1)^{pq} d(y \otimes x) \\ &= (-1)^{pq} dy \otimes x + (-1)^{pq+q} y \otimes dx \\ \rho d(x \otimes y) &= \rho[dx \otimes y + (-1)^p x \otimes dy] \\ &= (-1)^{(p-1)q} y \otimes dx + (-1)^{p+p(q-1)} dy \otimes x \\ &= (-1)^{pq-q} y \otimes dx + (-1)^{pq} dy \otimes x \end{aligned}$$

The final results of these two computations are the same, so $d\rho = \rho d$, as was to be shown.

It is natural to expect ρ to have the following property. ∎

Lemma 4-24. *Let* $\phi: E \to F$ *be a continuous map and* ϕ_1 *the induced homomorphism on singular chain groups. Then the following diagram commutes.*

$$\begin{array}{ccc} C(E) \otimes C(E) & \xrightarrow{\rho} & C(E) \otimes C(E) \\ \downarrow{\scriptstyle \phi_1 \otimes \phi_1} & & \downarrow{\scriptstyle \phi_1 \otimes \phi_1} \\ C(F) \otimes C(F) & \xrightarrow{\rho} & C(F) \otimes C(F) \end{array}$$

Proof. The proof of this is immediate.

Now compare the chain homomorphisms $g\Delta_1$ and $\rho g\Delta_1$. For each space E, $\rho g\Delta_1$ is a chain homomorphism of $C(E)$ into $C(E) \otimes C(E)$. If ϕ is a continuous map, the commutativity properties of ρ (Lemma 4-24), g (Lemma 4-11), and Δ_1 (induced by a continuous map) imply that the following diagram commutes.

$$\begin{array}{ccc} C(E) & \xrightarrow{\rho g \Delta_1} & C(E) \otimes C(E) \\ {\scriptstyle \varphi_1}\downarrow & & \downarrow{\scriptstyle \varphi_1 \otimes \varphi_1} \\ C(F) & \xrightarrow{\rho g \Delta_1} & C(F) \otimes C(F) \end{array}$$

A similar statement holds for $g\Delta_1$. Then, an acyclicity argument similar to that of Lemma 4-14 implies that, for any space, $\rho g\Delta_1$ is algebraically homotopic to $g\Delta_1$. This result can alternatively be stated as follows.

Lemma 4-25. *For any space E the following diagram is algebraic homotopy commutative.*

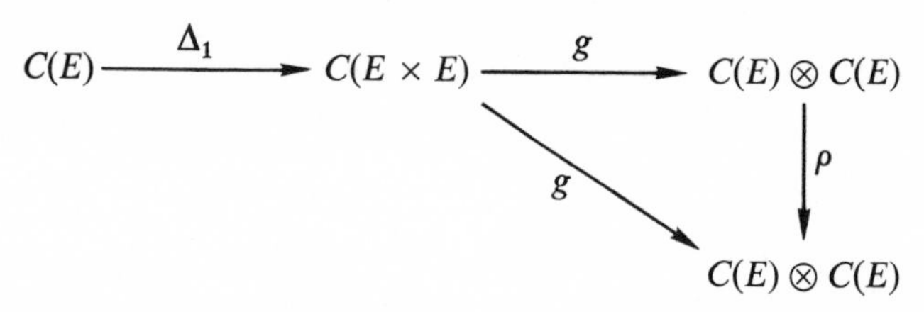

The use of the map ρ leads easily to the anticommutativity rule for the $\cup$ product.

Theorem 4-26. *The* $\cup$ *product*

$$\cup : H^*(E) \otimes H^*(E) H^* \to (E)$$

satisfies the equation

$$\bar{\alpha} \cup \bar{\beta} = (-1)^{pq} \bar{\beta} \cup \bar{\alpha}$$

where $\dim \bar{\alpha} = p$ *and* $\dim \bar{\beta} = q$.

Proof. By Lemma 4-25, the following diagram is homotopy commutative

$$\begin{array}{ccccccc} C^*(E) \otimes C^*(E) & \xrightarrow{\pi} & (C(E) \otimes C(E))^* & \xrightarrow{g'} & C^*(E \times E) & \xrightarrow{\Delta^1} & C^*(E) \\ & & \downarrow{\scriptstyle \rho'} & \nearrow{\scriptstyle g'} & & & \\ & & (C(E) \otimes C(E))^* & & & & \end{array} \qquad (4\text{-}58)$$

where Δ^1, g' and ρ' are the dual of Δ_1, g, and ρ and π is as in Section 4-6. Now take $\alpha \in C^p(E)$, $x \in C^q(E)$, $\beta \in C_p(E)$, $y \in C_q(E)$. Then

$$\begin{aligned}\rho'\pi(\alpha \otimes \beta)(y \otimes x) &= \pi(\alpha \otimes \beta)\rho(y \otimes x)\\ &= (-1)^{pq}\pi(\alpha \otimes \beta)(x \otimes y) \qquad \text{(Definition of } \rho)\\ &= (-1)^{pq}\alpha(x)\beta(y) \qquad \text{(Definition of } \pi)\\ &= (-1)^{pq}\beta(y)\alpha(x)\\ &= (-1)^{pq}\pi(\beta \otimes \alpha)(y \otimes x)\end{aligned}$$

Hence

$$\rho'\pi(\alpha \otimes \beta) = (-1)^{pq}\pi(\beta \otimes \alpha)$$

and also

$$\Delta^1 g'\rho'\pi(\alpha \otimes \beta) = (-1)^{pq}\Delta^1 g'\pi(\beta \otimes \alpha) \tag{4-59}$$

But the homotopy commutativity of (4-58) implies that, if α and β are cocycles representing elements $\bar{\alpha}$ and $\bar{\beta}$ of $H^p(E)$ and $H^q(E)$, then

$$\Delta^1 g'\rho'\pi(\alpha \otimes \beta) \sim \Delta^1 g'\pi(\alpha \otimes \beta)$$

Combine this with (4-59), take cohomology classes, and apply Theorem 4-17. It follows that

$$\bar{\alpha} \cup \bar{\beta} = (-1)^{pq}(\bar{\beta} \cup \bar{\alpha})$$

as required. ∎

4-15. THE COHOMOLOGY RING OF A PRODUCT SPACE

The object of this section is to try to express the $\cup$ product on a topological product $E \times F$ in terms of the $\cup$ products on the separate factors E and F. First, remember that for any space E the $\cup$ product is the composition of the homomorphisms

$$C^*(E) \otimes C^*(E) \xrightarrow{\pi} [C(E) \otimes C(E)]^* \xrightarrow{g'} C^*(E \times E) \xrightarrow{\Delta^1} C^*(E)$$

in the notation of Theorem 4-17. For the present purpose, it is more convenient to combine the homomorphisms g' and Δ^1. $\Delta^1 g'$ is the dual of $g\Delta_1$ where Δ_1 is the homomorphism induced on chains by the diagonal map and g is given by the formula

$$g(\sigma) = \sum_{i=0}^{r} (p\sigma F_i) \otimes (q\sigma L_{r-i})$$

where σ is a singular r simplex on $E \times E$ and p and q are the projections of $E \times E$ on the first and second factors, respectively. On the other hand, if σ is a singular r simplex on E then $\Delta_1 \sigma$ is the composition $\Delta\sigma$, given by $\Delta\sigma(x) = [\sigma(x), \sigma(x)]$ for any $x \in E$. By the formula for g applied to $\Delta\sigma$, since $p\Delta$ and $q\Delta$ are both identity maps, it follows that

$$g\,\Delta_1 \sigma = \sum_{i=0}^{r} (\sigma F_i) \otimes (\sigma L_{r-i})$$

for any singular r simplex σ on E. Now write $D_E = g\Delta_1$. Thus D_E is a chain homomorphism

$$D_E\colon C(E) \to C(E) \otimes C(E)$$

with the defining formula

$$D_E(\sigma) = \sum_{i=0}^{r} (\sigma F_i) \otimes (\sigma L_{r-i})$$

for any singular r simplex σ on E. If only one space is being studied D_E can simply be written as D. So the $\cup$ product now appears as the composition of the homomorphisms

$$C^*(E) \otimes C^*(E) \xrightarrow{\pi} [C(E) \otimes C(E)]^* \xrightarrow{D'_E} C^*(E)$$

where D'_E is the dual of D_E.

For a product space $E \times F$ the $\cup$ product is thus the composition of the cochain homomorphisms

$$C^*(E \times F) \otimes C^*(E \times F) \xrightarrow{\pi} [C(E \times F) \otimes C(E \times F)]^* \xrightarrow{D'_{E \times F}} C^*(E \times F) \tag{4-60}$$

It is known that, for the purposes of studying homology and cohomology, $C(E \times F)$ can be replaced by $C(E) \otimes C(F)$. Thus, it should be possible, if π and D' are replaced by the appropriate maps, to replace the sequence (4-60) by a sequence in which all the topological products are replaced by tensor products. More precisely, it should be possible to construct a diagram

$$\begin{array}{ccccc}
C^*(E \times F) & & [C(E \times F) & & \\
\otimes\, C^*(E \times F) & \xrightarrow{\pi} & \otimes\, C(E \times F)]^* & \xrightarrow{D'_{E \times F}} & C^*(E \times F) \\
\uparrow{\scriptstyle g' \otimes g'} & & \uparrow{\scriptstyle (g \times g)'} & & \uparrow{\scriptstyle g'} \\
[C(E) \otimes C(F)]^* & & [C(E) \otimes C(F) & & \\
\otimes\, [C(E) \otimes C(F)]^* & \xrightarrow{\phi} & \otimes\, C(E) \otimes C(F)]^* & \xrightarrow{\psi} & [C(E) \otimes C(F)]^*
\end{array} \tag{4-61}$$

which is commutative or at least homotopy commutative. Here g' is the dual of the chain homomorphism g of Section 4-9, and of course it has an algebraic homotopy inverse. Thus $g' \otimes g'$ and $(g \times g)'$ also have algebraic homotopy inverses, so as far as cohomology is concerned the two rows of (4-61) are essentially equivalent. Hopefully the maps φ and ψ can be defined in an algebraically natural way that will make computation possible.

The choice for ϕ is easy: it will be the homomorphism π of Section 4-6, that is, Eq. (4-23) with $K = L = C(E) \otimes C(F)$. The commutativity of π with cochain homomorphisms (Exercise 4-6) ensures then that the left-hand square of (4-61) is commutative.

On the other hand, the natural way of mapping $C(E) \otimes C(F)$ into $C(E) \otimes C(F) \otimes C(E) \otimes C(F)$ is to consider the map

$$D_E \otimes D_F\colon C(E) \otimes C(F) \to C(E) \otimes C(E) \otimes C(F) \otimes C(F)$$

and to compose it with the factor reversing map

$$\rho\colon C(E) \otimes C(E) \otimes C(F) \otimes C(F) \to C(E) \otimes C(F) \otimes C(E) \otimes C(F)$$

defined by

$$\rho(x \otimes y \otimes u \otimes v) = (-1)^{pq} x \otimes u \otimes y \otimes v$$

where p and q are the dimensions of y and u, respectively. By Lemma 4-23, ρ is a chain homomorphism. It will now be shown that, if ψ is taken as the dual of $\rho(D_E \otimes D_F)$; that is, if $\psi = (D_E \otimes D_F)'\rho'$, then the right half of (4-61) becomes homotopy commutative. It is more convenient to prove the homotopy commutativity before taking duals.

Lemma 4-27. *For any spaces E and F the following diagram is homotopy commutative.*

$$\begin{array}{ccc}
C(E \times F) & \xrightarrow{D_{E \times F}} & C(E \times F) \otimes C(E \times F) \\
\downarrow{\scriptstyle g} & & \downarrow{\scriptstyle g \otimes g} \\
C(E) \otimes C(F) & \xrightarrow{\rho(D_E \otimes D_F)} & C(E) \otimes C(F) \otimes C(E) \otimes C(F)
\end{array} \tag{4-62}$$

Proof. The proof is by an acyclicity argument of the type described in Section 4-10. Note that both $(g \otimes g)D_{E \times F}$ and $\rho(D_E \otimes D_F)g$ are chain homomorphisms of $C(E \times F)$ into $C(E) \otimes C(F) \otimes C(E) \otimes C(F)$ for any two spaces E and F. It is not hard to see that they commute with homomorphisms induced by continuous maps, and it is trivial that $(g \otimes g)D_{E \times F} = \rho(D_E \otimes D_F)g$ on dimension zero. The argument described in Section 4-10 must be modified somewhat since here the maps are from one complex to another (rather than from a complex into itself), but the basic idea is the same. In particular, the acyclic complexes that are used here are tensor products of singular chain complexes on Euclidean simplexes. The result is that $(g \otimes g)D_{E \times F}$ is algebraically homotopic to $\rho(D_E \times D_F)g$ for any two spaces E and F, as required. ∎

Then it follows from the dual of the diagram (4-62) that the right half of (4-61) is homotopy commutative if ψ is replaced by $(D_E \otimes D_F)'\rho'$. The result so far can be summed up.

Theorem 4-28. *The diagram* (4-61) *is homotopy commutative if* $\phi = \pi$ *and* $\psi = (D_E \otimes D_F)'\rho'$.

Now there is a homotopy commutative diagram

$$\begin{array}{ccccc} \begin{array}{c} C^*(E \times F) \\ \otimes\, C^*(E \times F) \end{array} & \xrightarrow{\pi} & \begin{array}{c} [C(E \times F) \\ \otimes\, C(E \times F)]^* \end{array} & \xrightarrow{D_{E \times F}'} & C^*(E \times F) \\ \uparrow{\scriptstyle g' \otimes g'} & & \uparrow{\scriptstyle (g \otimes g)'} & & \uparrow{\scriptstyle g'} \\ \begin{array}{c} [C(E) \otimes C(F)]^* \\ \otimes\, [C(E) \otimes C(F)]^* \end{array} & \xrightarrow{\pi} & \begin{array}{c} [C(E) \otimes C(F) \\ \otimes\, C(E) \otimes C(F)]^* \end{array} & \xrightarrow{(D_E \times D_F)'\rho'} & [C(E) \otimes C(F)]^* \end{array} \tag{4-63}$$

which has the property that the composition of the top line is the $\cup$ product on cochains of $E \times F$. Hence to calculate the product $\bar{\alpha}_0 \cup \bar{\beta}_0$ of two elements $\bar{\alpha}_0, \bar{\beta}_0$ in $H^*(E \times F)$, take representative cocycles α_0, β_0 and their, images α, β in $[C(E) \otimes C(F)]^*$ (under an algebraic homotopy inverse of g'), compute the action of the lower line of (4-63) on $\alpha \otimes \beta$, return to $C^*(E \times F)$ by means of g', and take the cohomology class of the result. The key to the computation is of course the calculation of the effect of the lower line of (4-63) on $\alpha \otimes \beta$. For convenience of notation, this is also denoted by $\cup$. Thus, $(D_E \otimes D_F)'\rho'\pi(\alpha \otimes \beta)$ is written as $\alpha \cup \beta$. This will now be computed.

An element of $[C(E) \otimes C(F)]^*$ is said to be homogeneous of degree (p, q) if it acts as zero on each group $C_h(E) \otimes C_k(F)$, except when $h = p$, $k = q$. Clearly $[C(E) \otimes C(F)]^*$ is generated by homogeneous elements for various p and q, so it is sufficient to calculate a product $\alpha \cup \beta$ of such elements.

So let $\alpha, \beta \in [C(E) \otimes C(F)]^*$ be homogeneous elements of degrees (p, q) and (r, s), respectively. Let σ, τ be singular simplexes of dimensions h and k

on E and F, respectively. Compute the operation of $\alpha \cup \beta$ on $\sigma \otimes \tau$.

$$\begin{aligned}(\alpha \cup \beta)(\sigma \otimes \tau) &= [(D_E \otimes D_F)'\rho'\pi(\alpha \otimes \beta)](\sigma \otimes \tau)\\ &= \pi(\alpha \otimes \beta)[\rho(D_E \otimes D_F)(\sigma \otimes \tau)]\\ &= \pi(\alpha \otimes \beta)[\rho(D_E\sigma \otimes D_F\sigma)]\\ &= \pi(\alpha \otimes \beta)\Big[\rho \sum_{\substack{i+j=h\\ l+m=k}} (\sigma F_i) \otimes (\sigma L_j) \otimes (\tau F_l) \otimes (\tau L_m)\Big]\\ &= \pi(\alpha \otimes \beta)\Big[\sum_{\substack{i+j=h\\ l+m=k}} (-1)^{lj}(\sigma F_i) \otimes (\tau F_l) \otimes (\sigma L_j) \otimes (\tau L_m)\Big]\end{aligned}$$

Here, since the degrees of α and β are (p, q) and (r, s), the only nonzero contribution are obtained from the terms with $i = p$, $l = q$ $j = r$, $m = s$. So from the definition of π,

$$(\alpha \cup \beta)(\sigma \otimes \tau) = (-1)^{qr}\alpha[(\sigma F_p) \otimes (\tau F_q)]\beta[(\sigma L_r) \otimes (\tau L_s)] \tag{4-64}$$

This completes the computation. Note that to get a nonzero result on the right the dimensions of σ and τ must satisfy $h = p + r$ and $k = q + s$; that is, $\alpha \cup \beta$ is of degree $(p + r, q + s)$.

4-16. SOME COMPUTATIONS

Some examples of the use of the formula (4-64) of the last section will now be given for the case in which E and F are simplicial complexes with given cell decompositions. The coefficient ring throughout this section will be the integers.

In the first place, since g: $C(E \times F) \to C(E) \otimes C(F)$ for any spaces E and F is a map with an algebraic homotopy inverse, it follows that, so far as the computation of homology and cohomology is concerned, the top line of the diagram (4-63) can be ignored. Cohomology groups for the product space can be computed from the cochain complex $[C(E) \otimes C(F)]^*$ and the $\cup$ product can be computed from the formula (4-64). Next, in the present case, E and F are supposed to be cell complexes with associated cellular chain complexes $\mathscr{C}(E)$ and $\mathscr{C}(F)$ in the usual notation, and there are inclusion maps i: $\mathscr{C}(E) \to C(E)$ and j: $\mathscr{C}(F) \to C(F)$ with algebraic homotopy inverses (Exercise 3-3). It follows, by Exercise 4-4, that the chain homomorphism

$$i \otimes j\colon \mathscr{C}(E) \otimes \mathscr{C}(F) \to C(E) \otimes C(F)$$

has an algebraic homotopy inverse. In fact, it is easy to see that $i \otimes j$ is also an inclusion map. It follows that the cohomology groups of $[C(E) \otimes C(F)]^*$ can be computed from the complex $[\mathscr{C}(E) \otimes \mathscr{C}(F)]^*$. On the other hand, the cellular chain groups $\mathscr{C}_p(E)$ and $\mathscr{C}_q(F)$ are free Abelian groups generated by the oriented cells on E and F. Write $\sigma_{p1}, \sigma_{p2}, \ldots$ and $\tau_{q1}, \tau_{q2}, \ldots$ for sets of generators of $\bar{\mathscr{C}}_p(E)$ and $\bar{\mathscr{C}}_q(F)$, respectively. The cochain groups $\bar{\mathscr{C}}^p(E)$ and $\bar{\mathscr{C}}^q(F)$ then have dual sets of generators, which are denoted by $\sigma^{p1}, \sigma^{p2}, \ldots$ and $\tau^{p1}, \tau^{p2}, \ldots$, respectively (these are cocells on E and F, respectively). It is not hard to see then that $\bar{\mathscr{C}}_p(E) \otimes \bar{\mathscr{C}}_q(F)$ is a free group generated by the products $\pi_{pi} \otimes \pi_{qj}$. Similarly, $\bar{\mathscr{C}}^p(E) \otimes \bar{\mathscr{C}}^q(F)$ is a free group generated by the $\sigma^{pi} \otimes \sigma^{qj}$. Then, if $\pi\colon \bar{\mathscr{C}}^*(E) \otimes \bar{\mathscr{C}}^*(F) \to [\bar{\mathscr{C}}(E) \otimes \bar{\mathscr{C}}(F)]^*$ is a homomorphism defined as in Section 4-6, $\pi(\sigma^{pi} \otimes \tau^{qj})(\sigma_{ph} \otimes \tau_{qk})$ is zero, except when $i = h$ and $j = k$, when it is equal to 1. That is to say π identifies the $\sigma^{pi} \otimes \tau^{qj}$ with a basis of the dual of $\bar{\mathscr{C}}_p(E) \otimes \bar{\mathscr{C}}_q(F)$, in fact, a basis dual to the $\sigma_{pi} \otimes \tau_{qj}$. In this case, π becomes an isomorphism (cf. Exercise 4-5) and is simply dropped from the notation. So the cohomology groups of $\bar{\mathscr{C}}(E) \otimes \bar{\mathscr{C}}(F)$ can be computed from the cochain complex $\bar{\mathscr{C}}^*(E) \otimes \bar{\mathscr{C}}^*(F)$. The problem now is to find the $\cup$ products of elements of the form $\sigma^{pi} \otimes \tau^{qj}$. Remember that $\sigma^{pi} \otimes \tau^{qj}$ is identified (by means of π) with an element of $[\bar{\mathscr{C}}(E) \otimes \bar{\mathscr{C}}(F)]^*$ and so is taken as the restriction to $\bar{\mathscr{C}}(E) \otimes \bar{\mathscr{C}}(F)$ of an element of $[C(E) \otimes C(F)]^*$. The product of two such elements can then be calculated from (4-64).

Take elements $\sigma^{pi} \otimes \tau^{qj}$ and $\sigma^{rh} \otimes \tau^{sk}$ in $\bar{\mathscr{C}}^*(E) \otimes \bar{\mathscr{C}}^*(F)$. These are of degree (p, q) and (r, s), respectively. Then (4-64) becomes

$$\begin{aligned}
&[(\sigma^{pi} \otimes \tau^{qj}) \cup (\sigma^{rh} \otimes \tau^{sk})](\sigma \otimes \tau) \\
&\quad = (-1)^{qr}(\sigma^{pi} \otimes \tau^{qj})[(\sigma F_p) \otimes (\tau F_q)](\sigma^{rh} \otimes \tau^{sk})[(\sigma L_r) \otimes (\tau L_s)] \\
&\quad = (-1)^{qr}\sigma^{pi}(\sigma F_p)\tau^{qj}(\tau F_q)\sigma^{rh}(\sigma L_r)\tau^{sk}(\tau L_s) \\
&\quad = (-1)^{qr}(\sigma^{pi} \cup \sigma^{rh})(\sigma)(\tau^{qj} \cup \tau^{sk})(\tau) \\
&\qquad\qquad\qquad \text{(these are cup products in } E \text{ and } F\text{, respectively)} \\
&\quad = (-1)^{qr}[(\sigma^{pi} \cup \sigma^{rh}) \otimes (\tau^{qj} \cup \tau^{sk})](\sigma \otimes \tau)
\end{aligned}$$

Again, at the last step π was used but it was dropped from the notation since it is an isomorphism on cellular cochains. Hence, the final result is

$$(\sigma^{pi} \otimes \tau^{qj}) \cup (\sigma^{rh} \otimes \tau^{sk}) = (-1)^{qr}(\sigma^{pi} \cup \sigma^{rh}) \otimes (\tau^{qj} \cup \tau^{sk}) \tag{4-65}$$

Examples

4-6. Consider the torus. It is the topological product $S^1 \times S^1$, where S^1 is the circle. S^1 has a cell decomposition with one 0 cell C_0, a point of S^1, and one 1 cell C_1, the complement of that point. The corresponding cell complex $\bar{\mathscr{C}}(S^1)$ is generated by a 0 cell c_0 on C_0 and a 1 cell c_1 on C_1. It is easy to see that the boundary relations are

$$dc_0 = dc_1 = 0$$

The associated cochain complex $\mathscr{C}(S^1)$ has a zero- and a one-dimensional group generated by the duals of c_0 and c_1, denoted by c^0 and c^1, with the coboundary relations

$$\delta c^0 = \delta c^1 = 0$$

Also note that the only nonzero $\cup$ products are (cf. Example 4-4)

$$c^0 \cup c^0 = c^0, \; c^0 \cup c^1 = c^1 \cup c^0 = c^1$$

According to remarks preceding this example, the cohomology groups of the torus are cohomology groups of the cochain complex $\mathscr{C}^*(S^1) \otimes \mathscr{C}^*(S^1)$. Here the zero-dimensional group is generated by $c^0 \times c^0$, the one-dimensional group by $c^0 \otimes c^1$ and $c^1 \otimes c^0$, and the two-dimensional group by $c^1 \otimes c^1$. The coboundary relations are

$$\delta(c^0 \otimes c^0) = \delta(c^0 \otimes c^1) = \delta(c^1 \otimes c^0) = \delta(c^1 \otimes c^1) = 0$$

It follows that $H^0(S^1 \times S^1) \cong H^2(S^1 \times S^1) \cong Z$ and $H^1(S^1 \times S^1)$ is free with two generators α and β represented by $c^0 \otimes c^1$ and $c^1 \otimes c^0$. Note that, geometrically, these correspond to two circles around the torus as in Example 4-3. Now calculate the $\cup$ products using (4-65). For example,

$$(c^0 \otimes c^0) \cup (c^0 \otimes c^0) = (c^0 \cup c^0) \otimes (c^0 \cup c^0) = c^0 \otimes c^0$$
$$(c^0 \otimes c^0) \cup (c^0 \otimes c^1) = (c^0 \cup c^0) \otimes (c^0 \cup c^1) = c^0 \otimes c^1$$

This, with similar computations for $c^1 \otimes c^0$ and $c^1 \otimes c^1$, shows that $c^0 \otimes c^0$ represents the identity of the cohomology ring.

$$(c^0 \otimes c^1) \cup (c^0 \otimes c^1) = (c^0 \cup c^0) \otimes (c^1 \cup c^1) = 0$$

Similarly $(c^1 \otimes c^0) \cup (c^1 \otimes c^0) = 0$.

$$(c^0 \otimes c^1) \cup (c^1 \otimes c^0) = -(c^0 \cup c^1) \otimes (c^1 \cup c^0) = -c^1 \otimes c^1$$

Similarly $(c^1 \otimes c^0) \cup (c^0 \otimes c^1) = c^1 \otimes c^1$.

Hence, the two generators α and β of $H^1(S^1 \times S^1)$ satisfy

$$\alpha \cup \alpha = \beta \cup \beta = 0$$
$$\alpha \cup \beta = -\beta \cup \alpha = \gamma$$

where γ is a generator of $H^2(S^1 \times S^1)$.

All other products are zero.

Note that this method reproduces the results of Example 4-3.

4-7. The object of this, and the following example is to show how the cohomology ring can distinguish between spaces that have the same cohomology groups. First, consider the topological product $P^3 \times P^2$ of the two- and three-dimensional projective spaces. In Example 2-9, a cell decomposition was obtained for P^n with one cell for each dimension from 0 up to n. Here denote by $\gamma_0, \gamma_1, \gamma_2$ oriented cells associated with the cell decomposition of P^2 and, by c_0, c_1, c_2, c_3, oriented cells associated with the decomposition of P^3. The boundary relations are (cf. Example 2-9)

$$d\gamma_0 = 0 \quad d\gamma_1 = 0 \quad d\gamma_2 = 2\gamma_1$$

$$dc_0 = 0 \quad dc_1 = 0 \quad dc_2 = 2c_1 \quad dc_3 = 0$$

The corresponding cochain complexes are generated by cocells γ^0, γ^1, γ^2 and c^0, c^1, c^2, c^3 and, by duality, the coboundary relations are

$$\delta\gamma^0 = 0 \quad \delta\gamma^1 = 2\gamma^2 \quad \delta\gamma^2 = 0$$

$$\delta c^0 = 0 \quad \delta c^1 = 2c^2 \quad \delta c^2 = 0 \quad \delta c^3 = 0$$

Then the cohomology groups of $P^3 \times P^2$ can be computed from the cochain complex $\mathscr{C}^*(P^3) \otimes \mathscr{C}^*(P^2)$. This complex is generated by the elements $c^i \otimes \gamma^j$ with coboundary relations computed from the coboundary formula (Section 4-6) along with the relations (4-66). The whole computation will not be carried out as only partial results are wanted in the comparison of this example with the next one.

The one- two-, and three-dimensional cochain groups of $\mathscr{C}^*(P^3) \otimes \mathscr{C}^*(P^2)$ are generated as follows, with coboundary relations as shown:

Dimension	Generators	Coboundaries
1	$c^0 \otimes \gamma^1$	$\delta(c^0 \otimes \gamma^1) = 2c^0 \otimes \gamma^2$
	$c^1 \otimes \gamma^0$	$\delta(c^1 \otimes \gamma^0) = 2c^2 \otimes \gamma^0$
2	$c^0 \otimes \gamma^2$	$\delta(c^0 \otimes \gamma^2) = 0$
	$c^1 \otimes \gamma^1$	$\delta(c^1 \otimes \gamma^1) = 2c^2 \otimes \gamma^1 - 2c^1 \otimes \gamma^2$
	$c^2 \otimes \gamma^0$	$\delta(c^2 \otimes \gamma^0) = 0$
3	$c^1 \otimes \gamma^2$	$\delta(c^1 \otimes \gamma^2) = 2c^2 \otimes \gamma^2$
	$c^2 \otimes \gamma^1$	$\delta(c^2 \otimes \gamma^1) = 2c^2 \otimes \gamma^2$
	$c^3 \otimes \gamma^0$	$\delta(c^3 \otimes \gamma^0) = 0$

The two-dimensional cocycle group is thus generated by $c^0 \otimes \gamma^2$ and $c^2 \otimes \gamma^0$, and since $2c^0 \otimes \gamma^2$ and $2c^2 \otimes \gamma^0$ are both coboundaries, it follows that $H^2(P^3 \times P^2) \cong Z_2 \oplus Z_2$.

The three-dimensional cocycle group is generated by $c^3 \otimes \gamma^0$ and $c^1 \otimes \gamma^2 - c^2 \otimes \gamma^1$. The cohomology class of the latter is of order 2 while that of the former is a free generator. Thus $H^3(P^3 \times P^2) \cong Z \oplus Z_2$. The important point is that the torsion part of this group, which will be denoted by $T^3(P^3 \times P^2)$, is isomorphic to Z_2.

Consider now the computation of the $\cup$ products. It was seen (Example 4-4) that in P^3, $c^1 \cup c^2 = c^2 \cup c^1 = c^3$ and γ^0 acts as the identity in P^2. Then, by (4-65),

$$(c^2 \otimes \gamma^0) \cup (c^1 \otimes \gamma^2 - c^2 \otimes \gamma^1) = c^3 \otimes \gamma^2$$

and this cocycle is not the coboundary of any four-dimensional cochain of $\mathscr{C}^*P(^3) \otimes \mathscr{C}^*(P^2)$. But $c^2 \otimes \gamma^0$ represents a generator of $H^2(P^3 \times P^2)$, and $c^1 \otimes \gamma^2 - c^2 \otimes \gamma^1$ represents a generator of the torsion group $T^3(P^3 \times P^2)$. Thus, there is a nonzero product $\alpha \cup \beta$ with α in $H^2(P^3 \times P^2)$ and β in $T^3(P^3 \times P^2)$.

4-8. Consider the space $(P^2 \vee S^2) \times P^2$. The first P^2 here has a cell decomposition with cells C_0, C_1, C_2 of dimensions 0, 1, 2 while S^3 has a cell decomposition with one 0 cell and one 3 cell Z_3. If the 0 cells of P^2 and S^3 are identified, a cell decomposition is obtained for $P^2 \vee S^3$ with cells C_0, C_1, C_2, Z_3. The associated cellular chain complex is then generated by oriented cells c_0, c_1, c_2, z_3 with boundary relations

$$dc_0 = 0 \quad dc_1 = 0 \qquad dc_2 = 2c_1 \quad dz_3 = 0$$

The corresponding cochain complex is generated by cocells c^0, c^1, c^2, z^3 with coboundary relations

$$\delta c^0 = 0 \quad \delta c^1 = 2c^2 \quad \delta c^2 = 0 \quad \delta z^3 = 0$$

Note that these are exactly the same as (4-66) with c^3 replaced by z^3. The second factor P^2 has a cell decomposition as in Example 4-7, with the associated cochain complex generated by γ^0, γ^1, γ^2 with the coboundary relations as in Example 4-7.

Clearly, the cochain complex $\mathscr{C}^*(P^2 \vee S^3) \otimes \mathscr{C}^*(P^2)$ is the same as $\mathscr{C}^*(P^3) \otimes \mathscr{C}^*(P^2)$ with c^3 replaced throughout by z^3, so the same cohomology groups will be obtained. Thus, the spaces $(P^2 \vee S^3) \times P^2$ and $P^3 \times P^2$ have the same cohomology groups.

Consider now the cohomology ring of $(P^2 \vee S^3) \times P^2$. In $P^2 \vee S^3$ the products $c^1 \cup c^2$ and $c^2 \cup c^1$ are 0 (*cf.* Examples 4-4 and 4-5). In other words, this result is not obtained from the corresponding statement of Example 4-7 simply by replacing c^3 by z^3. So consider the products of representatives $c^0 \otimes \gamma^2$ and $c^2 \otimes \gamma^0$ of generators of $H^2[(P^2 \vee S^3) \times P^2]$ with a representative $c^1 \otimes \gamma^2 - c^2 \otimes \gamma^1$ of a generator of $T^3[(P^2 \vee S^3) \times P^2]$. This time the products [using (4-65)] are both 0. Hence $\alpha \cup \beta = 0$ for any $\alpha \in H^2[(P^2 \vee S^3) \times P^2]$ and $\beta \in T^3[(P^2 \vee S^3) \times P^2]$. It follows that the cohomology ring of $(P^2 \vee S^3) \times P^2$ is not the same as that of $P^3 \times P^2$.

Exercise. 4-15. As another illustration of the point made by the preceding examples, compare the spaces $S^1 \vee S^1 \vee S^2$ and $S^1 \times S^1$ (the torus). Show that these have the same homology and cohomology groups but different cohomology rings. On the other hand, this is not such a satisfactory example, as the two spaces have different fundamental groups.

5

Čech Homology Theory—The Construction

Another form of homology theory is described in this chapter. This theory is the same as singular or simplicial homology theory for simplicial complexes, but presents some novel features of its own when applied to general spaces.

5-1. INTRODUCTION

So far, only one form of homology theory has been described for topological spaces in general, namely, singular homology theory. If the space is a simplicial complex, however, there is a special homology theory that is defined in terms of the simplicial structure, namely, the oriented simplicial homology theory. It has been shown that for simplicial complexes these two theories coincide.

Historically, the process occurred the other way around; that is, homology was considered first for simplicial complexes, or rather for topological spaces made into simplicial complexes by triangulation, and the definitions were formulated in terms of oriented simplexes. Two problems then arose. The first was to show that the resulting homology groups were topological invariants of the space and were independent of the manner in which the triangulation was done. The second was to extend the notion of homology to more general spaces, not necessarily triangulable. Both of these problems are solved by singular homology theory, for the singular groups can be defined for any space and, by their definition, they are automatically topological invariants. Then, since the singular groups are the same as the simplicial groups (Theorem 2-17) for a triangulable space, the topological invariance of the simplicial groups follows.

The second problem, the extension of homology from triangulable spaces to more general spaces, has been tackled in other ways. The usual procedure is to associate with the given space E a simplicial complex K and to compute the simplicial homology groups of K. The complex K is then made to tend to E in some sort of limiting process, and the homology groups of E are defined as some kind of limit of those of K. To illustrate this, the Čech homology groups of a space, or more generally, of a pair of spaces will be constructed. This construction is based on the notions of nerve of a covering and inverse limit, which are described in the following sections.

5-2. THE NERVE OF A COVERING

The idea here is to take a finite open covering of a space E and to associate with it a simplicial complex in such a way that the vertexes are in one-to-one correspondence with the sets of the covering; a collection of vertexes are to be the vertexes of a simplex if and only if the corresponding open sets have a nonempty intersection. To make this precise, a collection of vertexes, in a suitable product of real lines, is permanently associated with all the open sets of E, and the complexes constructed are all to be in Euclidean subspaces of this product. The definitions will now be set up formally.

Let E be a topological space. For each nonempty open set U in E, take a copy R_U of the real line and let ΠR_U be the topological product of these copies. Then, for a fixed open set V in E, take the point in ΠR_U whose projection on R_V is 1 and whose projection on each R_U, for $U \neq V$, is 0. Call this point V. The use of the same letter for the open set and for the point is convenient, and the context shows whether point or set is meant.

Now let $\mathscr{U}$ be a finite open covering of E, and construct the simplicial complex $K_{\mathscr{U}}$ as follows. The vertexes of $K_{\mathscr{U}}$ are the points U of $\Pi R_{\mathscr{U}}$ for which the corresponding sets U belong to the covering $\mathscr{U}$, and the complex $K_{\mathscr{U}}$ contains the simplex $[U_0 U_1 \cdots U_p]$ (the U_i being points of ΠR_U) if and only if the open sets $U_0, U_1, \ldots, U_p$ of E have a nonempty intersection. If this condition is not to contradict itself, it has to make each face of $[U_0 U_1 \cdots U_p]$ belong to $K_{\mathscr{U}}$ whenever $[U_0 U_1 \cdots U_p]$ itself does; clearly, it does, because if the intersection of all the sets $U_0, U_1, \ldots, U_p$ is nonempty, so is the intersection of any subcollection.

Note that, although $K_{\mathscr{U}}$ is constructed in the infinite product ΠR_U, it actually belongs to the finite product of R_U taken over those open sets U forming the covering $\mathscr{U}$. This finite product is a Euclidean space, so the complex $K_{\mathscr{U}}$ appears as a subspace of a Euclidean space in the usual way.

Definition 5-1. The complex $K_{\mathscr{U}}$ just constructed is called the *nerve* of the finite open covering $\mathscr{U}$.

Note that the definition has been formulated for finite coverings only. If the definition of a complex were allowed to include the case of infinitely many vertexes, then Definition 5-1 could be extended to infinite coverings. This extension is needed to get certain results in their full generality, but for the purpose of this introduction, finite coverings will suffice.

Examples

5-1. Let E be the circumference of a circle. Let $\mathscr{U}$ be the covering consisting of one open set, E itself. Then the complex $K_{\mathscr{U}}$ consists of a single point.

5-2. Let E be as it was in Example 5-1. This time let $\mathscr{U}$ consist of two sets, namely, two overlapping arcs, as in Fig. 15. $K_{\mathscr{U}}$ has two vertexes U_1 and U_2. The intersection $U_1 \cap U_2$ is not empty, so the segment $[U_1 U_2]$ belongs to $K_{\mathscr{U}}$. Thus, in this case, the complex $K_{\mathscr{U}}$ is a line segment.

5-3. Again, let E be the circumference of a circle but now let $\mathscr{U}$ consist of open arcs that overlap in pairs, as shown in Fig. 16. $K_{\mathscr{U}}$ has three vertexes, which are denoted by U_1, U_2, U_3 and which form the vertexes of a triangle, as in Fig. 17. The intersections $U_1 \cap U_2$, $U_2 \cap U_3$, $U_3 \cap U_1$ are non-empty, so the sides of the triangle belong to $K_{\mathscr{U}}$. On the other hand,

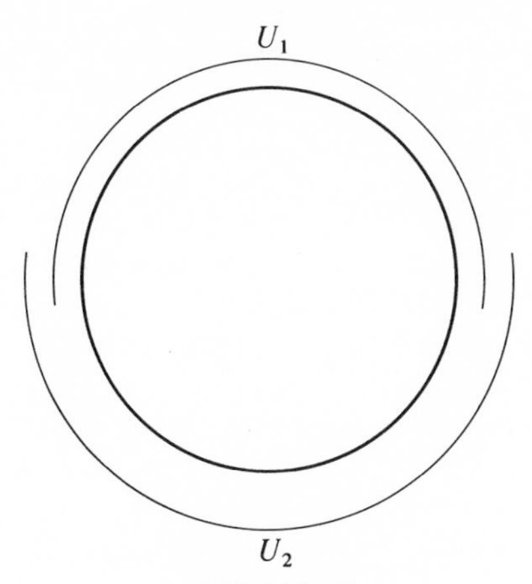

FIGURE 15

$U_1 \cap U_2 \cap U_3 = \varnothing$, so the triangle itself does not belong to $K_{\mathscr{U}}$. Thus, the complex $K_{\mathscr{U}}$ is the one-dimensional complex that consists of the boundary of a triangle. Note that, in this case, $K_{\mathscr{U}}$ is homeomorphic to the space E itself.

These and similar examples suggest that the nerve of a covering of E is an approximation of E which becomes closer as more open sets are used. This notion of approximation is brought out even more strikingly by the next example.

5-4. For each positive integer n take a circle in the (x, y) plane with center $(1/n, 0)$ and radius $1/n$, and let E be the union of these circles topologized as a subspace of the plane. Construct a covering $\mathscr{U}$ of E as follows. U_0 is the intersection of E with an open disk of center $(0, 0)$ and radius ε. U_0 contains all the circles in E of radius $1/n$ for sufficiently large n. If the circle in E of radius $1/n$ is not contained in U_0 (and this is so for a finite number of circles), then it can be covered by U_0 along with two overlapping open arcs U_{n1}, U_{n2}, each with $(0, 0)$ in its closure. U_0 and all the U_{ni} form a finite open covering of E. K_U has vertexes U_0, U_{n1}, U_{n2}, for a finite set of values of n, which correspond to those circles of E which are not contained in U_0. For each such n the sides $[U_0 U_{n1}]$, $[U_0 U_{n2}]$, $[U_{n1} U_{n2}]$ are in $K_{\mathscr{U}}$ and these are the only 1 simplexes of $K_{\mathscr{U}}$. $K_{\mathscr{U}}$ contains no simplexes of higher dimension. Thus, $K_{\mathscr{U}}$ is the union of the perimeters of finitely many triangles that have the vertex U_0 in common.

The smaller the open set U_0 is taken in this example, the more triangles there are in $K_{\mathscr{U}}$, and the more closely $K_{\mathscr{U}}$ resembles E. In this case, however, $K_{\mathscr{U}}$ can never actually be homeomorphic to E, since E is not triangulable ($H_1(E)$ has infinitely many generators).

It is natural to expect triangulable spaces to present special features in this connection. In fact, it will now be shown that, for a triangulable space E, a

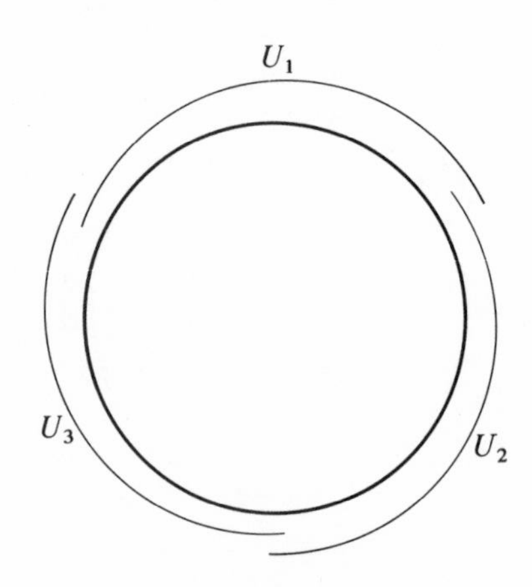

FIGURE 16

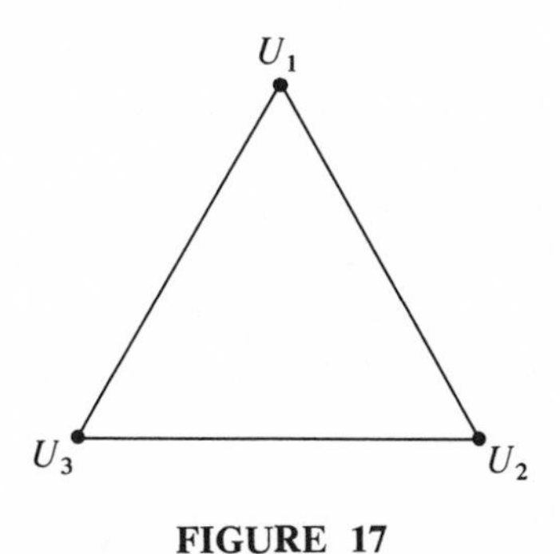

FIGURE 17

covering $\mathscr{U}$ can always be chosen so that $K_{\mathscr{U}}$ is homeomorphic to E. This point, of course, was illustrated by Example 5-3.

Let K be a simplicial complex embedded in some Euclidean space in such a way that all the vertexes are linearly independent. The vertexes of K then can be taken as vertexes of a simplex Σ, and K thus appears as a subcomplex of Σ. If x is a vertex of Σ, then Σ, when the face opposite x has been removed, is an open set in Σ, in its topology as a subspace of Euclidean space. Call this open set U'_x, and let $U_x = U'_x \cap K$. U_x is then an open neighborhood of x in the induced topology of K.

Definition 5-2. U_x is called the *star neighborhood* of x in K (cf. Exercise 2-7).

Note that U_x is the union of all simplexes of K having x as a vertex when, for each simplex, with the face opposite x has been removed. Or, alternatively, if the interior of a simplex is defined as the simplex with all its faces removed, U_x is the union of x and the interiors of all simplexes of K having x as a vertex.

Definition 5-3. Let $\mathscr{U}$ be the finite open covering of K which consists of all star neighborhoods of all the vertexes of K. Then, $\mathscr{U}$ is called the *star covering* of K.

Note that $\mathscr{U}$ depends on the representation of K as a complex. If the same space is subdivided into simplexes in a different way, the corresponding star covering is different.

Theorem 5-1. *Let K be a simplicial complex and $\mathscr{U}$ its star covering. Then there is a homeomorphism $f: K \to K_{\mathscr{U}}$ which sets the simplexes of K into one-to-one correspondence with those of $K_{\mathscr{U}}$, mapping each simplex of K linearly onto the corresponding simplex of $K_{\mathscr{U}}$.*

Proof. There is an open set U_x in $\mathscr{U}$ for each vertex x of K, and, as usual, the same symbol U_x is used to denote the corresponding vertex of the nerve $K_{\mathscr{U}}$ of $\mathscr{U}$. The map f is defined first on the vertexes of K by setting $f(x) = U_x$ for each vertex x. Next it must be checked that f can be extended to a simplicial map on K. Suppose, then, that $x, y, \ldots, z$ are the vertexes of a simplex of K. The star neighborhoods $U_x, U_y, \ldots, U_z$ have at least the interior of that simplex in common, so they certainly have a nonempty intersection. By Definition 5-1, therefore, the vertexes $U_x, U_y, \ldots, U_z$ of K_U are vertexes of a simplex in $K_{\mathscr{U}}$. It follows (cf. Definition 2-16) that f can be extended to a simplicial map of K into $K_{\mathscr{U}}$. On the other hand, there is an inverse map to f. Set $g(U_x) = x$ for each vertex U_x of $K_{\mathscr{U}}$. If $U_x, U_y, \ldots, U_z$ are vertexes of a simplex in $K_{\mathscr{U}}$, then the open sets $U_x, U_y, \ldots, U_z$ in K have a nonempty intersection and so must have the interior of a simplex in common (cf. the remark following Definition 5-2). Then, $x, y, \ldots, z$ are the vertexes of some face of this simplex, so g can be extended to a simplicial map of $K_{\mathscr{U}}$ into K. It is easy to see that f and g are inverse to each other, so f is the homeomorphism required by the statement of the theorem. ∎

The result just proved eventually leads to the proof that the Čech homology groups are the same as the simplicial.

Return now to the general discussion. Some machinery must be set up to enable relative homology to be defined. For this purpose, in addition to the nerves of coverings, subcomplexes of these nerves are needed, associated in a natural way with a given subspace of the space under discussion. So let (E, F) be a pair of spaces, let $\mathscr{U}$ be a finite open covering of E and let $K_{\mathscr{U}}$ be its nerve. A subcomplex $K'_{\mathscr{U}}$ of $K_{\mathscr{U}}$ is defined as follows: a simplex $[U_0 U_1 \cdots U_p]$ of $K_{\mathscr{U}}$ is in $K'_{\mathscr{U}}$ if and only if $U_0 \cap U_1 \cap \cdots \cap U_p \cap F \neq \varnothing$. If a simplex of $K_{\mathscr{U}}$ satisfies this condition, it is easy to see that its faces also do, so the set of all simplexes satisfying the condition forms a subcomplex of $K_{\mathscr{U}}$.

Definition 5-4. The subcomplex $K'_{\mathscr{U}}$ just constructed is called *the subcomplex of $K_{\mathscr{U}}$ associated with the subspace F.*

It should be noted that $K'_{\mathscr{U}}$ is not the same as the nerve $K_{\mathscr{U} \cap F}$ of the covering $\mathscr{U} \cap F$ induced by $\mathscr{U}$ on F, for two open sets defining different vertexes of $K_{\mathscr{U}}$ might have the same intersection with F and so define the same vertex of $K_{\mathscr{U} \cap F}$. They, nevertheless, would not define the same vertex of $K'_{\mathscr{U}}$. The complexes $K'_{\mathscr{U}}$ and $K_{\mathscr{U} \cap F}$ will be compared more precisely later (cf. remarks following Definition 5-7).

At this point, Theorem 5-1 can be extended to pairs of complexes. Let K be a simplicial complex, L a subcomplex, $\mathscr{U}$ the star open covering of K, and $\mathscr{U}'$ the star open covering of L (the latter being a covering of L by sets that are open in the subspace topology of L). In the first place, it is easy to see that

$\mathscr{U}'$ is the covering of L that is induced by the covering $\mathscr{U}$ of K, and in the second place, two different open sets of $\mathscr{U}$ intersect L in different open sets of $\mathscr{U}'$. So, in this case, the subcomplex $K'_{\mathscr{U}}$ of $K_{\mathscr{U}}$ associated with L, in the sense of Definition 5-4, can be identified with the nerve $K_{\mathscr{U}'}$ of $\mathscr{U}'$. It can easily be verified that the simplicial map of Theorem 5-1 maps L onto $K_{\mathscr{U}'}$. Thus, the following extension of Theorem 5-1 is obtained.

Theorem 5-2. *If (K, L) is a simplicial pair and $\mathscr{U}$ is the star open covering of K, there is a simplicial homeomorphism $f\colon K \to K_{\mathscr{U}}$ mapping L onto $K'_{\mathscr{U}}$. The map is set up by mapping each vertex x of K on the vertex U_x of $K_{\mathscr{U}}$, where U_x represents the open star neighborhood of x in K.*

5-3. HOMOLOGY GROUPS OF A COVERING

Homology groups will now be attached to the pair of spaces (E, F) corresponding to the finite open covering $\mathscr{U}$. These will be taken as the simplicial homology groups of $K_{\mathscr{U}}$ modulo $K'_{\mathscr{U}}$.

Definition 5-5. *The pth homology group of (E, F) associated with $\mathscr{U}$* is defined to be $\mathscr{H}_p(K_{\mathscr{U}}, K'_{\mathscr{U}})$. This group is denoted by $H_p(E, F; \mathscr{U})$.

Note that, if $F = \varnothing$, the subcomplex $K'_{\mathscr{U}}$ is also empty. In this case, $\mathscr{H}_p(K_{\mathscr{U}}, K'_{\mathscr{U}})$ reduces to $\mathscr{H}_p(K_{\mathscr{U}})$. This is denoted by $H_p(E; \mathscr{U})$.

Eventually, the Čech homology groups of the pair (E, F) will be defined by a limiting process applied to the groups $H_p(E, F; \mathscr{U})$. At the same time, induced homomorphisms must be defined. This will be done by first associating with a continuous map a homomorphism between the homology groups corresponding to certain coverings, and then applying a limiting process to these homomorphisms. The first part of this construction will be carried out now, the second in Section 5-6.

Let $f\colon X \to E$ be a continuous map and let $\mathscr{U}$ be a finite open covering of E. The set of all sets $f^{-1}(U)$ with U in $\mathscr{U}$ is a finite open covering $\mathscr{U}'$ of X. This covering is also denoted by $f^{-1}(\mathscr{U})$.

Definition 5-6. $\mathscr{U}' = f^{-1}(\mathscr{U})$ is the *covering of X induced by f and the covering $\mathscr{U}$ of E.*

Note that, if X is a subspace and f is the inclusion map, this notion coincides with the covering induced by intersection of the subspace with the sets of $\mathscr{U}$.

Let $K_{\mathscr{U}}$ be the nerve of $\mathscr{U}$ and $K_{\mathscr{U}'}$ that of $\mathscr{U}'$. A simplicial map of $K_{\mathscr{U}'}$ into $K_{\mathscr{U}}$ is now to be defined, as usual by defining it first on the vertexes and then extending by linearity. Let U' be a vertex of $K_{\mathscr{U}'}$. Then the set U' in X can be written as $f^{-1}(U)$ for some open set U in the covering $\mathscr{U}$. Note that U is not necessarily unique, since two different sets of $\mathscr{U}$ may have the same inverse images under f. However, fix a choice of U and then define the map $f_{\mathscr{U}}^{1}$ on the vertexes of $K_{\mathscr{U}'}$ by making it map the vertex U' of $K_{\mathscr{U}'}$ on the vertex U of $K_{\mathscr{U}}$.

To verify that $f_{\mathscr{U}}^{1}$ can be extended to a simplicial map, let $U_0', U_1', \ldots, U_p'$ be vertexes of $K_{\mathscr{U}'}$ and let their images under $f_{\mathscr{U}}^{1}$ be $U_0, U_1, \ldots, U_p$. Since the U_i' are vertexes of a simplex, the sets U_i' have nonempty intersection, so since $f(U_i') \subset U_i$ for each i, the sets U_i have nonempty intersection. It follows that the points U_i are the vertexes of a simplex in $K_{\mathscr{U}}$, so $f_{\mathscr{U}}^{1}$ can be extended to a simplicial map $f_{\mathscr{U}}^{1}: K_{\mathscr{U}'} \to K_{\mathscr{U}}$, as required.

It will presently be seen that the nonuniqueness of $f_{\mathscr{U}}^{1}$ makes no difference to the induced homomorphism on the homology groups. First, however consider maps of pairs of spaces.

Let f now be a continuous map of the pair (X, Y) into the pair (E, F) and, as before, let $K_{\mathscr{U}}, K_{\mathscr{U}'}$ be the nerves of $\mathscr{U}$ and $\mathscr{U}' = f^{-1}(\mathscr{U})$, respectively, where $\mathscr{U}$ is a finite open covering of E. Let $K_{\mathscr{U}}'$ be the subcomplex of $K_{\mathscr{U}}$ associated with F and $K_{\mathscr{U}'}'$ the subcomplex of $K_{\mathscr{U}'}$ associated with Y.

Then, if $U_0', U_1', \ldots, U_p'$ are vertexes of a simplex of $K_{\mathscr{U}'}$, the sets U_i' satisfy

$$U_0' \cap U_1' \cap \cdots \cap U_p' \cap Y \neq \varnothing$$

So, since $f(Y) \subset F$, the sets U_i satisfy

$$U_0 \cap U_1 \cap \cdots \cap U_p \cap F \neq \varnothing$$

It follows that the U_i are vertexes of a simplex in $K_{\mathscr{U}}'$. Hence the simplicial map $f_{\mathscr{U}}^{1}$ constructed above is a map of the pair $(K_{\mathscr{U}'}, K_{\mathscr{U}'}')$ into the pair $(K_{\mathscr{U}}, K_{\mathscr{U}}')$.

The following lemma leads to the proof that the possible ambiguity in the definition of $f_{\mathscr{U}}^{1}$ does not affect the induced homomorphisms on the homology groups.

Lemma 5-3. *Let $f_{\mathscr{U}}^{1}$ be constructed as above. For each set U' of $\mathscr{U}'$, make a second choice of set V in $\mathscr{U}$ such that $U' = f^{-1}(V)$, and define a second simplicial map $f_{\mathscr{U}}^{2}: (K_{\mathscr{U}'}, K_{\mathscr{U}'}') \to (K_{\mathscr{U}}, K_{\mathscr{U}}')$, which maps the vertex U' on the vertex V. Then $f_{\mathscr{U}}^{1}$ and $f_{\mathscr{U}}^{2}$ are contiguous, as maps of pairs.*

Proof. Let $U_0', U_1', \ldots, U_p'$ be vertexes of a simplex in $K_{\mathscr{U}'}$ and let $U_0, U_1, \ldots, U_p$ and $V_0, V_1, \ldots, V_p$ be their sets of images under $f_{\mathscr{U}}^{1}$ and $f_{\mathscr{U}}^{2}$, respectively. Thus, $f^{-1}(U_i) = f^{-1}(V_i) = U_i'$ for each i. Now

$$f^{-1}(U_0 \cap U_1 \cap \cdots \cap U_p \cap V_0 \cap V_1 \cap \cdots \cap V_p) \\ = U_0' \cap U_1' \cap \cdots \cap U_p' \neq \varnothing$$

so $U_0 \cap U_1 \cap \cdots \cap U_p \cap V_0 \cap V_1 \cap \cdots \cap V_p \neq \varnothing$. In other words, the U_i and V_i are all vertexes of some simplex of $K_{\mathscr{U}}$. Thus, for any simplex S of $K_{\mathscr{U}'}$, $f_{\mathscr{U}}{}^1(S)$ and $f_{\mathscr{U}}{}^2(S)$ are contained in some simplex of $K_{\mathscr{U}}$. In addition, suppose that the U_i' are vertexes of a simplex of $K'_{\mathscr{U}'}$. Then, in the above notation,

$$f^{-1}(U_0 \cap U_1 \cap \cdots \cap U_p \cap V_0 \cap V_1 \cap \cdots \cap V_p \cap F) \supset U_0' \cap U_1' \cap \cdots \cap U_p' \cap Y \neq \varnothing$$

So $U_0 \cap U_1 \cap \cdots \cap U_p \cap V_0 \cap \cdots \cap V_p \cap F \neq \varnothing$, which means that the U_i and V_i are vertexes of a simplex of $K'_{\mathscr{U}}$. Thus, if S is in $K'_{\mathscr{U}'}$, then $f_{\mathscr{U}}{}^1(S)$ and $f_{\mathscr{U}}{}^2(S)$ are contained in some simplex of $K'_{\mathscr{U}}$. It has been shown, therefore, that $f_{\mathscr{U}}{}^1$ and $f_{\mathscr{U}}{}^2$ are contiguous as maps of pairs. ∎

Exercise 3-2 can now be applied to the contiguous maps $f_{\mathscr{U}}{}^1$ and $f_{\mathscr{U}}{}^2$ to give the result that these simplicial maps induce the same homomorphism

$$f_{\mathscr{U}}: H_p(X, Y; \mathscr{U}') \to H_p(E, F; \mathscr{U})$$

Definition 5-7. $f_{\mathscr{U}}$ is called the *induced homomorphism associated with the covering $\mathscr{U}$ and the continuous map f.* The context always makes clear on what and into what $f_{\mathscr{U}}$ operates, so this does not have to be included in the terminology or description.

Of special interest is the inclusion map $i: X \to E$ of a subspace X of E into E. Here the covering $\mathscr{U}' = f^{-1}(\mathscr{U})$ is the covering $\mathscr{U} \cap X$ of X obtained by taking the intersections of X with the sets of $\mathscr{U}$. As already pointed out, the nerve of this covering $\mathscr{U}'$ of X is not necessarily the same as the subcomplex $K'_{\mathscr{U}}$ of $K_{\mathscr{U}}$ associated with the subspace X (cf. Definition 5-4). Using the notion of contiguous maps the two complexes $K_{\mathscr{U}'}$ and $K'_{\mathscr{U}}$ will now be compared. This is done in two senses, algebraic and geometric.

The comparisons will be made by showing that, associated with the simplicial map $i_{\mathscr{U}}{}^1: K_{\mathscr{U}'} \to K_{\mathscr{U}}$ induced as described above by the inclusion map $i: X \to E$, there is a simplicial map $j: K'_{\mathscr{U}} \to K_{\mathscr{U}'}$ such that both $i_{\mathscr{U}}{}^1 j$ and $j i_{\mathscr{U}}{}^1$ are contiguous to the appropriate identity maps. Stated another way, if, for a set U in the covering $\mathscr{U}$, the intersection $U' = U \cap X$ is not empty (so that U represents a vertex of $K'_{\mathscr{U}}$), let j map the vertex U of $K'_{\mathscr{U}}$ on the vertex U' of $K_{\mathscr{U}'}$. If $U_0, U_1, \ldots, U_p$ are vertexes of a simplex in $K'_{\mathscr{U}}$, it is clear that $j(U_0), j(U_1), \ldots, j(U_p)$ are the vertex of a simplex of $K_{\mathscr{U}'}$. Hence, j can be extended to a simplicial map of $K'_{\mathscr{U}}$ into $K_{\mathscr{U}'}$. Also, it is obvious that $j i_{\mathscr{U}}{}^1$ is the identity map of $K_{\mathscr{U}'}$ on itself. To examine the action of $i_{\mathscr{U}}{}^1 j$, let $U_0, U_1, \ldots, U_p$ be the vertexes of a simplex S in $K'_{\mathscr{U}}$ and let $U_0', U_1', \ldots, U_p'$ be their images under j. Thus, the set U_i' is $U_i \cap X$ for each i. On the other hand,

$i_{\mathscr{U}}^{1}(U_i')$ is a vertex V_i of $K_{\mathscr{U}}$ such that the set relation $V_i \cap X = U_i'$ holds. V_i is therefore a vertex of $K_{\mathscr{U}}'$ since U_i' is not empty. Also, however, the intersection

$$U_0 \cap U_1 \cap \cdots \cap U_p \cap V_0 \cap V_1 \cap \cdots \cap V_p \cap X = U_0' \cap U_1' \cap \cdots \cap U_p'$$

is not empty, so the vertexes $U_0, U_1, \ldots, U_p, V_0, V_1, \ldots, V_p$ are all vertexes of a simplex of $K_{\mathscr{U}}'$. Thus, $i_{\mathscr{U}}^{1}j(S) = [V_0 V_1 \cdots V_p]$ and $S = [U_0 U_1 \cdots U_p]$ are both contained in some simplex of $K_{\mathscr{U}}'$; that is, $i_{\mathscr{U}}^{1}j$ and the identity are contiguous.

From an application of Exercise 3-2 it follows that $i_{\mathscr{U}}^{1}j$ induces a chain homomorphism on the oriented chain complex $\mathscr{C}(K_{\mathscr{U}}')$, which is algebraically homotopic to the identity, while $ji_{\mathscr{U}}^{1}$, of course, induces the identity itself. Thus, $i_{\mathscr{U}}^{1}$ induces a chain homomorphism of $\mathscr{C}(K_{\mathscr{U}'})$ into $\mathscr{C}(K_{\mathscr{U}}')$, which has an algebraic homotopy inverse.

From the geometric point of view, (cf. Exercise 3-2) $i_{\mathscr{U}}^{1}j$ is homotopic to the identity map of $K_{\mathscr{U}}'$ on itself. Since $ji_{\mathscr{U}}^{1}$ is the identity, it follows that $i_{\mathscr{U}}^{1}$ and j are homotopy inverse to each other. This is sometimes expressed by saying that $K_{\mathscr{U}'}$ and $K_{\mathscr{U}}'$, although not homeomorphic, are of the same homotopy type. Clearly, two spaces that bear this relation to each other have the same homology groups in any theory in which the homotopy theorem holds.

Returning to the general theory, an important property of the homomorphisms $f_{\mathscr{U}}$ will now be obtained. This will be the main step in the later proof that in Čech homology theory the homomorphism induced by a composition of maps is the same as the composition of the individual induced homomorphisms.

Theorem 5-4. *Let*

$$f\colon (E, A) \to (F, B)$$

$$g\colon (F, B)(G, \to C)$$

be continuous maps and let $\mathscr{U}$ be a finite open covering of G. Write $\mathscr{U}' = g^{-1}(\mathscr{U})$. Then

$$(gf)_{\mathscr{U}} = g_{\mathscr{U}} f_{\mathscr{U}'}$$

Also, if $f\colon (E, A) \to (E, A)$ is the identity then $f_{\mathscr{U}}$ is the identity.

Proof. Write $h = gf$ and $\mathscr{U}'' = f^{-1}(\mathscr{U}')$. It has already been shown that, corresponding to f, g, h, simplicial maps can be set up; they are denoted by $f_{\mathscr{U}'}^{1}, g_{\mathscr{U}}^{1}, h_{\mathscr{U}}^{1}$. Hence, if U is a vertex of $K_{\mathscr{U}''}$, $f_{\mathscr{U}'}^{1}(U)$ is some vertex V of

$K_{\mathscr{U}'}$ such that the set relation $U = f^{-1}(V)$ holds, and $g_{\mathscr{U}}^{1}(V)$ is some vertex W of $K_{\mathscr{U}}$ such that $V = g^{-1}(W)$. Then, automatically, $U = h^{-1}(W)$, so W can be taken as defining $h_{\mathscr{U}}^{1}(U)$. Thus, the three simplicial maps can be set up so that

$$h_{\mathscr{U}}^{1} = g_{\mathscr{U}}^{1} f_{\mathscr{U}'}^{1}$$

Passing to the induced homomorphisms on the simplicial homology groups and an application of Theorem 1-5 (or 3-3) along with Section 2-8 yield the first part of the present theorem. The second part of the theorem is trivial. ∎

At this stage, homology groups have been associated with a pair of spaces, a finite open covering of the larger space, and of course, a fixed but unspecified coefficient group. Homology groups obtained in this way for different coverings will now be compared. Limits are eventually to be taken as the coverings become ever finer. And the first point here is to specify precisely what is meant by the phrase "becoming finer."

Definition 5-8. Let $\mathscr{U}$ and $\mathscr{V}$ be finite open coverings of a space E. $\mathscr{V}$ is called a *refinement* of $\mathscr{U}$ if every set V of $\mathscr{V}$ is contained in some set U of $\mathscr{U}$. This relation is written $\mathscr{V} > \mathscr{U}$.

It is easy to see that the relation just defined is a relation of partial order on the finite open coverings of E.

Example

5-5. In Example 5-2, it could be assumed that U_1 and U_2 are obtained by forming the unions of the sets in Example 5-3 two by two. The covering of Example 5-3 would then be a refinement of that in Example 5-2.

A second example is afforded by the following sequence of exercises, the object of which is to show that, if a simplicial complex is barycentrically subdivided, then the new star open covering is a refinement of the old.

Exercises. 5-1. Let K be a simplicial complex, L a subcomplex, and x a vertex of L. Prove that the star neighborhood of x in L is the intersection of L with the star neighborhood of x in K.

5-2. For any complex K let BK be its barycentric subdivision (cf. Exercise 2-6). Let $\mathscr{U}$ be the star open covering of K and $B\mathscr{U}$ that of BK. Remembering that any complex K can be taken as a subcomplex of a simplex S, use Exercise 5-1 to prove that, in order to show that $B\mathscr{U} > \mathscr{U}$, it is sufficient to prove the corresponding result replacing K by S.

5-3. Prove that if $\mathcal{U}$ is the star open covering of a simplex S then $B\mathcal{U} > \mathcal{U}$. To do this, remember that each simplex of BS is associated with an ascending sequence of faces of S (cf. Exercise 2-5), and this is such that, if x is a vertex of BS, x being the barycenter of a face S_0 of S, then the star of x in BS consists of simplexes associated with sequences containing S_0. It can then be shown that the open star of x in BS is contained in the open star, in S, of any vertex of S_0.

5-4. Having obtained the result of Exercise 5-3, take intersections with the subcomplex K of S, and hence show that if x is a vertex of BK and if x is the barycenter of the simplex S_0 of K, then the star neighborhood of x in BK is contained in the star neighborhood in K of any vertex of S_0. Also show that the vertexes of S_0 are the only vertexes of K whose star neighborhoods contain that of x in BK.

Consider now the nerves of two finite open coverings $\mathcal{U}$ and $\mathcal{V}$ such that $\mathcal{V} > \mathcal{U}$. It will be shown that a simplicial map of $K_{\mathcal{V}}$ into $K_{\mathcal{U}}$ can be defined in a more or less natural way. As in the case of the $f_{\mathcal{U}}{}^1$, this will be defined only up to contiguity, but it thus will define uniquely a map of the corresponding homology groups.

Given the coverings $\mathcal{U}$ and $\mathcal{V}$ of E such that $\mathcal{V} > \mathcal{U}$, let V be a vertex of $K_{\mathcal{V}}$. Since $\mathcal{V} > \mathcal{U}$, the set V is contained in some set of U of the covering $\mathcal{U}$. Define the map $f^1_{\mathcal{U}\mathcal{V}}$ on the vertexes of $K_{\mathcal{V}}$ by making it map the vertex V of $K_{\mathcal{V}}$ on the vertex U of $K_{\mathcal{U}}$. It will now be shown that $f^1_{\mathcal{U}\mathcal{V}}$ can be extended to a simplicial map. If $V_0, V_1, \ldots, V_p$ are vertexes of a simplex of $K_{\mathcal{V}}$ the sets V_i have a nonempty intersection, so the larger sets U_i, where $U_i = f^1_{\mathcal{U}\mathcal{V}}(V_i)$, have also a nonempty intersection. Thus, the U_i are vertexes of a simplex of $K_{\mathcal{U}}$, and so $f^1_{\mathcal{U}\mathcal{V}}$ can be extended as required.

The simplicial map $f^1_{\mathcal{U}\mathcal{V}}$ so defined depends on the choice of the set U of $\mathcal{U}$ containing a set V of $\mathcal{V}$. Suppose that a second simplicial map $f^2_{\mathcal{U}\mathcal{V}}: K_{\mathcal{V}} \to K_{\mathcal{U}}$ is defined in the same way. It will be shown that $f^1_{\mathcal{U}\mathcal{V}}$ and $f^2_{\mathcal{U}\mathcal{V}}$ are contiguous. Let $V_0, V_1, \ldots, V_p$ be the vertexes of a simplex S of $K_{\mathcal{V}}$ and let $f^1_{\mathcal{U}\mathcal{V}}(V_i) = U_i$, $f^2_{\mathcal{U}\mathcal{V}}(V_i) = W_i$. Thus the set relations $V_i \subset U_i$ and $V_i \subset W_i$ are satisfied. Since the V_i represent the vertexes of a simplex, the intersection of the V_i is not empty, so the intersection of the U_i and W_i is not empty. It follows that the vertexes U_i and W_i of $K_{\mathcal{U}}$ are all vertexes of a simplex in $K_{\mathcal{U}}$. Hence $f^1_{\mathcal{U}\mathcal{V}}(S)$ and $f^2_{\mathcal{U}\mathcal{V}}(S)$ are contained in some simplex of $K_{\mathcal{U}}$, so $f^1_{\mathcal{U}\mathcal{V}}$ and $f^2_{\mathcal{U}\mathcal{V}}$ are contiguous, as was to be shown.

Further let F be a subspace of E and let $K'_{\mathcal{U}}$ and $K'_{\mathcal{V}}$ be the subcomplexes of $K_{\mathcal{U}}$ and $K_{\mathcal{V}}$ associated with F. If the V_i are now the vertexes of a simplex of $K'_{\mathcal{V}}$, then the intersection $V_0 \cap V_1 \cap \cdots \cap V_p \cap F \neq \varnothing$, so also the intersection $U_0 \cap U_1 \cap \cdots \cap U_p \cap W_0 \cap W_1 \cap \cdots \cap W_p \cap F \neq \varnothing$. Hence, the U_i and W_i are vertexes of a simplex of $K'_{\mathcal{U}}$. Also, the intersections $U_0 \cap U_1 \cap \cdots \cap U_p \cap F$ and $W_0 \cap W_1 \cap \cdots \cap W_p \cap F$ are not empty. This shows that, in the first place, $f^1_{\mathcal{U}\mathcal{V}}$ and $f^2_{\mathcal{U}\mathcal{V}}$ are maps of the pair $(K_{\mathcal{V}}, K'_{\mathcal{V}})$ into the pair $(K_{\mathcal{U}}, K'_{\mathcal{U}})$, and in the second place, they are contiguous as maps of pairs.

From Exercise 3-2, it follows that $f^1_{\mathscr{U}\mathscr{V}}$ and $f^2_{\mathscr{U}\mathscr{V}}$ induce the same homomorphism.

$$f_{\mathscr{U}\mathscr{V}}: H_p(E, F; \mathscr{V}) \to H_p(E, F; \mathscr{U})$$

on the homology groups.

Definition 5-9. $f_{\mathscr{U}\mathscr{V}}$ is called the *homomorphism on the homology groups associated with the pair of coverings $\mathscr{V} > \mathscr{U}$.*

The next two theorems give fundamental properties of the $f_{\mathscr{U}\mathscr{V}}$ which are necessary for the construction of the Čech homology groups and the corresponding induced homomorphisms.

Theorem 5-5. (1) *For any finite open covering $\mathscr{U}$, $f_{\mathscr{U}\mathscr{U}}$ is the identity.*
(2) *If $\mathscr{U}$, $\mathscr{V}$, $\mathscr{W}$ are finite open coverings of E such that $\mathscr{W} < \mathscr{V} < \mathscr{W}$, then*

$$f_{\mathscr{U}\mathscr{V}} f_{\mathscr{V}\mathscr{W}} = f_{\mathscr{U}\mathscr{W}}$$

Proof. The first part is obvious, since the simplicial map $f^1_{\mathscr{U}\mathscr{U}}$ can be made the identity.

To prove the second part, take a vertex W of the nerve $K_{\mathscr{W}}$ of $\mathscr{W}$. Since $\mathscr{W} > \mathscr{V}$, there is a set V in $\mathscr{V}$ such that $W \subset V$. Let $f^1_{\mathscr{V}\mathscr{W}}$ map the vertex W on the vertex V of $K_{\mathscr{V}}$. Since $\mathscr{V} > \mathscr{U}$, there is a set U of $\mathscr{U}$ such that $V \subset U$. So let $f^1_{\mathscr{U}\mathscr{V}}$ map the vertex V of $K_{\mathscr{V}}$ on the vertex U of $K_{\mathscr{U}}$. Now the set U of $\mathscr{U}$ contains the set W of $\mathscr{W}$, so the simplicial map $f^1_{\mathscr{U}\mathscr{W}}$ can be taken (remember that this choice is arbitrary) to map the vertex W of $K_{\mathscr{W}}$ on the vertex U of $K_{\mathscr{U}}$. If this is done for each vertex of $K_{\mathscr{W}}$, the composition formula

$$f^1_{\mathscr{U}\mathscr{V}}\ f^1_{\mathscr{U}\mathscr{W}} = f^1_{\mathscr{U}\mathscr{W}}$$

will be satisfied on the vertexes of $K_{\mathscr{W}}$ and so, when extended by linearity, will be satisfied by the simplicial maps on all of $K_{\mathscr{W}}$. Passing to the induced homomorphisms on the simplicial groups, and using Theorem 1-5, and Section 2-8, the second part of the present theorem follows. ∎

The next theorem gives a connection between the $f_{\mathscr{U}}$ of Definition 5-7 and the $f_{\mathscr{U}\mathscr{V}}$ just introduced.

Theorem 5-6. *Let $f: (X, Y) \to (E, F)$ be a continuous map, let $\mathscr{U}$, $\mathscr{V}$ be finite open coverings of E, and let $\mathscr{U}' = f^{-1}(\mathscr{U})$, $\mathscr{V}' = f^{-1}(\mathscr{V})$. Then the* *following diagram commutes whenever $\mathscr{V} > \mathscr{U}$.*

$$\begin{array}{ccc} H_p(X, Y; \mathscr{U}') & \xrightarrow{f_{\mathscr{U}}} & H_p(E, F; \mathscr{U}) \\ \uparrow f_{\mathscr{U}'\mathscr{V}'} & & \uparrow f_{\mathscr{U}\mathscr{V}} \\ H_p(X, Y; \mathscr{V}') & \xrightarrow{f_{\mathscr{V}}} & H_p(E, F; \mathscr{V}) \end{array} \tag{5-1}$$

Proof. If $\mathscr{V} > \mathscr{U}$, $\mathscr{V}' > \mathscr{U}'$ automatically, so both $f_{\mathscr{U}\mathscr{V}}$ and $f_{\mathscr{U}'\mathscr{V}'}$ are defined. Take $V' \in \mathscr{V}'$. $V' = f^{-1}(V)$ for some V in $\mathscr{V}$. V is contained in some U of $\mathscr{U}$, so $V' \subset U' = f^{-1}(U)$. Now define the various simplicial maps as follows:

$f_{\mathscr{V}}{}^1$ maps V' on V

$f^1_{\mathscr{U}\mathscr{V}}$ maps V on U

$f^1_{\mathscr{U}'\mathscr{V}'}$ maps V' on U'

$f_{\mathscr{U}}{}^1$ maps U' on U

This makes the formula

$$f^1_{\mathscr{U}\mathscr{V}} f_{\mathscr{V}}{}^1 = f_{\mathscr{U}}{}^1 f^1_{\mathscr{U}'\mathscr{V}'} \tag{5-2}$$

hold for the operation of both sides on the vertexes of $K_{\mathscr{V}'}$. At this point $f_{\mathscr{U}}{}^1$ and $f^1_{\mathscr{U}\mathscr{V}}$ are not completely defined. On the vertexes of $K_{\mathscr{U}'}$ and $K_{\mathscr{V}}$ which are not indcluded in the above list, these maps may be defined in accordance with Definitions 5-7 and 5-9, subject to no further conditions. When extended by linearity the formula (5-2) holds on the whole of $K_{\mathscr{V}'}$, and the stated result is detained by taking the induced homomorphisms on homology groups. ∎

The next step is to show that the family of groups $H_p(E, F; \mathscr{U})$, along with the homomorphisms $f_{\mathscr{U}\mathscr{V}}$, admits the construction of a certain type of limit process. A digression will be made in the next section to describe this process, which can then be applied to the construction of the Čech homology groups.

5-4. INVERSE LIMITS

The limit process described here will be applied only to groups for the moment, although it can appear in other contexts (cf. Section 6-5 for the case of topological spaces). Whatever the context, the notion depends on directed sets.

Definition 5-10. A *directed set* is a partially ordered set A, the order relation being written as $>$, with the additional condition that for every pair of elements α, β of A, there is an element γ of A such that $\gamma > \alpha$ and $\gamma > \beta$.

Examples

5-6. A totally ordered set automatically satisfies the additional condition and so is a directed set.

5-7. Let E be any set and A the set of all its subsets. Order A by defining $\alpha > \beta$ to mean that α is contained in β. Then, for any α and β in A, take $\gamma = \alpha \cap \beta$, and it can be seen at once that the condition of Definition 5-10 is satisfied.

5-8. The set A of Example 5-7 can also be ordered by making $\alpha > \beta$ mean that α contains β. Again, A is a directed set, for if α, β are in A the condition of Definition 5-10 can be satisfied by taking $\gamma = \alpha \cup \beta$.

Throughout this section, only additive Abelian groups will be discussed, and so this qualification will be dropped temporarily and such an object will simply be called a group.

Definition 5-11. An *inverse system of groups* is a set of groups G_α indexed by a directed set A such that, for every pair of indexes α, β with $\beta > \alpha$, there is a homomorphism $\pi_{\alpha\beta}: G_\beta \to G_\alpha$, and these homomorphisms satisfy the conditions

(1) $\pi_{\alpha\alpha}$ is the identity for all α in A;
(2) $\pi_{\alpha\beta}\pi_{\beta\gamma} = \pi_{\alpha\gamma}$ whenever $\gamma > \beta > \alpha$.

Such a system may be denoted by $\{G_\alpha, \pi_{\alpha\beta}, A\}$. If the directed set or the set of homomorphisms can be understood from the context, this will be abbreviated to $\{G_\alpha, \pi_{\alpha\beta}\}$, or simply $\{G_\alpha\}$.

Examples

5-9. Let G be a group and A the set of all its subgroups. Order A by making $K > H$ mean $K \subset H$. A is directed, for if H and K are in A, then $H \cap K$ satisfies $H \cap K \subset H$ and $H \cap K \subset K$. Write $G_H = G/H$, and whenever $K > H$, define the homomorphism

$$\pi_{HK}: G_K \to G_H$$

to be the natural homomorphism

$$G/K \to \frac{G/K}{H/K} \cong G/H$$

It is easy to see that conditions (1) and (2) of Definition 5-11 are satisfied, so $\{G_H, \pi_{HK}, A\}$ is an inverse system.

5-10. A similar construction is possible if A, instead of being the set of all subgroups of G, is some family of subgroups such that, whenever H and K are in A, there is an L in A satisfying $L \subset H$ and $L \subset K$.

Let $\{G_\alpha, \pi_{\alpha\beta}, A\}$ be an inverse system of groups. An element of the Cartesian product ΠG_α of all the G_α is specified by giving its coordinates (g_α), where $g_\alpha \in G_\alpha$ for each α. Let G be the set of all elements of ΠG_α whose coordinates satisfy the relation

$$g_\alpha = \pi_{\alpha\beta} g_\beta$$

whenever $\beta > \alpha$.

Let g and g', with coordinates (g_α) and (g'_α), respectively, be two elements of the set G. Then, if $\beta > \alpha$,

$$g_\alpha - g'_\alpha = \pi_{\alpha\beta}(g_\beta) - \pi_{\alpha\beta}(g'_\beta) = \pi_{\alpha\beta}(g_\beta - g'_\beta)$$

so the difference $g - g'$ is in G. Thus, G is a subgroup of ΠG_α.

Definition 5-12. The group G just constructed is called *the inverse limit of the system* $\{G_\alpha, \pi_{\alpha\beta}, A\}$. The notation for this is

$$G = \lim \{G_\alpha, \pi_{\alpha\beta}, A\}$$

If no confusion is likely to arise this notation may be abbreviated to

$$G = \lim_A G_\alpha$$

or simply $G = \lim G_\alpha$.

Note that, if A consists of one element α, then $\lim G_\alpha = G_\alpha$, trivially.

There are natural homomorphisms $\pi_\alpha \colon G \to G_\alpha$ for each α, for each element g of G is in ΠG_α and so is specified by its coordinates (g_α). Then, $\pi_\alpha(g)$ is defined as g_α. It is easy to see that π_α is a homomorphism.

Definition 5-13. π_α is called the *projection of* $\lim\{G_\alpha, \pi_{\alpha\beta}, A\}$ *into* G_α.

Note that π_α is simply the restriction to G of the projection of the Cartesian product into G_α.

The π_α and the $\pi_{\alpha\beta}$ are related as follows. Take any g in $G = \lim_A G_\alpha$, with coordinates (g_α). If $\beta > \alpha$, $g_\alpha = \pi_{\alpha\beta} g_\beta$. In terms of the projections, this equation can be written as

$$\pi_\alpha(g) = \pi_{\alpha\beta}\,\pi_\beta(g)$$

Since this holds for all g in G, it follows that the equation

$$\pi_\alpha = \pi_{\alpha\beta}\,\pi_\beta$$

holds whenever $\beta > \alpha$.

The notion of projection can be generalized. Let B be a subset of A. In any case, B is partially ordered by the ordering relation given in A. Suppose, in addition, that B is directed by this ordering relation. Then from the inverse system $\{G_\alpha, \pi_{\alpha\beta}, A\}$ pick those G_α whose indexes are in B. The relations between the $\pi_{\alpha\beta}$ remain true if the indexes are restricted to B and in this way, a new inverse system $\{G_\alpha, \pi_{\alpha\beta}, B\}$ is obtained. Thus, it is possible to form the two inverse limits $\lim_A G_\alpha$ and $\lim_B G_\alpha$ of the original system and the restricted system, respectively.

If $g \in \lim_A G_\alpha$, let $\pi_B(g)$ be the element of $\Pi\, G_\alpha$ whose coordinates are the coordinates of g with indexes in B. Since the coordinates (g_α) of g satisfy the relations $g_\alpha = \pi_{\alpha\beta}\, g_\beta$ for all α and β in A such that $\beta > \alpha$, these relations still hold with the indexes restricted to B. Hence $\pi_B(g)$ is an element of $\lim_B G_\alpha$. Thus, a map

$$\pi_B\colon \lim_A G_\alpha \to \lim_B G_\alpha$$

has been constructed. It is easy to see that this is a homomorphism.

Definition 5-14. g_B is the *projection map of* $\lim_A G_\alpha$ *into* $\lim_B G_\alpha$.

The projection map of Definition 5-13 is a special case of π_B, corresponding to the case in which B is the single element α.

Clearly, a further generalization is possible. If B and C are subsets of A, both directed under the order relation of A and if $C \subset B$, then the projection π_C as defined in Definition 5-14 can be restricted to $\lim_B G_\alpha$ to give a homomorphism $\pi_{CB}\colon \lim_B G \to \lim_C G_\alpha$.

Exercise. 5-5. If $B \supset C \supset D$ are subsets of A directed by the order relation of A prove that, in the notation just introduced,

$$\pi_{DB} = \pi_{DC}\pi_{CB}$$

Note that there are special cases of this relation which may be obtained by taking D to be one element or B to be equal to A.

Examples

5-11. A simple example of the notion of inverse limit is afforded by considering an infinite Cartesian product (infinite because the discussion becomes trivial in the finite case). Let $\{G^i\}$ be a family of groups indexed by the elements of a set F. Let A be the set of finite subsets of F partially ordered by inclusion. In other words, $\alpha > \beta$ means that α contains β. A is directed since the union of two sets in A is greater than each of them. If the finite set α consists of $i, j, \ldots, k$, let G_α be the finite product $G^i \times G^j \times \cdots \times G^k$. If $\beta > \alpha$, then G_α is a partial product of $G_\beta \cdot \pi_{\alpha\beta}$ is defined as the projection of G_β onto G_α. It is not hard to prove that $\{G_\alpha, \pi_{\alpha\beta}, A\}$ is an inverse system of groups.

It will now be shown that $G = \lim_A G_\alpha$ can be identified with $G' = \Pi G^i$, the product being taken over all the i in F. In the first place, an element x of G is an element of ΠG_α, and so it has coordinates (x_α) with $x_\alpha \in G_\alpha$. These coordinates satisfy the compatibility relations

$$x_\alpha = \pi_{\alpha\beta}(x_\beta)$$

for $\beta > \alpha$. On the other hand, x_α and x_β have coordinates as elements of the finite products G_α and G_β, and the compatibility relations mean that, for any factor G^i appearing in the finite product G_β, and so also in G_α, the G^i coordinate of x_β is the same as that of x_α. More generally, if α, β are any elements of A and $\gamma = \alpha \cup \beta$, so that $\gamma > \alpha$ and $\gamma > \beta$, then the above remark implies that, if G^i is any factor common to the products G_α and G_β, then the coordinates in G^i of x_α and x_β are the same as that of x_γ and so are equal.

Now take any element x of $G = \lim_A G_\alpha$ and define $f(x) \in \Pi G^i$ as follows. For any i in F, take some α in A that contains i as a member and let x^i be the G^i coordinate of x_α in the product G_α. By the remarks just made, x^i is independent of the α chosen. Now let $f(x)$ be the element of ΠG^i whose G^i coordinate is x^i, for each i. It is an easy matter to verify that f is a homomorphism and, in fact, is an isomorphism onto.

Also, in this example it is easy to check that the projection $\pi_\alpha\colon \lim_A G_\alpha \to G_\alpha$ coincides with the projection of the infinite product ΠG^i onto the finite product G_α in the usual sense of projection in Cartesian products.

This example provides an intuitive motivation for the use of the word "limit" in this context, since it makes an infinite product appear as the limit of finite partial products.

5-12. As a further example of inverse limits, consider the inverse system of Example 5-9. Here a group G is given and A is a family of subgroups such that, when H and K are in A, there is an L in A with $L \subset H \cap K$. The set A is directed by the order relation defined by making $H > K$ mean $H \subset K$. The inverse system $\{G_H, \pi_{HK}, A\}$ is constructed with $G_H = G/H$ and π_{HK}, $K > H$, equal to the natural map $G/K \to G/H$.

Let $G' = \lim_A G_H$. An element g of G' can be given by its coordinates

(g_H), where $g_H \in G_H$; that is, g_H is a coset of H in G. For two coordinates g_H and g_K of g, with $K > H$, the relation $g_H = \pi_{HK} g_K$ holds. This says that the coset g_K of K in G is contained in the coset g_H of H in G. Thus, an element of G' is a family of cosets g_H such that those cosets that correspond to later indices H in the ordering of A, that is, to smaller subgroups, are contained in those corresponding to earlier suffixes. If the intersection of such a family of cosets is nonempty, this suggests characterizing the element of the inverse limit by this intersection: this, in turn, suggests comparing G' with G/G_0 where G_0 is the intersection of all subgroups in the family A. In fact, it will now be shown that G/G_0 can always be identified with a subgroup of G', and that under suitable conditions (namely, requiring the nonempty intersection of the cosets corresponding to an element of G'), G/G_0 is isomorphic to G'.

For this purpose, a map f of G/G_0 into G' will be constructed. First, for each H in A, let f_H be the natural homomorphism

$$G/G_0 \to \frac{(G/G_0)}{(H/G_0)} \cong G/H = G_H$$

To show that the f_H fit together to give a map into G', note that, if $K > H$, the images $f_H(g)$ and $f_K(g)$ of any element g of G/G_0 satisfy the relations $f_H(g) = \pi_{HK} f_K(g)$. Thus, the images of g under the f_H are the coordinates of an element of G'. Denote this element by $f(g)$. In this way, the map f: $G/G_0 \to G'$ is constructed. It is easy to check that f is a homomorphism. It will now be shown to have zero kernel.

Let g be in the kernel of f. Thus, $f_H(g) = 0$ for all H in A; that is, g, regarded as a coset of G_0 in G, is contained in the zero coset of H in G for each H. To state this yet another way, it is contained in the group H for each H in A, so it is in the intersection of all these H, namely, G_0. Thus, g is the zero element of G/G_0, so f has zero kernel, as was to be shown.

It has been shown, therefore, that $\lim_A G_H$ is an extension of G/G_0; that is, it contains an isomorphic image of G/G_0. The next step is to see under what conditions the map f is onto. If this is to happen, each element (g_H) of $\lim_A G_H = G'$ (given here by its coordinates) must have an inverse image under f. This means that there must be an element g of G/G_0 mapping on g_H for each H in A. This, in turn, means that the coset g of G_0 in G must be contained in the coset g_H of H in G for each H in A. Then the intersection of the g_H, regarded as subsets of G, certainly must be nonempty as a necessary condition for f to be onto. This condition is also sufficient. Suppose that, in fact, for each element (g_H) of G', the intersection of the g_H, as subsets of G, is not empty, and take $g_0 \in \bigcap g_H$. Then, if g_1 is any element of G_0, g_1 is also in each H, so $g_0 + g_1$ is in the same coset modulo H as g_0, namely, g_H, for each H in A. This implies that $g_0 + g_1$ is contained in $\bigcap g_H$, that is, allowing g_1 to vary throughout G_0, the whole coset $g = g_0 + G_0$ is contained in $\bigcap g_H$ and $f(g) = (g_H)$. It has thus been shown that f is onto.

Summing up, it has been proved that f is onto, if and only if $\bigcap g_H \neq \varnothing$ for the set of coordinates (g_H) of each element of G'. Note that, in any case, this argument shows that the image of f consists of exactly those elements of G' whose coordinates satisfy this condition.

Note that Example 5-11 can be regarded as a special case of Example 5-12. If $G = \Pi G^i$, let A be the family of subgroups that are products of infinitely many of the G^i. If H and K are two such subgroups, then the product of all the factors common to H and K is a member of A contained in both H and K. So A is a family such as that described in Example 5-12. $G_H = G/H$ for H in A is a product of finitely many of the G^i and so can be identified with one of the G_α of Example 5-11. Also, the π_{HK} can be identified with the $\pi_{\alpha\beta}$ for suitable α and β. The result of Example 5-11, namely, that $\lim_A G_\alpha$ is G, will now be proved using the discussion of Example 5-12.

To begin with, G_0 is, in this case, the intersection of all the H in A. Any nonzero element of a group H in A has at least one nonzero coordinate, in G^i, for example, so it is not in the group H' of A which is the product of all the factors of H except G^i. Since H' is in A, this means that the selected nonzero element of H is not in G_0. It follows that G_0 is zero. Hence, f is an isomorphism of G into $G' = \lim_A G_H$. To that check the condition for f to be onto is satisfied here, first note that a coset of H in G is obtained by fixing a finite number of coordinates, those corresponding to factors not in H and allowing the others to vary freely in the corresponding G^i. If (g_H) is an element of G', the compatibility conditions between the g_H imply that, for any index i, all cosets g_H that involve fixing the G^i coordinate fix it in the same way. Thus, an element of each G^i is picked out, and the collection of all these elements are the coordinates of an element of $G = \Pi G^i$ common to all the g_H. The condition for f to be onto is thus satisfied, so $G \cong G'$, as required.

5-13. Finally, consider the following topological example. Take the space of Example 5-4 and the sequence of coverings given in that example. If K_n is the nerve of the nth covering, $H_1(K_n)$ is the Cartesian product of n copies of Z. If $m > n$, the mth covering is a refinement of the nth and there is a simplicial map of K_m into K_n set up as for Definition 5-9 which maps $m - n$ of the triangles on a point. The induced homomorphism of $H_1(K_m)$ into $H_1(K_n)$ maps $m - n$ of the factors of Z^m on zero. This is exactly the situation of Example 5-11 with G replaced by a product of denumerably many copies of Z. So the inverse limit of the $H_1(K_n)$, indexed by the natural number n, is the product of denumerably many factors each equal to Z.

In this way, $H_1(E)$, which is known to be the product of a denumerable set of copies of Z, is obtained as the inverse limit of the one-dimensional homology groups of nerves of certain coverings of E. This gives some sort of motivation for the definition of the Čech groups, which is given in Section 5-6, although there is no guarantee that in the general case the result coincides with the singular homology groups.

5-5. LIMITS OF HOMOMORPHISMS

It has been indicated already that the Čech homology groups are defined as inverse limits of systems of simplicial homology groups of nerves of coverings. At the same time, induced homomorphisms have to be constructed. This will be done by means of a limiting process, which will now be described.

Let $\{G_\alpha, \pi_{\alpha\beta}, A\}$ and $\{H_\alpha, \kappa_{\alpha\beta}, B\}$ be two given inverse systems of groups. Let $\phi: B \to A$ be an order-preserving map; that is, if α, β are in B and $\alpha > \beta$, then $\phi(\alpha) > \phi(\beta)$. To simplify notation, write a prime on a letter to denote the effect of the map ϕ; thus $\alpha' = \phi(\alpha)$, and so on. Now, for each α in B, let a homomorphism $f_\alpha: G_{\alpha'} \to H_\alpha$ be given such that, for every $\beta > \alpha$ in B, the following diagram is commutative.

$$\begin{array}{ccc} G_{\alpha'} & \xrightarrow{f_\alpha} & H_\alpha \\ {\scriptstyle \pi_{\alpha'\beta'}}\uparrow & & \uparrow{\scriptstyle \kappa_{\alpha\beta}} \\ G_{\beta'} & \xrightarrow{f_\beta} & H_\beta \end{array} \tag{5-3}$$

The condition $\beta > \alpha$ ensures that $\beta' > \alpha'$, so the vertical maps are both defined.

Definition 5-15. Such a family of homomorphisms $\{f_\alpha\}$ is called an *inverse system of homomorphisms of the system* $\{G_\alpha, \pi_{\alpha\beta}, A\}$ *into the system* $\{H_\alpha, \kappa_{\alpha\beta}, B\}$, *corresponding to the map* $\phi: B \to A$.

The motivation for this definition will be clear when the construction of the induced homomorphisms is carried out in Čech theory. There a continuous map $g: (X, Y) \to (E, F)$ is given, and the two inverse systems are $\{H_p(X, Y; \mathscr{U}), f_{\mathscr{U}\mathscr{V}}\}$ and $\{H_p(E, F; \mathscr{U}), f_{\mathscr{U}\mathscr{V}}\}$ in which the indexing sets are the finite open coverings of X and E, respectively. The map $\phi: B \to A$ assigns to each finite open covering $\mathscr{U}$ of E its inverse image $g^{-1}(\mathscr{U})$, and then the inverse system of homomorphisms consists of homomorphisms

$$g_{\mathscr{U}}: H_p(X, Y; g^{-1}(\mathscr{U})) \to H_p(E, F; \mathscr{U})$$

induced by g as described in Definition 5-7.

To show that an inverse system of homomorphisms, as it is defined in Definition 5-15, leads in a natural way to a homomorphism f of $\lim_A G_\alpha$ into $\lim_B H_\alpha$, take each element $g = (g_\alpha)$ of $\lim_A G_\alpha$ and construct its image under f. In other words, define each coordinate of the image, for each β in B. Take g in $\lim_A G_\alpha$ with coordinates (g_α), $\alpha \in A$, and consider the set of elements $h_\alpha = f_\alpha(g_{\alpha'})$ for each α in B; here $\alpha' = \phi(\alpha)$. For each α in B, this is an element of H_α, and for $\beta > \alpha$,

$$\begin{aligned}\kappa_{\alpha\beta}(h_\beta) &= \kappa_{\alpha\beta} f_\beta(g_{\beta'}) \\ &= f_\alpha \pi_{\alpha'\beta'}(g_{\beta'}) \quad \text{[by the commutativity of the diagram (5-3)]} \\ &= f_\alpha(g_{\alpha'}) \quad \text{(by the compatibility relations on the } \pi_{\alpha'\beta'})\end{aligned}$$

In other words,

$$\kappa_{\alpha\beta}(h_\beta) = h_\alpha$$

so the h_α are coordinates of an element h in $\lim_B H_\alpha$. Write

$$h = f(g)$$

It is easy to check that the map f so defined is a homomorphism.

Definition 5-16. The homomorphism $f\colon \lim_A G_\alpha \to \lim_B H$ constructed, as it was above, from the inverse system of f_α is called *the inverse limit of the* f_α and is denoted by $\lim_B f_\alpha$ (note that the f_α are, in fact, indexed by the set B).

Lemma 5-7. *Let* $\{G_\alpha, \pi_{\alpha\beta}, A\}$, $\{H_\alpha, \kappa_{\alpha\beta}, B\}$, $\{K_\alpha, \lambda_{\alpha\beta}, C\}$ *be three inverse systems of groups. Let* $\phi\colon B \to A$ *and* $\psi\colon C \to B$ *be order-preserving maps and corresponding to them let inverse systems of homomorphisms*

$$f_\alpha\colon G_{\alpha'} \to H_\alpha\, [a \in B,\, \alpha' = \phi(\alpha)]$$

$$h_\beta\colon H_{\beta'} \to K_\beta\, [\beta \in C,\, \beta' = \psi(\beta)]$$

be given. Then the system of homomorphisms

$$h_\alpha f_{\alpha'}\colon G_{\alpha''} \to K_\alpha$$

with $\alpha \in C$, $\alpha' = \psi(\alpha)$, $\alpha'' = \phi\psi(\alpha)$ *is an inverse system of homomorphisms corresponding to the order-preserving map* $\phi\psi\colon C \to A$ *and*

$$\lim_C h_\alpha f_{\alpha'} = (\lim_C h_\alpha)(\lim_B f_\alpha)$$

Proof. Consider the diagram

$$\begin{array}{ccccc} G_{\alpha''} & \xrightarrow{f_{\alpha'}} & H_{\alpha'} & \xrightarrow{h_\alpha} & K_\alpha \\ \uparrow{\scriptstyle \pi_{\alpha''\beta''}} & & \uparrow{\scriptstyle \kappa_{\alpha'\beta'}} & & \uparrow{\scriptstyle \lambda_{\alpha\beta}} \\ G_{\beta''} & \xrightarrow{f_{\beta'}} & H_{\beta'} & \xrightarrow{h_\beta} & K_\beta \end{array}$$

for $\beta > \alpha$. The two halves are commutative by the definition of inverse systems of homomorphisms applied to the h_α, $\alpha \in C$, and the f_α, $\alpha \in B$, in particular, for the $f_{\alpha'}$ with $\alpha' \in \psi(C)$. Hence, the whole diagram is commutative, so the $h_\alpha f_{\alpha'}$ form an inverse system of homomorphisms.

To complete the proof, let $g \in \lim_A G_\alpha$ and let (g_α) be its coordinates. Then, if $f = \lim_B f_{\alpha'}$ the coordinates of $f(g)$ are $[f_\alpha(g_{\alpha'})]$ where $\alpha \in B$ and $\alpha' = \phi(\alpha)$. If $h = \lim_C h_\alpha$, the coordinates of $h[f(g)]$ are $\{h_\alpha[f_{\alpha'}(g_{\alpha''})]\}$ with α in C, $\alpha' = \psi(\alpha)$ and $\alpha'' = \phi\psi(\alpha)$. On the other hand, the α coordinate of $(\lim_C h_\alpha f_{\alpha'})(g)$ is $h_\alpha f_{\alpha'}(g_{\alpha''})$ with $\alpha \in C$, $a'' = \phi\psi(\alpha)$, which is the same as the α coordinate of $h[f(g)]$. Hence

$$\lim_C(h_\alpha f_{\alpha'}) = (\lim_C h_\alpha)(\lim_B f_\alpha)$$

as was to be proved. ∎

Exercises. 5-6. Let $\{G_\alpha, \pi_{\alpha\beta}, A\}$ be an inverse system and let $f_\alpha: G_\alpha \to G_\alpha$ be the identity for each α. Show that the f_α form an inverse system of homomorphisms corresponding to the identity map of A on itself, and show that $\lim_A f_\alpha$ is the identity.

5-7. Let $f_\alpha: G_{\alpha'} \to H_\alpha$ be the zero homomorphism for each $\alpha \in B$, when $\{G_\alpha, \pi_{A\beta}, A\}$ and $\{H_\alpha, \kappa_{\alpha\beta}, B\}$ are given inverse systems and ϕ is any order-preserving map of B into A, $\phi(\alpha) = \alpha'$. Show that the f_α form an inverse system corresponding to ϕ and that $\lim_B f_\alpha$ is the zero homomorphism.

5-8. Let $\{f_\alpha\}$ be an inverse system of homomorphisms of $\{G_\alpha, \pi_{A\beta}, A\}$ into $\{H_\alpha, \kappa_{\alpha\beta}, B\}$ corresponding to some map $\phi: B \to A$. Let $\pi_\alpha: \lim_A G_\alpha \to G_\alpha$ and $\kappa_\alpha: \lim_B H_\alpha \to H_\alpha$ be projections like those in Definition 5-13. Prove that $\kappa_\alpha(\lim_B f_\alpha) = f_\alpha \pi_{\alpha'}$ for each π in B, $\alpha' = \phi(\alpha)$.

More generally, if B' is a subset of B and A' a subset of A such that $\phi(B') \subset A'$, prove that the following diagram is commutative.

$$\begin{array}{ccc}
\lim_A G_\alpha & \xrightarrow{\lim_B f_\alpha} & \lim_B H_\alpha \\
{\scriptstyle \pi_{A'}}\downarrow & & \downarrow{\scriptstyle \pi_{B'}} \\
\lim_{A'} G_\alpha & \xrightarrow{\lim_{B'} f_\alpha} & \lim_{B'} H_\alpha
\end{array}$$

5-9. Let $\{G_\alpha, \pi_{A\beta}, A\}$ be an inverse system in which all the $\pi_{\alpha\beta}$ are isomorphisms onto. Thus, the G_α are all isomorphic to some group G. Show that $\lim G_\alpha = G$.

(*Hint*: Form an inverse system of homorphisms of the G_α onto the inverse system that has just one group G.)

5-6. *CONSTRUCTION OF THE ČECH HOMOLOGY GROUPS*

All the elements needed for the construction of the Čech homology groups have now been defined and it only remains to assemble them. Let E be a topological space and let F be a subspace. Let $\Gamma(E)$ be the set of finite open coverings of E ordered by refinement. $\Gamma(E)$ is a directed set. If $\mathscr{U}$ and $\mathscr{V}$ are in $\Gamma(E)$, let $\mathscr{W}$ be the covering consisting of sets $U \cap V$ with U in $\mathscr{U}$ and

V in $\mathscr{V}$. Then $\mathscr{W}$ is a refinement of both $\mathscr{U}$ and $\mathscr{V}$. It follows from Theorem 5-5 that $\{H_p(E, F; \mathscr{U}), f_{\mathscr{U}\mathscr{V}}, \Gamma(E)\}$ is an inverse system, so the following definition makes sense.

Definition 5-17. *The pth Čech homology group of the pair* (E, F) *is defined as the inverse limit of the system*

$$\{H_p(E, F; \mathscr{U}), f_{\mathscr{U}\mathscr{V}}, \Gamma(E)\}$$

and is denoted by $\check{H}_p(E, F)$. Briefly,

$$\check{H}_p(E, F) = \lim_{\Gamma(E)} H_p(E, F; \mathscr{U})$$

In particular, if F is the empty set, $\check{H}_p(E, F)$ is written simply as $\check{H}_p(E)$. No coefficient groups are mentioned here, but it is understood that some fixed coefficient group is used in constructing all the $H_p(E, F; \mathscr{U})$ and that it is the coefficient group for $\check{H}_p(E, F)$. It is not included in the notation unless some special emphasis is needed.

Note that this definition is rather different in character from the definition of singular homology. Here the elements of the homology group $\check{H}_p(E, F)$ are defined directly, rather than in terms of cycles and boundaries in a complex. In fact, the algebraic properties of the inverse limit process make it impossible, in general, to construct a complex of which the $\check{H}_p(E, F)$ are the homology groups. This also means that there is no associated cochain complex. In fact, the corresponding Čech type of cohomology must be constructed by taking cohomology groups of nerves of coverings and taking a different kind of limit.

All this makes it a bit difficult to visualize geometrically the meaning of a element of $\check{H}_p(E, F)$. The following may help to give some sort of picture of what is happening. Let E be a space and, for the moment, take F as the empty set. Let $\mathscr{U}$ be a finite open covering of E, and consider a sequence $U_0, U_1, \ldots, U_p$ of sets in $\mathscr{U}$ such that $U_i \cap U_{i+1} \neq \varnothing$ for each i and also $U_p \cap U_0 \neq \varnothing$. Then, the $((U_i U_{i+1}))$ and $((U_p U_0))$ are oriented one simplexes on the nerve of $\mathscr{U}$, and

$$\alpha_{\mathscr{U}} = \sum_{i=0}^{p-1} ((U_i U_{i+1})) + ((U_p U_0))$$

is an oriented one cycle on that nerve. Its homology class is thus an element $\bar{\alpha}_{\mathscr{U}}$ of $H_1(E; U)$. $\alpha_{\mathscr{U}}$ can be thought of geometrically in E as a closed chain formed by the open sets $U_0, U_1, \ldots, U_p$ (cf. Fig. 18). Now let $\mathscr{V}$ be a refinement of $\mathscr{U}$ and let $V_0, V_1, \ldots, V_q$ be open sets of $\mathscr{V}$ giving rise, as

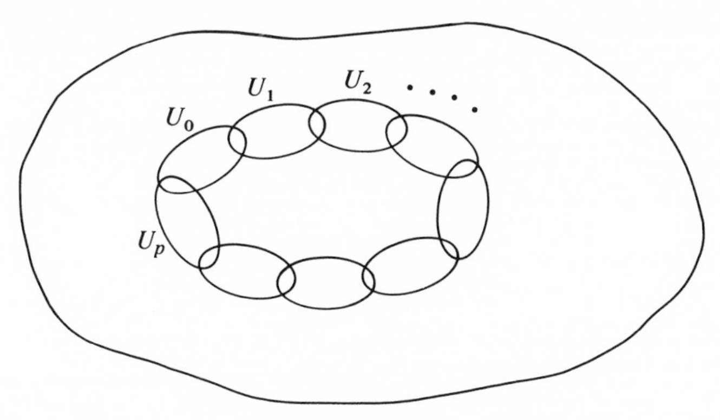

FIGURE 18

above, to a cycle $\alpha_{\mathscr{V}}$ on the nerve of $\mathscr{V}$ with homology class $\bar{\alpha}_{\mathscr{V}}$ in $H_1(E; \mathscr{V})$. Suppose that, when the simplicial map $f^1_{\mathscr{U}\mathscr{V}}: K_{\mathscr{V}} \to K_{\mathscr{U}}$ is constructed as it was in Section 5-3, each V_i belonging to $\alpha_{\mathscr{V}}$ is mapped on one of the U_j (i.e., is contained in it, in set theoretic language in E) that belongs to $\alpha_{\mathscr{U}}$ in such a way that $\bar{\alpha}_{\mathscr{U}} = f_{\mathscr{U}\mathscr{V}}\,\bar{\alpha}_{\mathscr{V}}$. Geometrically, this is a process of obtaining the chain of sets in E corresponding to $\alpha_{\mathscr{V}}$ by refining that corresponding to $\alpha_{\mathscr{U}}$ (cf. Fig. 19). The inverse limit process can be thought of as taking the limit of the $\bar{\alpha}_{\mathscr{U}}$ as the covering $\mathscr{U}$ is refined indefinitely. Intuitively, the corresponding chains of sets in E (representing the cycles $\alpha_{\mathscr{U}}$) become, as it were, strings of finer and finer beads, so that they appear to close down on a closed curve in E (cf. Fig. 20). Note, however, that this closed curve is in the nature of a limiting state, and in fact, there may not actually be any closed curve in E corresponding to an element of $\check{H}_1(E)$. This is in contradistinction to the situation in singular homology, where a singular cycle can always be associated with the curve carrying the singular simplexes that make it up.

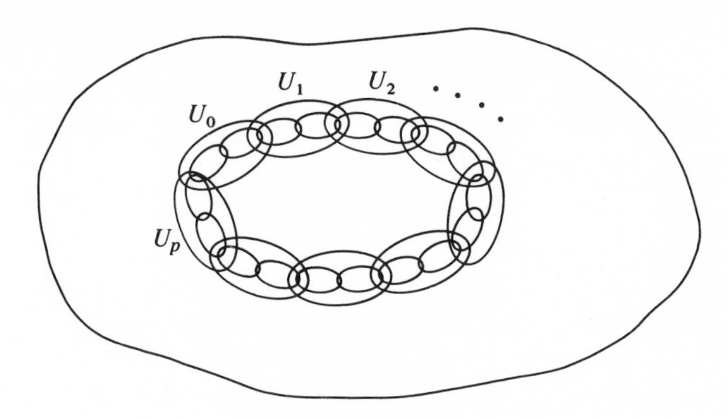

FIGURE 19

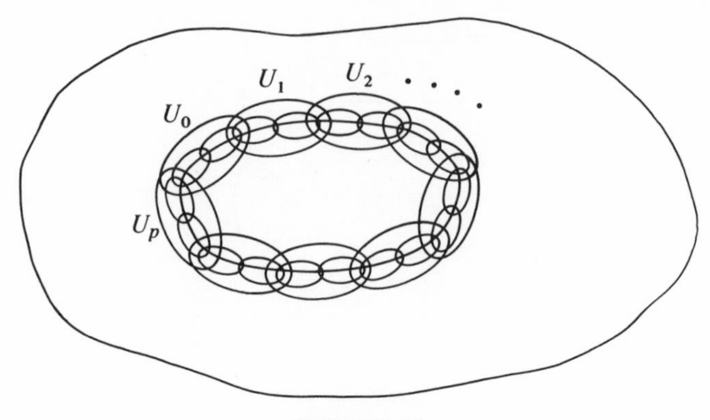

FIGURE 20

Example

5-14. Let E be the subspace of the plane consisting of the segments $(0, y)$ and $(2/\pi, y)$ for all y such that $1 \geqq y \geqq -2$, the segment $(x, -2)$ for all x such that $0 \leqq x \leqq 2/\pi$, and the part of the graph of $y = \sin(1/x)$ for which $0 < x \leqq 2/\pi$, as shown in Fig. 21. Here there is no continuous curve going all the way round the figure. In fact, any singular one simplex is in the complement of a neighborhood of the origin 0. The same applies to any singular 1 chain, so any 1 cycle is on a set homeomorphic to an interval and so is homologous to zero. Hence $H_1(E) = 0$.

On the other hand, in the manner of the preceding intuitive discussion, arbitrarily fine but finite chains of open sets can be constructed in E, making exactly one circuit of the figure. The point is that one of the open sets must contain 0, and then a finite chain is sufficient to get around the rest of the figure. Thus, it is plausible that $\check{H}_1(E)$ should be infinite cyclic. This will be proved later.

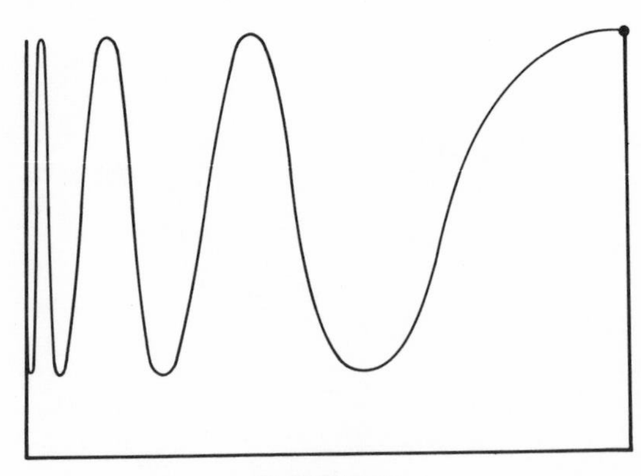

FIGURE 21

Induced homomorphisms corresponding to continuous maps will now be constructed, using the theory of Section 5-5. Let f be a continuous map of the pair (X, Y) into the pair (E, F). Let $\Gamma(E)$ and $\Gamma(X)$ be the sets of finite open coverings of E and X, respectively, and let $\phi: \Gamma(E) \to (X)$ be defined by setting $\phi(\mathscr{U}) = f^{-1}(\mathscr{U})$; that is, the covering of X which consists of sets $f^{-1}(U)$ with U in $\mathscr{U}$. This map ϕ is order preserving. Write $\mathscr{U}' = \phi(\mathscr{U})$ and let

$$f_{\mathscr{U}}: H_p(X, Y; \mathscr{U}') \to H_p(E, F; \mathscr{U})$$

be as it was constructed in Section 5-3. Theorem 5-6 now implies that the $f_{\mathscr{U}}$, with $\mathscr{U}$ in $\Gamma(E)$, form an inverse system of homomorphisms of the system $\{H_p(X, Y; \mathscr{U}), f_{\mathscr{U}\mathscr{V}}, \Gamma(X)\}$ into the system $\{H_p(E, F; \mathscr{U}), f_{\mathscr{U}\mathscr{V}}, \Gamma(E)\}$ corresponding to the map ϕ. It follows that a limit can be constructed such as the one in Section 5-5, namely

$$\lim_{\Gamma(E)} f_{\mathscr{U}} : \lim_{\Gamma(X)} H_p(X, Y; \mathscr{U}) \to \lim_{\Gamma(E)} H_p(E, F; \mathscr{U})$$

By Definition 5-15, this is, in fact, a homomorphism

$$f_*: \check{H}_p(X, Y) \to \check{H}_p(E, F)$$

Definition 5-18. f_* is called the *induced homomorphism on the pth Čech homology group corresponding to the continuous map f.* Note that there is an f_* for each dimension p.

The following theorem shows that the induced homomorphisms so defined behave in a similar way to those of the singular homology theory.

Theorem 5-8. *If $f: (X, Y) \to (E, F)$ and $g: (E, F) \to (A, B)$ are continuous maps of the pairs indicated, then the relation*

$$(gf)_* = g_* f_*$$

holds between the induced homomorphisms on the Čech groups for each dimension. Also, if $i: (E, F) \to (E, F)$ is the identity, then i_* is the identity homomorphism for each dimension.

Proof. The first part follows from the combination of the first part of Theorem 5-4 with Lemma 5-7. The second part follows from the combination of the second part of Theorem 5-4 with Exercise 5-6. ∎

Corollary. *The Čech homology groups of a pair are topological invariants.*

Proof. The proof is exactly like that of Theorem 1-6, using Theorem 5-8 instead of Theorem 1-5.

The construction of the Čech homology groups thus gives a new set of topological invariants associated with a topological space, and these groups are already seen to satisfy the property (2) of Section 2-1. Also, if E is a single point, it has only one finite open covering, whose nerve is a single point. The following is then trivial.

Theorem 5-9. *If E is a single point, $\check{H}_p(E) = 0$ for $p > 0$ and $\check{H}_0(E) = \mathcal{G}$, the coefficient group.*

In other words, the dimension theorem is satisfied. Whether or not the other properties listed in Section 2-1 are satisfied by the Čech homology groups will be investigated presently. In the meantime, a digression will be made to examine the case in which the spaces under consideration are simplicial complexes.

5-7. ČECH HOMOLOGY FOR SIMPLICIAL PAIRS

It will be shown in this section that, if (E, F) is a simplicial pair, then the singular group $H_p(E, F)$ is isomorphic to the Čech group $\check{H}_p(E, F)$ for each p, both being isomorphic to $\mathcal{H}_p(E, F)$. It will appear later that the singular and Čech groups are not necessarily isomorphic for nontriangulable spaces. In fact, it has already been indicated that, for the space E of Example 5-14, $H_1(E)$ and $\check{H}_1(E)$ are different; this point will be examined and proved later.

The approach here will be to show directly that $\check{H}_p(E, F)$ is isomorphic to $\mathcal{H}_p(E, F)$ for a triangulable pair. This will be done by showing that in forming the inverse limit which defines $\check{H}_p(E, F)$ it is sufficient to use a subset of the set of all finite open coverings, namely, the star coverings of barycentric subdivisions of E. The reduction to a subset of the indexing set (in this instance the finite open coverings) is carried out by using the notion of cofinal subsets of a directed set. This general notion will now be described.

Definition 5-19. Let A be a directed set and let A' be a subset, also directed by the ordering relation of A. A' is called *cofinal in A* if, for every α in A, there is a β in A' such that $\beta > \alpha$.

The idea of this definition is that, in a suitable sense, arbitrarily large elements of A can be obtained by taking elements of the subset A' only.

Examples

5-15. If A is the set of integers with the usual order, any subset that is unbounded above is cofinal in A.

5-16. Let E be a simplicial complex and A the set of finite open coverings of E. Let A' be the set of coverings of the form $B^r\mathscr{U}$, where this denotes the star open covering of the rth barycentric subdivision of E. Then A' is cofinal in A. To see this, it is sufficient to show that any finite open covering $\mathscr{V}$ of E has a refinement of the form $B^r\mathscr{U}$ for some r. But it is known that, on the one hand, if $\mathscr{V}$ is a given finite open covering, there is a positive ε such that any set of diameter less than ε is in some set of $\mathscr{V}$ and, on the other hand, all the open sets of $B^r\mathscr{U}$ are of diameter less than a preassigned ε for sufficiently large r, so the required result follows.

5-17. Let A be the set of all open sets in the plane, and order them by making $U > V$ mean that $U \subset V$. Let A' be the set of all open disks. Then A' is cofinal in A. Clearly, this can be generalized to any topological space with A' taken as a basis for the topology.

Cofinal sets are important because an inverse limit over a directed set is the same as that over a cofinal subset. This is somewhat analogous to the statement that, if a sequence of numbers has a limit, then any subsequence has the same limit. The sequence is indexed by the natural numbers and the subsequence by an unbounded subset, which (cf. Example 5-15) is cofinal.

Theorem 5-10. *Let $\{G_\alpha, \pi_{\alpha\beta}, A\}$ be an inverse system of groups and let A' be a cofinal subset of the indexing set A. Then the projection homomorphism (cf. Definition 5-14)*

$$\pi_{A'}\colon \lim_A G_\alpha \to \lim_{A'} G_\alpha$$

is an isomorphism onto.

Proof. First, to show that $\pi_{A'}$ is onto, take g in $\lim_{A'} G_\alpha$ and let its coordinates be (g_α), a coordinate for each α in A'. An element h in $\lim_A G_\alpha$ has to be found such that $\pi_{A'}(h) = g$; that is, h has to be an element with coordinates (h_α) for each α in A, such that when α is in A', $h_\alpha = g_\alpha$. To construct h, start by taking any α in A. By the cofinal property of A', there is a β in A' such that $\beta > \alpha$; g_β is already given. Define $h_\alpha = \pi_{\alpha\beta} g_\beta$. (It is clear from the compatibility relations that the h_α should satisfy that they must be defined in this way.) In particular, if α is in A', take $\beta = \alpha$ so that $h_\alpha = g_\alpha$.

It will now be shown that the h_α are coordinates of an element of $\lim_A G_\alpha$. Take α and α' in A. As before, elements β and β' in A' have been chosen so that $\beta > \alpha$ and $\beta' > \alpha'$, and h_α and $h_{\alpha'}$ are defined by

$$h_\alpha = \pi_{\alpha\beta} g_\beta$$

$$h_{\alpha'} = \pi_{\alpha'\beta'} g_{\beta'}$$

Now suppose $\alpha' > \alpha$, and let γ be an element of A' such that $\gamma > \beta$ and $\gamma > \beta'$. Then

$$\pi_{\alpha\alpha'} h_{\alpha'} = \pi_{\alpha\alpha'} \pi_{\alpha'\beta'} g_{\beta'} = \pi_{\alpha\alpha'} \pi_{\alpha'\beta'} \pi_{\beta'\gamma} g_\gamma$$

In the second equation, the compatibility relations for the coordinates of g are used. Also,

$$\pi_{\alpha\alpha'} \pi_{\alpha'\beta'} \pi_{\beta'\gamma} g_\gamma = \pi_{\alpha\gamma} g_\gamma = \pi_{\alpha\beta} \pi_{\beta\gamma} g_\gamma$$
$$= \pi_{\alpha\beta} g_\beta$$

where again the compatibility relations for the coordinates of g are used. But $\pi_{\alpha\beta} g_\beta = h_\alpha$, so, combining all these steps,

$$\pi_{\alpha\alpha'} h_{\alpha'} = h_\alpha$$

for all α and α' in A such that $\alpha' > \alpha$. Hence, the h_α are coordinates of an element h of $\lim_A G_\alpha$, and if α is in A', $h_\alpha = g_\alpha$ by definition, so $\pi_{A'}(h) = g$. Thus, any element g of $\lim_{A'} G_\alpha$ is in the image of $\pi_{A'}$, so this map is onto.

To see that the kernel of $\pi_{A'}$ is zero, let h be in this kernel, so that $\pi_{A'}(h) = 0$. Thus h has coordinates $h_\alpha = 0$ for all α in A'. Take β in A. Then there is an α in A' such that $\beta > \alpha$, since A' is cofinal in A. By the compatibility relations, $h_\beta = \pi_{\beta\alpha} h_\alpha$ and $h_\alpha = 0$. Hence, $h_\beta = 0$ for all β in A, so $h = 0$, as was to be shown. ∎

Exercise. 5-10. Show in addition that the limit of an inverse system of homomorphisms over a cofinal subset is the same as the limit over the whole index set.

It has been seen (cf. Example 5-16) that, if E is a simplicial complex, the set $\Gamma_0(E)$ of star open coverings of the successive barycentric subdivisions of E is cofinal in the set $\Gamma(E)$ of finite open coverings of E. So the theorem just proved implies that, to show that $\check{H}_p(E, F) \cong \mathscr{H}_p(E, F)$ for the simplicial pair (E, F), it is sufficient to show that

$$\lim\{H_p(E, F; \mathscr{U})\} \cong \mathscr{H}_p(E, F)$$

where the limit is taken over the coverings of the family $\Gamma_0(E)$ only. This can be done by showing that, for each $\mathscr{U}$ in $\Gamma_0(E)$, $H_p(E, F: \mathscr{U})$ can be identified with $\mathscr{H}_p(E, F)$ and that, when this is done, all the maps $f_{\mathscr{U}\mathscr{V}}$ become identities. The required result follows from Exercise 5-9.

Note that, by Theorem 5-2, a simplicial pair (E, F) can be identified with the pair $(K_\mathscr{U}, K'_\mathscr{U})$ where $K_\mathscr{U}$ is the nerve of the star open covering and $K'_\mathscr{U}$ is the subcomplex associated with the subspace F. This can be applied to each barycentric subdivision of E, so it follows that, if $B^r\mathscr{U}$ is the star open covering of the rth barycontric subdivision of E, then

$$H_p(E, F; B^r\mathscr{U}) \cong \mathscr{H}_p(B^rE, B^rF)$$

Of course, at this point, singular homology theory may be applied to show very quickly that the right-hand side of the last equation is $H_p(E, F)$ for all r and that all the $f_{\mathscr{U}\mathscr{V}}$ in the inverse system $\{\mathscr{H}_p(E, F; \mathscr{U}), f_{\mathscr{U}\mathscr{V}}, \Gamma_0(E)\}$ are identities. In this and similar situations, however, it is more satisfactory to preserve some historical perspective in discussing Čech homology and to use only simplicial homology groups of the nerves of coverings, making no reference to their identity with the singular groups.

The homomorphisms

$$f_{\mathscr{V}\mathscr{W}}: H_p(E, F; \mathscr{U}^r) \to H_p(E, F; \mathscr{V})$$

for coverings $\mathscr{V}$ and $\mathscr{W}$ of the family $\Gamma_0(E)$ will now be examined and all shown to be isomorphisms onto. To do this, it is sufficient to take the case where $\mathscr{V} = B^r\mathscr{U}$, $\mathscr{W} = B^{r+1}\mathscr{U}$. Thus, $\mathscr{V}$ is the star open covering of B^rE, while $\mathscr{W}$ is the star open covering of the barycentric subdivision of this complex. The map $f_{\mathscr{V}\mathscr{W}}$ is defined by first constructing a simplicial map $f^1_{\mathscr{V}\mathscr{W}}$ (cf. Definition 5-9) of $(K_{\mathscr{W}}, K'_{\mathscr{W}})$ into $(K_{\mathscr{V}}, K'_{\mathscr{V}})$. A vertex W of $K_{\mathscr{W}}$ is mapped on some vertex V of $K_{\mathscr{V}}$ such that the open set V of E contains the open set W of Z. If $K_{\mathscr{W}}$ and $K_{\mathscr{V}}$ are identified with $B^{r+1}E$ and B^rE, respectively, as they were in Theorem 5-2, V and W are identified with vertexes x, y of $B^{r+1}E$ and B^rE, respectively, such that W is the star of x and V is the star of y. To say that $W \subset V$ means that y is some vertex of the smallest simplex in B^rE containing x, that is, some vertex of the simplex of which x is the barycenter (cf. Exercise 5-4). With this identification, the construction of $f^1_{\mathscr{V}\mathscr{W}}$ can be stated as follows: $f^1_{\mathscr{V}\mathscr{W}}$ maps each vertex x of $B^{r+1}E$ on some vertex of the smallest simplex of B^rE containing x.

Reference to Section 3-7 shows that this map induces a homomorphism on the simplicial chain groups which has an algebraic homotopy inverse, so the induced map $f_{\mathscr{V}\mathscr{W}}$ on the simplicial homology groups is an isomorphism onto, as was to be shown.

This result has been proved with $\mathscr{V}$ and $\mathscr{W}$ as the star open coverings of consecutive barycentric subdivisions of E, but application of the transitivity relations between the $f_{\mathscr{V}\mathscr{W}}$ shows that $f_{\mathscr{V}\mathscr{W}}$ is an isomorphism onto for any pair of coverings of the family $\Gamma_0(E)$ such that $W > \mathscr{V}$. In particular, if $\mathscr{V}$ is taken as the star open covering of E, it follows that each group $H_p(E, F; B^r\mathscr{U})$ is isomorphic to $H_p(E, F)$. Then, from the application of Exercise 5-9, it follows that

$$\lim H_p(E, F; \mathscr{V}) \cong \mathscr{H}_p(E, F)$$

where the limit is taken over coverings in $\Gamma_0(E)$. Combination of this with the cofinality of $\Gamma_0(E)$ in $\Gamma(E)$, proves the following theorem.

Theorem 5-11. *If (E, F) is a simplicial pair then*

$$\check{H}_p(E, F) \cong \mathscr{H}_p(E, F)$$

for each p and for any coefficient group.

6

Further Properties of Čech Homology

The basic properties of the Čech homology groups remain to be studied. These bear resemblances to the results of Chapter 1, but some quite novel features appear here.

6-1. INTRODUCTION

It follows from the discussion so far that Čech homology satisfies the general requirements for a homology theory as set out in Section 5-1. Namely, it coincides with the simplicial theory on simplicial complexes and the Čech groups are topological invariants. It is of interest now to compare Čech theory with the other general homology theory already studied, namely, the singular homology theory, and to see whether the other fundamental properties of singular theory are satisfied by the Čech theory. It has already been seen that the behavior of induced homomorphisms is similar in the two theories (Theorems 1-5 and 5-8). Also, the dimension theorem is satisfied by both theories (Theorems 1-7 and 5-9). The excision, homotopy, and exactness theorems will now be examined in relation to Čech homology theory. The excision theorem will be shown to hold in a slightly weaker form than in singular homology theory. The homotopy theorem will be proved for compact spaces only. This restriction is due only to the fact that Čech theory has been set up using finite coverings only. If arbitrary coverings had been used (with infinite complexes as nerves), the result would have been obtained in general (cf. [1]). The exactness theorem fails in Čech theory, or at least survives only in a much weakened form.

On the other hand, Čech theory enjoys a property that has no place at all in the singular theory, namely the continuity property, to be described in Section 6-5.

6-2. THE HOMOTOPY THEOREM FOR ČECH HOMOLOGY

Turning now to the homotopy theorem, attention will be restricted to compact spaces only. As with the singular homology theory, the essential part of the proof is to obtain the theorem in the special case of the two maps f and g of a space E into $E \times I$ which is defined by $f(p) = (p, 0)$ and $g(p) = (p, 1)$. If F is a subspace of E, it will be shown that the induced homomorphisms f_* and g_* of $\check{H}_p(E, F)$ into $\check{H}_p(E \times I, F \times I)$ are the same.

In defining f_* and g_* by limiting processes, coverings of $E \times I$ must be used. Some difficulty arises here because an arbitrary finite open covering of $E \times I$ displays no regularity with respect to the product structure. So the first step in the proof will be to pick out a cofinal set of finite open coverings of $E \times I$ whose nerves have a product structure. It will then follow that the simplicial maps induced by f and g on the nerves will be homotopic.

Definition 6-1. A finite open covering of $E \times I$ is called a *product covering* if it consists of sets $U_i \times I_j$ where the U_i form a covering $\mathcal{U}$ of E and the I_j form a covering $\mathcal{I}$ of I. Such a covering is denoted by $\mathcal{U} \times \mathcal{I}$.

Lemma 6-1. *If E is compact, the product coverings of $E \times I$ are cofinal in the set of all finite open coverings of $E \times I$.*

Proof. Let $\mathcal{U}$ be a given finite open covering of $E \times I$. Let t be a point of I. Each point (x, t) in $E \times \{t\}$ is in some set U of $\mathcal{U}$, so each has a rectangular neighborhood (i.e., a neighborhood of the form $V \times W$, where V is a neighborhood of x in E and W is a neighborhood of t in I) contained in $\mathcal{U}$. $E \times \{t\}$ is contained in the union of these rectangular neighborhoods, and so, since it is a compact set, it is contained in the union of finitely many of them. Since only a finite number of rectangular neighborhoods is now involved, they may be adjusted so that they all have the same second factor. Thus, there is a neighborhood I_t of t in I and a finite open covering $\mathcal{U}_t$ of E such that each set $U \times I_t$ for U in $\mathcal{U}_t$ is contained in some set of $\mathcal{U}$. Of course, $E \times \{t\}$ is contained in the union of the $U \times I_t$ for U in U_t. Note that when these conditions are satisfied by a covering $\mathcal{U}_t$, they are automatically satisfied by any refinement of $\mathscr{U}_t$. Next, the I_t form an open covering of I, so I, being compact, is contained in a finite collection $\mathscr{I}$ of the I_t. Let $\mathscr{V}$ be a common refinement of the corresponding finitely many $\mathscr{U}_t$. The collection $\mathscr{V} \times \mathscr{I}$ is then a finite open covering of $E \times I$ and, by the remark made at the end of the last paragraph, it is a refinement of $\mathscr{U}$. This completes the proof of the lemma. ∎

Note that if $\mathscr{V} \times \mathscr{I}$ is a refinement of $\mathscr{U}$, so is $\mathscr{V}' \times \mathscr{I}'$, where $\mathscr{V}'$ and $\mathscr{I}'$ are refinements of $\mathscr{V}$, $\mathscr{I}$ in E and I, respectively. This remark will now be used to strengthen Lemma 6-1 by putting $\mathscr{I}$ into a specially simple form.

Lemma 6-2. *Any finite open covering of I has a refinement consisting of open intervals $I_0, I_1, \ldots, I_n$ such that $I_j \cap I_i = \varnothing$, except when $j = i + 1$ or $i - 1$.*

Proof. For any x in I and any open set U containing x, there is an open interval V such that $x \in V \subset U$; so any finite open covering of I has a refinement that is a covering by open intervals. By the compactness of I, the latter covering can be assumed to be finite. So it may as well be assumed that the given covering $\mathscr{I}$ is a covering by open intervals. The lemma will now be proved by induction. Suppose that $I_0, I_1, \ldots, I_r$ have already been constructed so that each I_i is an open interval contained in some set of $\mathscr{I}$ so that $I_i \cap I_j = \varnothing$, except when $j = i + 1$. Also assume that the I_i are arranged in order from left to right. Let x be the right-hand boundary point of I_r. x is contained in some I'_{r+1} of the covering I. Shrink the part of I'_{r+1} to the left of x to get a smaller interval I_{r+1} contained in I'_{r+1} such that $I_{r+1} \cap I_{r-1} = \varnothing$. Then, the $I_i, i = 1, 2, \ldots, r + 1$ satisfy the intersection conditions of the lemma. Since the right-hand end points of the I_i form a strictly increasing sequence and since they are picked from the right-hand end points of the sets in $\mathscr{I}$, the sequence of I_i covers I in a finite number of terms. Also, the choice of the I_i at each stage ensures that the covering of I by the I_i is a refinement of $\mathscr{I}$, so the proof is completed. ∎

For convenience, a covering of I of the type obtained in this lemma is called simple. Then, combining the above lemmas produces the following result.

Lemma 6-3. *If E is compact, any finite open covering of $E \times I$ has a refinement of the form $\mathscr{V} \times \mathscr{I}$, where $\mathscr{V}$ is a finite open covering of E and $\mathscr{I}$ is a simple finite open covering of I.*

The main step in the argument is the proof that the nerve of a covering $\mathscr{V} \times \mathscr{I}$, where $\mathscr{I}$ is simple, contains the product $K_{\mathscr{V}} \times I$, where $K_{\mathscr{V}}$ is the nerve of $\mathscr{V}$.

Remember that, if K is a simplicial complex, $K \times I$ can be triangulated (cf. Exercise 2-3); that is, if $K \times I$ is embedded in N space so that $K \times \{0\}$ and $K \times \{1\}$ are in the hyperplanes $x_N = 0$ and $x_N = 1$, respectively, points $(p, 0)$ and $(p, 1)$ differing only in their last coordinates, then $K \times I$ is expressed as the union of simplexes of the type $[x_0 x_1 \cdots x_i y_i y_{i+1} \cdots y_p]$ where the x_i are the vertexes of a simplex of K, identified with $K \times \{0\}$, while the y_i are the corresponding vertexes of the corresponding simplex in $K \times \{1\}$. Note that the triangulation is by no means unique but depends on the ordering of the vertexes of K. The same principle can be used to obtain a more elaborate triangulation for $K \times I$. Namely let

$$0 = t_0 < t_1 < \cdots < t_n = 1$$

be $(n + 1)$ points of I, embed $K \times I$ in N space, as before, for each vertex x of K, let $x^{(i)}$ denote the point (x, t_i), which is assumed to be in the hyperplane $x_N = t_i$. Then if $x_0, x_1, \ldots, x_\alpha$ are vertexes of a simplex in K, all the vertexes in K having been ordered initially in some way, form all the simplexes of the type $[x_0^{(i)} x_1^{(i)} \cdots x_j^{(i)} x_j^{(i+1)} \cdots x_r^{(i+1)}]$. For each i the set of such simplexes forms a triangulation, as before, of the product of K and the interval $[t_i t_{i+1}]$, and these triangulations clearly fit together to give a triangulation of the whole of $K \times I$. This is called an *n layered triangulation* of $K \times I$.

Now let f, g be maps of a space E into $E \times I$ defined by

$$f(p) = (p, 0)$$

$$g(p) = (p, 1)$$

Let $\mathscr{U} \times \mathscr{I}$ be a product finite open covering of $E \times I$ where $\mathscr{I}$ is a simple covering of I, and define simplicial maps f^1 and g^1 of $K_{\mathscr{U}}$ (the nerve of $\mathscr{U}$) into $K_{\mathscr{U} \times \mathscr{I}}$ by making f^1 map the vertex U of $K_{\mathscr{U}}$ on the vertex $U \times I_1$ of $K_{\mathscr{U} \times \mathscr{I}}$ while g^1 maps U on $U \times I_n$. Note that these are simplicial maps corresponding to f and g, respectively, in the manner described in Section 5-3. Finally, let f^2 and g^2 be simplicial maps of $K_{\mathscr{U}}$ into $K_{\mathscr{U}} \times I$ which are defined by setting $f^2(x) = (x, 0)$ and $g^2(x) = (x, 1)$ for each x in $K_{\mathscr{U}}$. Here it is understood that $K_{\mathscr{U}} \times I$ is given a certain n-layered triangulation, but clearly, f^2 and g^2 are independent of this.

All this notation makes it possible to take one of the main steps in the proof of the homotopy theorem. The idea is to replace the given homotopic maps by homotopic simplicial maps, which in turn induce algebraically homotopic chain homomorphisms on the simplicial chains.

Lemma 6-4. *In the above notation there is a simplicial map H of an n-layered triangulation of $K_{\mathscr{U}} \times I$ into $K_{\mathscr{U} \times \mathscr{I}}$ such that the following diagrams are commutative.*

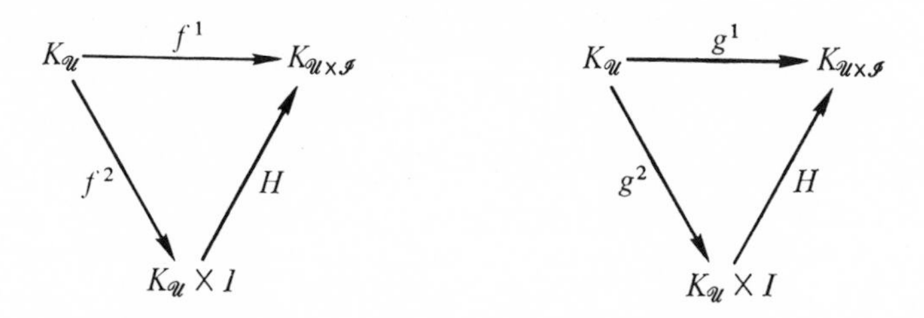

Proof. Denote the sets of $\mathscr{U}$ by U, and as usual, use the same notation for the corresponding vertexes of $K_{\mathscr{U}}$. Then, taking points $0 = t_0 < t_1 < \cdots < t_n = 1$ on I, the vertexes of an n-layered triangulation of $K_{\mathscr{U}} \times I$ at the level $x_N = t_j$ can be denoted by the symbols $U_i^{(j)}$, where i runs over the indexing set of $\mathscr{U}$. On the other hand, the sets of the covering $\mathscr{U} \times \mathscr{I}$ are of the form $U_i \times I_j$, and these same symbols are used for the corresponding vertexes of $K_{\mathscr{U}\times\mathscr{I}}$.

The required simplicial map H of the n-layered triangulation of $K_{\mathscr{U}} \times I$ into $K_{\mathscr{U}\times\mathscr{I}}$ will be constructed by first mapping the vertex $U_i^{(j)}$ of $K_{\mathscr{U}} \times I$ on the vertex $U_i \times I_j$ of $K_{\mathscr{U}\times\mathscr{I}}$ for each i and j. To check that H, so defined on the vertexes, can be extended to a simplicial map it must be shown that whenever a collection of vertexes of $K_{\mathscr{U}} \times I$ are vertexes of a simplex, so are their images. Now a set of vertexes of a simplex of $K_{\mathscr{U}} \times I$ must be of the form

$$U_0^{(i)}, U_1^{(i)}, \ldots, U_j^{(i)}, U_j^{(i+1)}, \ldots, U_r^{(i+1)}$$

where $U_0, U_1, \ldots, U_r$ are vertexes of a simplex of $K_{\mathscr{U}}$. Thus $U_0 \cap U_1 \cap \cdots \cap U_r \neq \varnothing$ in E. The images of these vertexes under H then correspond to the sets

$$U_0 \times I_i, U_1 \times I_i, \ldots, U_j \times I_i, U_j \times I_{i+1}, \ldots, U_r \times I_{i+1}$$

and the intersection of these sets is

$$(U_0 \cap U_1 \cap \cdots \cap U_r) \times (I_i \cap I_{i+1})$$

in $E \times I$. The first factor has just been seen to be nonempty and the second is nonempty by the assumption of the simplicity of $\mathscr{I}$. Thus the vertexes

$$U_0 \times I_i, U_1 \times I_i, \ldots, U_j \times I_i, U_j \times I_{i+1}, \ldots, U_r \times I_{i+1}$$

are the vertexes of a simplex in $K_{\mathscr{U}\times\mathscr{I}}$. This is exactly what had to be proved.

To complete the proof of the lemma, the commutativity of the two diagrams must be shown. To do this it is, of course, sufficient to check what happens to the vertexes of $K_{\mathscr{U}}$. If U is such a vertex, then $f^1(U)$ is the vertex $U \times I_1$

of $K_{\mathscr{U}\times\mathscr{I}}$. On the other hand, $f^2(U)$ is the vertex $U \times \{0\}$ of $K_{\mathscr{U}} \times I$, and H maps this on the vertex $U \times I_1$ of $K_{\mathscr{U}\times\mathscr{I}}$. Thus f^1 agrees with Hf^2 on the vertexes of $K_{\mathscr{U}}$, so the two maps, being simplicial, are equal. Hence the first diagram is commuative, and a similar argument can be applied to the second. ∎

This lemma is actually needed in the following relative form.

Lemma 6-5. *Let F be a subspace of E. Let $K'_{\mathscr{U}}$ be the subcomplex of $K_{\mathscr{U}}$ associated with F and let $K'_{\mathscr{U}\times\mathscr{I}}$ be the subcomplex of $K_{\mathscr{U}\times\mathscr{I}}$ associated with $F \times I$. Then the simplicial maps of the last lemma are maps of pairs such that the following diagrams are commutative.*

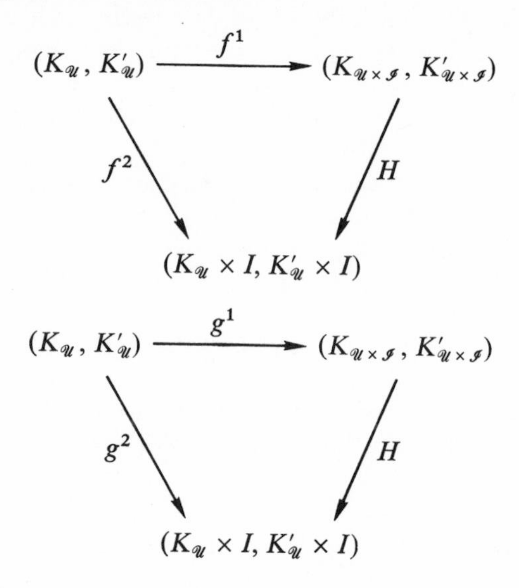

Proof. The commutativity has in fact been proved already, and it only remains to check that the various maps are maps of pairs as indicated. This is an easy verification, which is left to the reader. ∎

Refer back now to Example 3-12 (Section 3-5) to see that f^2 and g^2 induce the same homomorphisms on the simplicial homology groups. By composing these with the homomorphism induced by H, it follows that the homomorphisms induced by $f^1 = Hf^2$ and $g^1 = Hg^2$ are the same. This can be stated as follows.

Lemma 6-6. *The homomorphisms $f_{\mathscr{U}\times\mathscr{I}}$ and $g_{\mathscr{U}\times\mathscr{I}}$ of $\mathscr{H}_p(K_{\mathscr{U}}, K'_{\mathscr{U}}) = H_p(E, F; \mathscr{U})$ into $\mathscr{H}_p(K_{\mathscr{U}\times\mathscr{I}}, K'_{\mathscr{U}\times\mathscr{I}}) = H_p(E \times I, F \times I; \mathscr{U} \times \mathscr{I})$ are induced by f and g and, as in Section 5-3, are equal.*

Note that this could also be established by using the homotopy theorem of singular homology theory, but as usual, this is avoided so that Čech theory depends only on simplicial homology theory.

The proof of the homotopy theorem can now be carried out.

Theorem 6-7. *Let f and g be maps of (E, F) into $(E \times I, F \times I)$ defined by setting $f(x) = (x, 0)$ and $g(x) = (x, 1)$ for all x in E. Then the induced homomorphisms f_* and g_* of $\check{H}_p(E, F)$ into $\check{H}_p(E \times I, F \times I)$ are equal for each p.*

Proof. It has just been shown that, if $\mathscr{U} \times \mathscr{I}$ is a product open covering of $E \times I$ with $\mathscr{I}$ simple, then the homomorphisms

$$f_{\mathscr{U} \times \mathscr{I}}: H_p(E, F; \mathscr{U}) \to H_p(E \times I, F \times I; \mathscr{U} \times \mathscr{I})$$

$$g_{\mathscr{U} \times \mathscr{I}}: H_p(E, F; \mathscr{U}) \to H_p(E \times I, F \times I; \mathscr{U} \times \mathscr{I})$$

are equal. Remember that the coverings $\mathscr{U} \times \mathscr{I}$ of this type are cofinal in the set of all finite open coverings of $E \times I$ (Lemma 6-3). It then follows from Exercise 5-10 that

$$\lim f_{\mathscr{U} \times \mathscr{I}} = f_*$$

$$\lim g_{\mathscr{U} \times \mathscr{I}} = g_*$$

taking limits over these product coverings. Then, since $f_{\mathscr{U} \times \mathscr{I}} = g_{\mathscr{U} \times \mathscr{I}}$ for each $\mathscr{U} \times \bar{B}$, it follows that $f_* = g_*$, as was to be shown. ∎

6-3. THE EXCISION THEOREM IN ČECH HOMOLOGY

The excision theorem takes the following form in Čech homology theory.

Theorem 6-8. *Let (E, F) be a pair of spaces and let A be an open set in E such that $\bar{A} \subset \mathring{F}$. Then the homomorphism*

$$i_*: \check{H}_p(E - A, F - A) \to \check{H}_p(E, F)$$

induced by inclusion is an isomorphism onto for each p.

Note the difference between this and the corresponding result in the singular homology theory (Theorem 1-23). There A was any set such that $\bar{A} \subset \mathring{F}$, while here it must be open.

Proof. Remember that in singular homology theory the essential step in proving the excision theorem was the cutting down of the sizes of the singular simplexes by barycentric subdivision. The corresponding step here is the restriction to sufficiently fine coverings of E; namely, the two sets $E - \bar{A}$ and $\mathring{F}$ form a finite open covering $\mathscr{U}_0$ of E, and given any finite open covering $\mathscr{U}$ of E, there is a common refinement $\mathscr{V}$ of $\mathscr{U}$ and $\mathscr{U}_0$. In other words, every finite open covering of E has a refinement that is a refinement of $\mathscr{U}_0$. Thus the finite open coverings that are refinements of $\mathscr{U}_0$ are cofinal in the set of all finite open coverings. Hence, in the construction of $\check{H}_p(E, F)$ it is sufficient (by Theorem 5-10) to take the inverse limit over finite open coverings that are refinements of $\mathscr{U}_0$. Note that for such a covering every open set that meets A is contained entirely in F. At the same time, consider the family of finite open coverings of $E - A$. Each open set of $E - A$ is (by definition of the induced topology) of the form $i^{-1}(U)$ for some open set U of E, and if a collection of open sets $i^{-1}(U)$ forms a covering $\mathscr{U}'$ of $E - A$, then the union of the corresponding open sets in E contains $E - A$. If A, an open set, is adjoined to this collection, a finite open covering $\mathscr{U}$ of E is obtained that has $\mathscr{U}'$ as its inverse image under i. Thus every finite open covering $\mathscr{U}'$ of $E - A$ is of the form $i^{-1}(\mathscr{U})$, where $\mathscr{U}$ is a finite open covering of E. On the other hand, $\mathscr{U}$ has a refinement $\mathscr{V}$, which is a refinement of $\mathscr{U}_0$, so $\mathscr{U}' = i^{-1}(\mathscr{U})$ has a refinement $\mathscr{V}' = i^{-1}(\mathscr{V})$, which is a refinement of $\mathscr{U}_0' = i^{-1}(\mathscr{U}_0)$. Thus, in constructing $\check{H}_p(E - A, F - A)$, it is sufficient to take the limit over finite open open coverings which are refinements of $\mathscr{U}_0'$.

In the remainder of this proof, then, it can be assumed that the only coverings of E appearing are refinements of $\mathscr{U}_0$ and the only coverings of $E - A$ are refinements of $\mathscr{U}_0'$. Write $\Gamma_0(E)$ for the family of finite open coverings of E which are refinements of $\mathscr{U}_0$, $\Gamma_0(E - A)$ for the family of coverings of $E - A$ which are refinements of $\mathscr{U}_0'$.

The homomorphism i_* is constructed as $\lim i_U$ where the limit is taken over $\Gamma(E)$ and the $i_{\mathscr{U}}$ are as in Definition 5-7. The $i_{\mathscr{U}}$ form an inverse system of homomorphisms of $\{H_p(E - A,\ F - \mathscr{A};\ \mathscr{U}),\ f_{\mathscr{U}\mathscr{V}},\ \Gamma(E - A)\}$ into $\{H_p(E,\ F;\ \mathscr{U}),\ f_{\mathscr{U}\mathscr{V}},\ \Gamma(E)\}$ corresponding to the map ϕ of $\Gamma(E)$ into $\Gamma(E - A)$ defined by setting $\phi(\mathscr{U}) = \mathscr{U}' = i^{-1}(\mathscr{U})$. But $\Gamma_0(E)$ is cofinal in $\Gamma(E)$, and $\Gamma_0(E - A)$ in $\Gamma(E - A)$, so by Exercise 5-10, it follows that i_* is $\lim i_U$ where the limit is taken over $\Gamma_0(E)$. Further, if Exercise 5-10 is used to show that i_* is an isomorphism onto, it is sufficient to show that $i_{\mathscr{U}}$ is an isomorphism onto for each $\mathscr{U}$ in $\Gamma_0(E)$.

To complete the proof of the excision theorem, consider the homomorphism

$$i_{\mathscr{U}}\colon H_p(E - A, F - A; \mathscr{U}') \to H_p(E, F; \mathscr{U})$$

constructed as in Section 5-3, where $\mathscr{U}$ is in $\Gamma_0(E)$ and $\mathscr{U}' = i^{-1}(\mathscr{U})$. As usual, write $K_{\mathscr{U}}$ for the nerve of $\mathscr{U}$, $K_{\mathscr{U}}'$ for the subcomplex associated with $\mathscr{F}$, $K_{\mathscr{U}'}$ for the nerve of $\mathscr{U}'$, and $K_{\mathscr{U}'}'$ for the subcomplex associated with

$F - A$. As in the construction of induced homomorphisms in general, $i_{\mathscr{U}}$ will be induced by a simplicial map $i_{\mathscr{U}}{}^1$, which will now be examined.

Let U' be a vertex of $K_{\mathscr{U}'}$ corresponding to an open set U' of $E - A$ such that U' is, in fact, an open set of E that does not meet A. Then U' appears as a vertex of $K_{\mathscr{U}}$, and regarded as a vertex of $K_{\mathscr{U}}$, it is denoted by U. This sets up a one-to-one correspondence between certain vertexes of $K_{\mathscr{U}'}$ and certain vertexes of $K_{\mathscr{U}}$. Also, it is clear that a collection of such vertexes of $K_{\mathscr{U}'}$ are vertexes of a simplex in $K_{\mathscr{U}'}$ if and only if the corresponding vertexes of $K_{\mathscr{U}}$ are the vertexes of a simplex in $K_{\mathscr{U}}$. Pick out all the simplexes of $K_{\mathscr{U}'}$ whose vertexes are of the type described. Their union clearly forms a subcomplex $K_{\mathscr{U}'}$. Similarly, the simplexes of $K_{\mathscr{U}}$, all of whose vertexes are open sets in E which do not meet A, form a subcomplex $K''_{\mathscr{U}}$ of $K_{\mathscr{U}}$. Let $i_{\mathscr{U}}{}^1$ be defined on $K''_{\mathscr{U}'}$ by making it map each vertex U' on the vertex U of $K_{\mathscr{U}}$ which corresponds to it as just described. Then, extended linearly to all of $K_{\mathscr{U}'}$, $i_{\mathscr{U}}{}^1$ clearly maps this complex homeomorphically onto $K''_{\mathscr{U}}$. It remains now to define $i_{\mathscr{U}}{}^1$ on the rest of $K_{\mathscr{U}'}$. Any vertex of $K_{\mathscr{U}'}$ not in $K''_{\mathscr{U}'}$ is of the form U' where $U' = i^{-1}(U)$ and U is an open set of E meeting A. Since $\mathscr{U}$ is a refinement of $\mathscr{U}_0$, U is contained in F, so U is a vertex of $K'_{\mathscr{U}}$ and U' is a vertex of $K'_{\mathscr{U}'}$. For each such U', a U is picked such that $U' = i^{-1}(U)$ and then $i_{\mathscr{U}}{}^1(U')$ is defined to be U. Also, any simplex of $K_{\mathscr{U}'}$ (or $K_{\mathscr{U}}$) having U' (respectively U) as a vertex must be in the subcomplex $K'_{\mathscr{U}'}$ (respectively $K'_{\mathscr{U}}$).

Thus completely defined, $i_{\mathscr{U}}{}^1$ operates in such a way that $K'_{\mathscr{U}'}$ is mapped into $K'_{\mathscr{U}}$, while $K''_{\mathscr{U}'}$ is mapped homeomorphically onto $K''_{\mathscr{U}}$, the homeomorphism setting up a one-to-one correspondence between the simplexes of $K''_{\mathscr{U}'}$ and $K''_{\mathscr{U}}$ and mapping each simplex of $K''_{\mathscr{U}'}$ linearly on its image. Note also that $K_{\mathscr{U}'} = K'_{\mathscr{U}'} \cup K''_{\mathscr{U}'}$ and $K_{\mathscr{U}} = K'_{\mathscr{U}} \cup K''_{\mathscr{U}}$.

Consider then the following commutative diagram:

$$\begin{array}{ccc}
(K_{\mathscr{U}'}, K'_{\mathscr{U}'}) & \xrightarrow{i_{\mathscr{U}}{}^1} & (K_{\mathscr{U}}, K'_{\mathscr{U}}) \\
\uparrow & & \uparrow \\
(K''_{\mathscr{U}'}, K''_{\mathscr{U}'} \cap K'_{\mathscr{U}'}) & \xrightarrow{i_{\mathscr{U}}{}^1} & (K''_{\mathscr{U}}, K''_{\mathscr{U}} \cap K'_{\mathscr{U}})
\end{array}$$

In the lower line, $i_{\mathscr{U}}{}^1$ is actually the restriction of the upper $i_{\mathscr{U}}{}^1$ to $K''_{\mathscr{U}'}$, and the two vertical maps are inclusions. It has just been arranged that the lower line is an isomorphism of the two simplicial pairs, so the corresponding induced homomorphism of the simplicial homology groups is an isomorphism onto. On the other hand, by Exercise 2-13, the two vertical maps both induce isomorphisms onto of the simplicial homology groups, so by the commutativity of the diagram of the induced homomorphisms, $i_{\mathscr{U}}$ is an isomorphism onto, as was to be proved. ∎

6-4. THE PARTIAL EXACTNESS THEOREM

Since the Čech homology groups agree with the simplicial groups and so with the singular groups on triangulable pairs, it follows that the exactness theorem (Theorem 1-9) is satisfied by the Čech groups of such pairs. However, the Čech homology sequence of a pair of spaces is not, in general, exact but only partially so, in the sense that the image of each map is contained in the kernel of the next. Thus the composition of any two consecutive maps in the sequence is zero. It is convenient to have a name for a sequence with this property.

Definition 6-2. If the sequence

$$\to G_q \xrightarrow{f_q} G_{q-1} \to$$

of groups and homomorphisms is such that $f_{q-1}f_q = 0$ for each q, the sequence is said to be *of order two.*

The partial exactness theorem to be proved here is a purely algebraic matter concerning inverse limits of sequences. In other words, it will be shown that the inverse limit of a system of sequences of order two is also of order two.

Let A be a directed set to be used as indexing set. For each integer i, let $\{G_\alpha{}^i, \pi^i_{\alpha\beta}, A\}$ be an inverse system, and for each i, let f^i be a homomorphism of the system $\{G_\alpha{}^i, \pi^i_{\alpha\beta}, A\}$ into the system $\{G^{i-1}_\alpha, \pi^{i-1}_{\alpha\beta}, A\}$. Thus each f^i is a family of homomorphisms f^i such that the diagram

$$\begin{array}{ccc} G_\alpha{}^i & \xrightarrow{f_\alpha{}^i} & G^{i-1}_\alpha \\ {\scriptstyle \pi^i_{\alpha\beta}}\uparrow & & \uparrow{\scriptstyle \pi^{i-1}_{\alpha\beta}} \\ G_\beta{}^i & \xrightarrow{f_\beta{}^i} & G^{i-1}_\beta \end{array}$$

is commutative for all α and β in A such that $\beta > \alpha$.

Think of this in an alternative way. For fixed α, the $G_\alpha{}^i$ and $f_\alpha{}^i$ form a sequence of groups and homomorphisms and the $\pi^i_{\alpha\beta}$ constitute a family of homomorphisms from the β sequence to the α sequence, making all the above diagrams commutative. Thus the situation may be thought of either as a sequence of inverse systems or as an inverse system of sequences.

Now let $G^i = \lim G_\alpha{}^i$, and let $f^i = \lim f_\alpha{}^i$. The latter exists since the commutativity conditions of Section 5-5 are satisfied. Thus a limit sequence

$$\to G^i \xrightarrow{f^i} G^{i-1} \to$$

is obtained.

Lemma 6-9. *If, for each α, the sequence of $G_\alpha{}^i$ with homomorphisms $f_\alpha{}^i$ is of order two, then the limit sequence of G^i with homomorphisms f^i is also of order two.*

Proof. Take x in G^i. x has coordinates $\{x_\alpha\}$ with x_α in $G_\alpha{}^i$. $f^{i-1}f^i$ is the limit of the inverse system of homomorphisms $f_\alpha^{i-1}f_\alpha{}^i$(cf. Lemma 5-7) and, by the definition of limit homomorphism, $f^{i-1}f^i(x)$ has coordinates $f_\alpha^{i-1}f_\alpha{}^i(x_\alpha)$. For each α, however, this is zero by hypothesis, so $f^{i-1}f^i = 0$ for each i, as required. ∎

However, it is not possible, in general, to strengthen Lemma 6-9 by replacing the property of being of order two by that of being exact. In fact, even in the simple case of a system of three-term sequences

$$G_\alpha{}^1 \xrightarrow{f_\alpha} G_\alpha{}^0 \to 0$$

with each f_α onto, it is not necessarily true that the inverse limit of the f_α is onto. The difficulty is that, although each coordinate x_α of an element x of $G^0 = \lim G_\alpha{}^0$ has an inverse image y_α in $G_\alpha{}^1$, it may not be possible to choose the y_α coherently so that they are coordinates of some y in $\lim G_\alpha{}^1$.

Example

6-1. To illustrate the point just mentioned, take the natural numbers as the indexing set A and let each $G_\alpha{}^1$ be the group of integers Z, let each $G_\alpha{}^0$ be Z_2, the integers modulo 2, and let each f_α be the natural homomorphism $Z \to Z_2$ (reduction modulo 2). For each pair α, β of integers where $\beta > \alpha$, let $\pi_{\alpha\beta}^1$ be defined by $\pi_{\alpha\beta}^1(x) = 3^{\beta-\alpha}x$, while $\pi_{\alpha\beta}^0$ is the identity. The diagram

$$\begin{array}{ccc} Z & \xrightarrow{f_\alpha} & Z_2 \\ {\scriptstyle \pi_{\alpha\beta}^1}\uparrow & & \uparrow{\scriptstyle \pi_{\alpha\beta}^0} \\ Z & \xrightarrow{f_\beta} & Z_2 \end{array}$$

is commutative for each pair α, β. Here $\lim \{Z_2, \pi_{\alpha\beta}^0, A\} = Z_2$. But if $x \in \lim \{Z, \pi_{\alpha\beta}^1, A\}$ then x must have coordinates $\{x_1, x_2, \ldots\}$ where $x_\alpha = 3x_{\alpha+1}$. Each x_α is hence divisible by every power of 3, and this is possible only if each x_α is 0. Hence $\lim \{Z, \pi_{\alpha\beta}^1, A\} = 0$, so $\lim f_\alpha$ is zero and is certainly not onto.

Here the difficulty is just that described before; that is, taking the element 1 in $Z_2 = \lim \{Z_2, \pi_{\alpha\beta}^0, A\}$, it has coordinates $\{1, 1, \ldots\}$. The inverse image of each coordinate under the appropriate f_α is an odd integer x_α, but it is not possible to choose the x_α so that $x_\alpha = 3x_{\alpha+1}$ for each x_α.

Consider again the topology. It will be shown that any pair of spaces has, in the Čech theory, a homology sequence of order two. This is stated more precisely as follows.

Theorem 6-10. *Let (E, F) be a pair of spaces and let $i: F \to E$ and $j: E \to (E, F)$ be the inclusion maps (in the second case E is identified with the pair $(E, \varnothing)$). Then there is a homomorphism ∂ such that the sequence*

$$\to \check{H}_p(F) \xrightarrow{i_*} \check{H}_p(E) \xrightarrow{j_*} \check{H}_p(E, F) \xrightarrow{\partial} \check{H}_{p-1}(F) \to \qquad (6\text{-}1)$$

is of order two, and in addition, ∂ commutes with homomorphisms induced by continuous maps.

Note. The statement on commutativity means that, if $f: (E, F) \to (E', F')$ is a continuous map and if the homomorphism ∂ is constructed for the pair (E', F') as well as for the pair (E, F), then the diagram

$$\begin{array}{ccc} \check{H}_p(E, F) & \xrightarrow{\partial} & \check{H}_{p-1}(F) \\ {\scriptstyle f^*}\downarrow & & {\scriptstyle f^*}\downarrow \\ \check{H}_p(E', F') & \xrightarrow{\partial} & \check{H}_{p-1}(F') \end{array}$$

is commutative for each p.

The homomorphism ∂ is called the *boundary homomorphism* of Čech homology theory.

Proof. Let $\mathscr{U}$ be a finite open covering of E and let $\mathscr{U}' = i^{-1}(\mathscr{U})$ be the induced covering of the subspace F. Denote by $K_{\mathscr{U}}, K_{\mathscr{U}'}, K'_{\mathscr{U}}$ the nerve of $\mathscr{U}$, the nerve of $\mathscr{U}'$, and the subcomplex of $K_{\mathscr{U}}$ associated with F, respectively. For the pair $(K_{\mathscr{U}}, K'_{\mathscr{U}})$ there is an exact homology sequence (Exercise 3-4)

$$\to \mathscr{H}_p(K'_{\mathscr{U}}) \xrightarrow{i'_{\mathscr{U}}} \mathscr{H}_p(K_{\mathscr{U}}) \xrightarrow{j_{\mathscr{U}}} \mathscr{H}_p(K_{\mathscr{U}}, K'_{\mathscr{U}}) \xrightarrow{\partial'_{\mathscr{U}}} \mathscr{H}_{p-1}(K'_{\mathscr{U}}) \to \qquad (6\text{-}2)$$

Here $j_{\mathscr{U}}$ is associated with the continuous map $j: E \to (E, F)$, as in Denfition 5-7, so when the limit is taken over the set of infinite open coverings of E, $\lim j_{\mathscr{U}}$ is j_*. $i'_{\mathscr{U}}$ is the homomorphism on homology groups induced by the inclusion map $h: K'_{\mathscr{U}} \to K_{\mathscr{U}}$. On the other hand (cf. Section 5-3), it has been seen that there is a commutative diagram

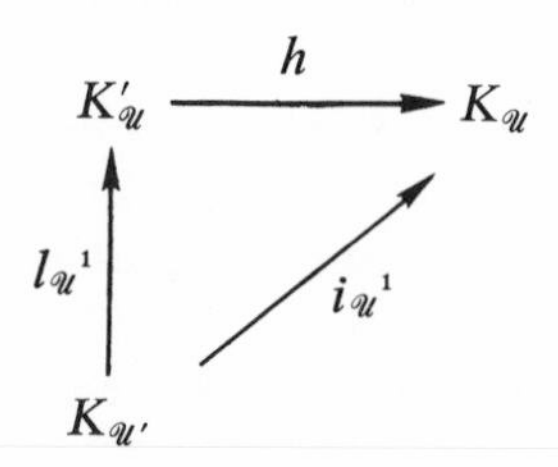

where $i_{\mathscr{U}}^1$ is a simplicial map induced, as in Section 5-3, by the inclusion map $i\colon F \to E$ and $l_{\mathscr{U}}^1$ is a simplicial map that induces on simplicial chains a homomorphism with an algebraic homotopy inverse. Hence, in the diagram of induced homomorphisms on homology groups,

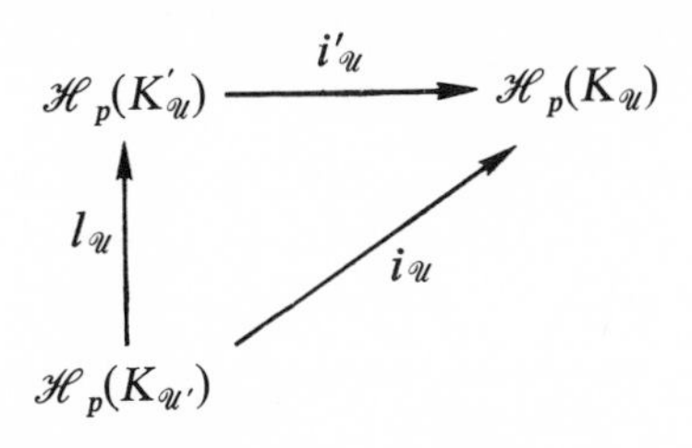

the map $l_{\mathscr{U}}$ is an isomorphism onto.

Attach this diagram to the sequence (6-1) to obtain the diagram

$$\begin{array}{ccccccccc}
\mathscr{H}_p(K'_{\mathscr{U}}) & \xrightarrow{i'_{\mathscr{U}}} & \mathscr{H}_p(K_{\mathscr{U}}) & \xrightarrow{j_{\mathscr{U}}} & \mathscr{H}_p(K_{\mathscr{U}}, K'_{\mathscr{U}}) & \xrightarrow{\partial'_{\mathscr{U}}} & \mathscr{H}_{p-1}(K'_{\mathscr{U}}) & \xrightarrow{i'_{\mathscr{U}}} & \mathscr{H}_{p-1}(K_{\mathscr{U}}) \\
\uparrow l_{\mathscr{U}} & \nearrow i_{\mathscr{U}} & & & & & \uparrow l_{\mathscr{U}} & \nearrow i_{\mathscr{U}} & \\
\mathscr{H}_p(K_{\mathscr{U}'}) & & & & & & \mathscr{H}_{p-1}(K_{\mathscr{U}'}) & &
\end{array} \tag{6-3}$$

It will now be checked that all the homomorphisms in this diagram belong to systems satisfying the commutativity conditions of Definition 5-15 so that the inverse limits can be taken. Of course, this is known from general considerations for $i_{\mathscr{U}}$ and $j_{\mathscr{U}}$, so it only remains to consider $\partial'_{\mathscr{U}}$ and $l_{\mathscr{U}}$. The conditions for the $i_{\mathscr{U}}$ then follow from the fact that $i'_{\mathscr{U}} = i_{\mathscr{U}} l_{\mathscr{U}}^{-1}$.

Consider the $l_{\mathscr{U}}$ first. Let $\mathscr{V}$ be a refinement of $\mathscr{U}$ and let $\mathscr{V}'$ be the covering of F induced by $\mathscr{V}$. $l_{\mathscr{U}}$ and $l_{\mathscr{V}}$ are respectively induced by simplicial maps $l_{\mathscr{U}}^1$ and $l_{\mathscr{V}}^1$. In the diagram

$$\begin{array}{ccc}
K'_{\mathscr{V}} & \xrightarrow{f^1_{\mathscr{U}\mathscr{V}}} & K'_{\mathscr{U}} \\
l_{\mathscr{V}}^1 \uparrow & & \uparrow l_{\mathscr{U}}^1 \\
K_{\mathscr{V}'} & \xrightarrow[f^1_{\mathscr{U}'\mathscr{V}'}]{} & K_{\mathscr{U}'}
\end{array} \tag{6-4}$$

$f^1_{\mathscr{U}\mathscr{V}}$ is as in Section 5-3, but it is restricted to the subcomplex $K'_{\mathscr{V}}$. $f^1_{\mathscr{U}'\mathscr{V}'}$ corresponds similarly to the pair of coverings $\mathscr{U}'$ and $\mathscr{V}'$ of F. If $\mathscr{V}'$ is a vertex of $K_{\mathscr{V}'}$ then $l_{\mathscr{V}}^1(V')$ is a vertex V of $K'_{\mathscr{V}}$ such that, in terms of the corresponding open sets, $V' = V \cap F$. $f^1_{\mathscr{U}\mathscr{V}}(V)$ is a vertex U of $K'_{\mathscr{U}}$ such that $V \subset U$. Note then that, if $U' = U \cap F$, V' is contained in U', so $f^1_{\mathscr{U}'\mathscr{V}'}(V')$

can be taken to be U', and $l^1_{\mathscr{U}}(U') = U$. This means that (6-4) is commutative, so the diagram of induced homomorphisms on homology groups is commutative. This is the condition of Definition 5-15 for the $l_{\mathscr{U}}$ to form an inverse system of homomorphisms.

Now consider the $\partial'_{\mathscr{U}}$. Let $\mathscr{V}$ be a refinement of $\mathscr{U}$ and consider the diagram

$$\begin{array}{ccc} \mathscr{H}_p(K_{\mathscr{U}}, K'_{\mathscr{U}}) & \xrightarrow{\partial'_{\mathscr{U}}} & \mathscr{H}_{p-1}(K'_{\mathscr{U}}) \\ \uparrow f_{\mathscr{U}\mathscr{V}} & & \uparrow f_{\mathscr{U}\mathscr{V}} \\ \mathscr{H}_p(K'_{\mathscr{V}}, K'_{\mathscr{V}}) & \xrightarrow{\partial'_{\mathscr{V}}} & \mathscr{H}_{p-1}(K'_{\mathscr{V}}) \end{array} \tag{6-5}$$

Here the $f_{\mathscr{U}\mathscr{V}}$ on the left, as in Section 5-3, is induced by a simplicial map $f^1_{\mathscr{U}\mathscr{V}}$, while the $f_{\mathscr{U}\mathscr{V}}$ on the right is induced by the restriction of $f^1_{\mathscr{U}\mathscr{V}}$ to $K'_{\mathscr{V}}$. The commutativity of (6-5) is then simply a special case of Theorem 1-8.

So, as stated above, all the homomorphisms in the diagram (6-3) belong to inverse systems of homomorphisms indexed by the finite open coverings of E. Take the inverse limits of these systems over the finite open coverings of E, writing

$$j_* = \lim j_{\mathscr{U}},\ i_* = \lim i_{\mathscr{U}},\ i'_* = \lim i'_{\mathscr{U}}$$
$$l = \lim l_{\mathscr{U}},\ \partial' = \lim \partial'_{\mathscr{U}},\ \check{H}'_p(F) = \lim \mathscr{H}_p(K'_{\mathscr{U}})$$

and obtain the commutative diagram (commutativity in the triangles follows from Lemma 5-7):

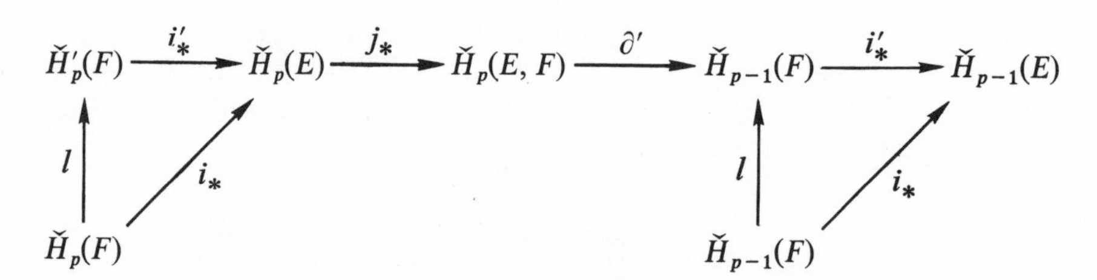

By Lemma 6-9, the sequence along the top of this diagram is of order 2. l is the inverse limit of isomorphisms onto and so is an isomorphism onto. Hence a homomorphism ∂ can be defined as $l^{-1}\partial$. Clearly, then, the sequence (6-1) is of order two.

To complete the proof, it must be shown that ∂ commutes with homomorphisms induced by continuous maps. To do this let

$$f\colon (E, F) \to (X, Y)$$

be a continuous map of the pair (E, F) into the pair (X, Y). Let $\mathscr{U}$ be a finite open covering of X and let $\mathscr{V} = f^{-1}(\mathscr{U})$ be the induced finite open covering of E. Let $\mathscr{U}'$ be the covering of Y induced by $\mathscr{U}$ and $\mathscr{V}'$ the covering of F induced by $\mathscr{V}$. Denote by $K_{\mathscr{U}}, K_{\mathscr{V}}, K_{\mathscr{U}'}, K_{\mathscr{V}'}$ the nerves of $\mathscr{U}$, $\mathscr{V}$, $\mathscr{U}'$, and $\mathscr{V}'$, respectively, and let $K'_{\mathscr{U}}$ be the subcomplex of $K_{\mathscr{U}}$ associated with Y, $K'_{\mathscr{V}}$ the subcomplex of $K_{\mathscr{V}}$ associated with F.

As in Section 5-3, f induces a simplicial map $f_{\mathscr{U}}^1$ of the pair $(K_{\mathscr{V}}, K'_{\mathscr{V}})$ into the pair $(K_{\mathscr{U}}, K'_{\mathscr{U}})$. This induces a homomorphism

$$f_{\mathscr{U}} : \mathscr{H}_p(K_{\mathscr{V}}, K'_{\mathscr{V}}) \to \mathscr{H}_p(K_{\mathscr{U}}, K'_{\mathscr{U}})$$

and the restriction of $f_{\mathscr{U}}^1$ to $K'_{\mathscr{V}}$, which is a simplicial map into $K'_{\mathscr{U}}$, induces a homomorphism

$$f'_{\mathscr{U}} : \mathscr{H}_p(K'_{\mathscr{V}}) \to \mathscr{H}_p(K'_{\mathscr{U}})$$

By Theorem 1-8, the diagram

$$\begin{array}{ccc} \mathscr{H}_p(K_{\mathscr{V}}, K'_{\mathscr{V}}) & \xrightarrow{f_{\mathscr{U}}} & \mathscr{H}_p(K_{\mathscr{U}}, K'_{\mathscr{U}}) \\ \partial'_{\mathscr{V}} \downarrow & & \downarrow \partial'_{\mathscr{U}} \\ \mathscr{H}_{p-1}(K'_{\mathscr{V}}) & \xrightarrow{f'_{\mathscr{U}}} & \mathscr{H}_{p-1}(K'_{\mathscr{U}}) \end{array} \tag{6-6}$$

is commutative.

Consider now the following diagram:

$$\begin{array}{ccc} K'_{\mathscr{V}} & \xrightarrow{f_{\mathscr{U}}^1} & K'_{\mathscr{U}} \\ l_{\mathscr{V}}^1 \uparrow & & \uparrow l_{\mathscr{U}}^1 \\ K_{\mathscr{V}'} & \xrightarrow{g_{\mathscr{U}'}^1} & K_{\mathscr{U}'} \end{array} \tag{6-7}$$

Here g is written for the restriction of f to F and $g_{\mathscr{U}'}^1$ is a simplicial map induced by g, as in Section 5-3. $f_{\mathscr{U}}^1$ is actually the restriction of the former $f_{\mathscr{U}}^1$ to the subcomplex $K'_{\mathscr{V}}$ of $K_{\mathscr{V}}$. The maps $l_{\mathscr{U}}^1$ and $l_{\mathscr{V}}^1$ are associated with the coverings $\mathscr{U}$ and $\mathscr{V}$ as in the earlier part of this proof. It will now be checked that this diagram induces a commutative diagram of homomorphisms on the homology groups. Since the maps in the diagram (6-7) are all simplicial, it is sufficient, in studying them, to examine how they act on the vertexes.

Take a vertex V' and $K_{\mathscr{V}'}$. $l_{\mathscr{V}}^1(V')$ is a vertex V of $K'_{\mathscr{V}} \subset K_{\mathscr{V}}$ such that the open sets V and V' of E and F, respectively, satisfy $V \cap F = V'$. $f_{\mathscr{U}}^1 l_{\mathscr{V}}^1(V')$ is a vertex U of $K'_{\mathscr{U}} \subset K_{\mathscr{U}}$ such that $V = f^{-1}(U)$. Now take the vertex U'

of $K_{\mathscr{U}'}$ such that $U' = U \cap Y$. The simplicial map k of $K'_{\mathscr{U}}$ into $K_{\mathscr{U}'}$ which maps U on U' induces a chain homomorphism that is an algebraic homotopy inverse to $l_{\mathscr{U}}{}^1$. On the other hand, $f^{-1}(U') = f^{-1}(U) \cap f^{-1}(Y) = V \cap f^{-1}(Y)$, so $g^{-1}(U') = V \cap F = V'$. $g^1_{\mathscr{U}'}(V')$ can be taken as U' (here the arbitrariness of the simplicial map $g^1_{\mathscr{U}'}$ is being exploited, cf. Lemma 5-3), and it follows that

$$g^1_{\mathscr{U}'}(V') = kf_{\mathscr{U}}{}^1 l_{\mathscr{V}}{}^1(V')$$

for any vertex V' of $K_{\mathscr{V}'}$. Since all the maps here are simplicial, this means that $g^1_{\mathscr{U}'} = kf_{\mathscr{U}}{}^1 l_{\mathscr{V}}{}^1$. And since k and $l_{\mathscr{U}}{}^1$ induce chain homomorphisms that are algebraically homotopic inverse to each other, it follows that $l_{\mathscr{U}}{}^1 g_{\mathscr{U}}{}^1$, and $f_{\mathscr{U}}{}^1 l_{\mathscr{V}}{}^1$ induce algebraically homotopic homomorphisms. Hence the induced diagram of homomorphisms on homology groups is commutative, namely:

$$\begin{array}{ccc} \mathscr{H}_{p-1}(K'_{\mathscr{V}}) & \xrightarrow{f'_{\mathscr{U}}} & \mathscr{H}_{p-1}(K'_{\mathscr{U}}) \\ {\scriptstyle l_{\mathscr{V}}}\uparrow & & \uparrow{\scriptstyle l_{\mathscr{U}}} \\ \mathscr{H}_{p-1}(K_{\mathscr{V}'}) & \xrightarrow{g_{\mathscr{U}'}} & \mathscr{H}_{p-1}(K_{\mathscr{U}'}) \end{array} \tag{6-8}$$

Arguing as in the case of the $i'_{\mathscr{U}}$ in the earlier part of this proof shows that the $f'_{\mathscr{U}}$ form an inverse system of homomorphisms, indexed by the finite open coverings of X. It has already been checked that the $\partial'_{\mathscr{U}}$, $\partial'_{\mathscr{V}}$, $l_{\mathscr{U}}$, $l_{\mathscr{V}}$ and, of course, as in general, the $f_{\mathscr{U}}$ and $g_{\mathscr{U}'}$ form inverse systems. Hence the inverse limits of all the groups and homomorphisms in the diagrams (6-6) and (6-8) can be taken, giving, by Lemma 5-7, the commutative diagrams

$$\begin{array}{ccc} \check{H}_p(E, F) & \xrightarrow{f_*} & \check{H}_p(X, Y) \\ {\scriptstyle \partial'}\downarrow & & \downarrow{\scriptstyle \partial'} \\ \check{H}'_{p-1}(F) & \xrightarrow{f'_*} & \check{H}'_{p-1}(Y) \end{array} \qquad \begin{array}{ccc} \check{H}'_{p-1}(F) & \xrightarrow{f'_*} & \check{H}'_{p-1}(Y) \\ {\scriptstyle l}\uparrow & & \uparrow{\scriptstyle l} \\ \check{H}_{p-1}(F) & \xrightarrow{g_*} & \check{H}_{p-1}(Y) \end{array}$$

where $f'_* = \lim f'_{\mathscr{U}}$, over the finite open coverings of X, and $\check{H}'_{p-1}(F)$, $\check{H}'_{p-1}(Y)$ are $\lim \mathscr{H}'_{p-1}(K_{\mathscr{V}'})$ and $\lim \mathscr{H}_{p-1}(K_{\mathscr{U}'})$, taken over the finite open coverings of F and Y, respectively. Combine these diagrams, remembering for each pair of spaces that ∂ is defined as the composition of ∂' with the appropriate l^{-1}, to obtain the commutative diagram

$$\begin{array}{ccc} \check{H}_p(E, F) & \xrightarrow{f_*} & \check{H}_p(X, Y) \\ \downarrow{\scriptstyle \partial} & & \downarrow{\scriptstyle \partial} \\ \check{H}_{p-1}(F) & \xrightarrow{g_*} & \check{H}_{p-1}(Y) \end{array}$$

as required. ∎

6-5. THE CONTINUITY THEOREM

The last section is somewhat negative in that the exactness theorem, which holds for singular homology theory, survives only in a rather mutilated form in Čech homology theory. On the other hand, the Čech theory has a distinguishing feature of its own that is not shared by singular theory, namely, the continuity property. Here a limit process for systems of spaces will be defined and continuity will consist in the fact that, considering compact spaces only, the Čech homology groups of the limit space are the inverse limits of the Čech homology groups of the spaces of the system.

Inverse systems and limits will now be defined for spaces in much the same manner as they were for groups.

Definition 6-3. Let A be a directed set and let $\{E_\alpha\}$ be a collection of topological spaces indexed by A such that for each $\beta > \alpha$ there is a continuous map $\pi_{\alpha\beta}\colon E_\beta \to E_\alpha$ satisfying the conditions

$$\pi_{\alpha\alpha} = \text{identity, all } \alpha \in A$$

$$\pi_{\alpha\beta}\pi_{\beta\gamma} = \pi_{\alpha\gamma} \text{ whenever } \gamma > \beta > \alpha$$

Then $\{E_\alpha, \pi_{\alpha\beta}, A\}$ is an *inverse system of topological spaces.*

Let E be the subset of the Cartesian product ΠE_α consisting of points with coordinates $\{x_\alpha\}$ such that $x_\alpha = \pi_{\alpha\beta}(x_\beta)$ whenever $\beta > \alpha$. ΠE_α is given the product topology and E the induced topology as a subspace.

Definition 6-4. The space E just constructed is called the *inverse limit* of the system $\{E_\alpha, \pi_{\alpha\beta}, A\}$ and is denoted by $\lim \{E_\alpha, \pi_{\alpha\beta}, A\}$ or $\lim_A E_\alpha$, or simply $\lim E_\alpha$.

The *projection map*

$$\pi_\beta\colon \lim E_\alpha \to E_\beta$$

is defined by mapping the point $\{x_\alpha\}$ on its β coordinates x_β. Note that this is the restriction of the projection map of the topological product on its β factor and so is automatically continuous. As in the case of groups, the relation

$$\pi_\alpha = \pi_{\alpha\beta}\pi_\beta$$

holds.

Exercise. 6-1. Let E be a topological space and let $\{E_\alpha\}$ be a family of subspaces indexed by the directed set A with the property that if $\alpha > \beta$ then $E_\alpha \subset E_\beta$. Prove that if $\pi_{\beta\alpha}\colon E_\alpha \to E_\beta$ is the inclusion map, $\alpha > \beta$, then $\{E_\alpha, \pi_{\alpha\beta}, A\}$ is an inverse system of spaces, and prove that $\lim E_\alpha = \cap E_\alpha$ the intersection being taken over all the indexes in A.

If all the E_α are Hausdorff spaces, then E is a closed subset of the product ΠE_α. This can be seen by noting first that, in this case ΠE_α is Hausdorff. The two maps f and g defined on ΠE_α by

$$f(x) = x_\alpha$$

$$g(x) = \pi_{\alpha\beta}(x_\beta),\ \beta > \alpha$$

where x_α and x_β are the α and β coordinates of x, are continuous maps into E_α. Then the set where $f = g$ is closed; that is, the subset of ΠE_γ, where $x_\alpha = \pi_{\alpha\beta}(x_\beta)$, for fixed $\beta > \alpha$, is closed. E is the intersection of all sets of this type for all pairs $\beta < \alpha$, so E is closed, as required.

It follows, in particular, that if each E_α is compact so that ΠE_α is compact, then E is also compact.

Of course, it is conceivable that E might be empty. It will now be shown that this does not happen in the case of interest in this section, that is, when all the E_α are compact.

Suppose then that all the E_α are compact, and for each pair α, β in A where $\beta > \alpha$, let $E_{\alpha\beta}$ be the subset of ΠE_γ consisting of all points $x = \{x_\gamma\}$ such that $x_\alpha = \pi_{\alpha\beta}(x_\beta)$. As pointed out before, $E_{\alpha\beta}$ is a closed set in ΠE_γ.

Take any finite collection $E_{\alpha_1\beta_1}, E_{\alpha_2\beta_2}, \ldots, E_{\alpha_k\beta_k}$ of sets of the type just described. Since A is directed, there is a γ in A such that $\gamma > \beta_i$ for each i. Take any x_γ in E_γ and define

$$x_{\alpha_i} = \pi_{\alpha_i\gamma}(x_\gamma)$$

$$x_{\beta_i} = \pi_{\beta_i\gamma}(x_\gamma)$$

for $i = 1, 2, \ldots, k$. Let x have the coordinates x_{α_i} and x_{β_i} in the appropriate spaces, the remaining coordinates being unrestricted. Then, since

$$\pi_{\alpha_i\beta_i}(x_{\beta_i}) = \pi_{\alpha_i\beta_i}\pi_{\beta_i\gamma}(x_\gamma) = \pi_{\alpha_i\gamma}(x_\gamma) = x_{\alpha_i}$$

the point x is in $E_{\alpha_i\beta_i}$ for each i. Thus the intersection of any finite collection of the closed sets $E_{\alpha\beta}$ in ΠE_γ is not empty. Since ΠE_γ is compact, it follows that the intersection of all the $E_{\alpha\beta}$ is nonempty. The intersection of all the $E_{\alpha\beta}$ is E, so it has been shown that, if all the E_α are compact, the inverse limit E is not empty.

The following lemma will be useful later, and its proof is a further example of the type of reasoning just used to prove that lim E_α is not empty when the E_α are compact.

Lemma 6-11. *If all the E_α are compact then*

$$\pi_\alpha(E) = \bigcap \pi_{\alpha\beta}(E_\beta)$$

the intersection being taken over all $\beta > \alpha$.

Proof. Since $\pi_\alpha = \pi_{\alpha\beta}\pi_\beta$, it is clear that

$$\pi_\alpha(E) \subset \pi_{\alpha\beta}(E_\beta)$$

for each $\beta > \alpha$. Hence

$$\pi_\alpha(E) \subset \bigcap_{\beta > \alpha} \pi_{\alpha\beta}(E_\beta)$$

Conversely, suppose that

$$x_\alpha \in \bigcap_{\beta > \alpha} \pi_{\alpha\beta}(E_\beta)$$

Thus, for each $\beta > \alpha$, $x_\alpha \in \pi_{\alpha\beta}(E_\beta)$, so $\pi_{\alpha\beta}^{-1}(x_\alpha)$ is nonempty for each $\beta > \alpha$. It is to be shown that $x_\alpha \in \pi_\alpha(E)$. In other words, there is an element in E whose α coordinate is x_α. Now E is the intersection of the sets $E_{\beta\gamma}$ (in the notation just used), so this is the same as showing that, if F is the set of points of ΠE_γ with α coordinate x_α, then F has a nonempty intersection with the collection of $E_{\beta\gamma}$. Since F is a product of compact spaces (namely, all the E_γ except E_α itself), it is compact, so it is sufficient to show that F has a nonempty intersection with any finite collection of the $E_{\beta\gamma}$.

So consider the finite collection $E_{\alpha_1\beta_1}, E_{\alpha_2\beta_2}, \ldots, E_{\alpha_r\beta_r}$ and choose γ in A such that $\gamma > \alpha_i$ and $\gamma > \beta_i$ for each i. As just pointed out, $\pi_{\alpha\gamma}^{-1}(x_\alpha)$ is not empty. So take $x_\gamma \in \pi_{\alpha\gamma}^{-1}(x_\alpha)$ and define

$$x_{\alpha_i} = \pi_{\alpha_i\gamma}(x_\gamma)$$
$$x_{\beta_i} = \pi_{\beta_i\gamma}(x_\gamma)$$

for each i. Let x be any element of ΠE_β whose α_i and β_i coordinates are x_{α_i} and x_{β_i}, respectively, for $i = 1, 2, \ldots, r$, and whose α coordinate is x_α. There is a possible clash of definition here if α is equal to one of the α_i or β_i. Suppose $\alpha = \alpha_i$. x_{α_i} is defined as $\pi_{\alpha_i\gamma}(x_\gamma) = \pi_{\alpha\gamma}(x_\gamma)$. On the other hand, $x_\gamma \in \pi_{\alpha\gamma}^{-1}(x_\alpha)$, so $\pi_{\alpha\gamma}(x_\gamma) = x_\alpha$; that is, if $\alpha = \alpha_i$ then $x_\alpha = x_{\alpha_i}$. Similarly, if $\alpha = \beta_i$ then $x_\alpha = x_{\beta_i}$, so no contradiction is involved in the definition of x. Clearly, $x \in F$. Also

$$\begin{aligned}\pi_{\alpha_i\beta_i}(x_{\beta_i}) &= \pi_{\alpha_i\beta_i}\pi_{\beta_i\gamma}(x_\gamma)\\ &= \pi_{\alpha_i\gamma}(x_\gamma)\\ &= x_{\alpha_i}\end{aligned}$$

for each i. Hence $x \in E_{\alpha_i\beta_i}$ for each i. This means that the intersection of F and the collection of $E_{\alpha_i\beta_i}$ is not empty, as was to be shown. Then, as remarked before, this implies that the given element x_α of $\bigcap_{\beta>\alpha}\pi_{\alpha\beta}(E_\beta)$ is in $\pi_\alpha(E)$, and this completes the proof of the lemma. ∎

There is an important feature of the topology of the space $E = \lim E_\alpha$ which will be noted before the continuity theorem is discussed, namely, the existence of a particularly simple basis for the open sets. In the first place, a basis in E is induced by a basis in ΠE_α. Hence, if $x \in E$ and U is an open set in E containing x, there is an open set V in E such that $x \in V \subset U$, where V consists of points $y = \{y_\alpha\}$ such that y_{α_i} is in an open set U_i in E_{α_i} for a finite collection of α_i, the remaining y_α being unrestricted, except by the coordinate conditions in the definition of E. But A is a directed set, so there is a γ in A such that $\gamma > \alpha_i$ for each i. The continuity of each $\pi_{\alpha_i\gamma}$ implies that there is an open neighborhood U_0 of y_γ such that $\pi_{\alpha_i\gamma}(U_0) \subset U_i$ for each i. Hence V contains the open set V_0 defined by the condition $y_\gamma \in U_0$, along with the coordinate conditions defining E, and $x \in V_0 \subset U$. It follows that sets of the type V_0 form a basis for the topology of E.

In the formulation of the continuity theorem, systems of pairs of spaces also appear. This means that, in Definition 6-3, each E_α contains a subspace F_α and each map $\pi_{\alpha\beta}$ is a map of the pair (E_β, F_β) into (E_α, F_α). The inverse system of pairs is denoted by $\{(E_\alpha, F_\alpha), \pi_{\alpha\beta}, A\}$. It is clear that, in this situation, the F_α also form an inverse system $\{F_\alpha, \pi_{\alpha\beta}, A\}$. The topological product ΠE_α contains ΠF_α as a subspace, and the intersection of $E = \lim E_\alpha$ with ΠF_α clearly coincides with $F = \lim F_\alpha$.

Definition 6-5. The pair (E, F) is called the *inverse limit of the system* $\{(E_\alpha, F_\alpha), \pi_{\alpha\beta}, A\}$.

The continuity theorem now can be formulated. Let $\{(E_\alpha, F_\alpha), \pi_{\alpha\beta}, A\}$ be an inverse system of pairs of spaces and, for each α in A, construct $\check{H}_p(E_\alpha, F_\alpha)$. If $\beta > \alpha$, the continuous map $\pi_{\alpha\beta}$ induces a homomorphism

$$\pi_{\alpha\beta*}\colon \check{H}_p(E_\beta, F_\beta) \to \check{H}_p(E_\alpha, F_\alpha)$$

By Theorem 5-8, these homomorphisms satisfy the conditions

$$\pi_{\alpha\alpha*} = \text{identity}$$

$$\pi_{\alpha\gamma*} = \pi_{\alpha\beta*}\pi_{\beta\gamma*}(\gamma > \beta > \alpha)$$

Thus $\{\check{H}_p(E_\alpha, F_\alpha), \pi_{\alpha\beta*}, A\}$ is an inverse system of groups whose inverse limit $\lim \check{H}_p(E_\alpha, F_\alpha)$ can be constructed.

Also, there is a homomorphism of $\check{H}_p(E, F)$ into $\lim \check{H}_p(E_\alpha, F_\alpha)$ which can be constructed in a natural way, (E, F) being $\lim (E_\alpha, F_\alpha)$; namely for each α there is a projection (cf. Definition 6-4)

$$\pi_\alpha: (E, F) \to (E_\alpha, F_\alpha)$$

that satisfies the composition relation

$$\pi_{\alpha\beta}\pi_\beta = \pi_\alpha$$

for each pair of indexes such that $\beta > \alpha$. The induced homomorphisms

$$\pi_{\alpha*}: \check{H}_p(E, F) \to \check{H}_p(E_\alpha, F_\alpha)$$

thus satisfy

$$\pi_{\alpha\beta*}\pi_{\beta*} = \pi_{\alpha*}$$

(cf. Theorem 5-8). So the $\pi_{\alpha*}$ form an inverse system of homomorphisms from $\check{H}_p(E, F)$ into $\{H_p(E_\alpha, F_\alpha), \pi_{\alpha\beta*}, A\}$ (Definition 5-15). Hence (cf. Definition 5-16) there is a limit homomorphism $\pi = \lim \pi_{\alpha*}$,

$$\pi: \check{H}_p(E, F) \to \lim \check{H}_p(E_\alpha, F_\alpha)$$

The object of the continuity theorem is to show that, if all the spaces concerned are compact, then π is an isomorphism onto. The proof is rather long and complicated and will be broken down into a series of lemmas. The idea of the proof will be sketched first.

To show that π is onto, an inverse image must be constructed for any given element $\theta \in \lim \check{H}_p(E_\alpha, F_\alpha)$. Now θ can be represented by its coordinates $\{\theta_\alpha\}$, where $\theta \in \check{H}_p(E_\alpha, F_\alpha)$, so finding an inverse image for θ under π is equivalent to finding a common inverse image for each θ_α under the corresponding $\pi_{\alpha*}$. In other words, an element $\phi \in \check{H}_p(E, F)$ is to be found such that

$$\pi_{\alpha*}(\phi) = \theta_\alpha$$

for each $\alpha \in A$.

On the other hand, $\check{H}_p(E, F)$, the $\check{H}_p(E_\alpha, F_\alpha)$ are themselves inverse limits, and $\pi_{\alpha*}$ is an inverse limit of homomorphisms $\pi_{\alpha\mathscr{U}}$ over the finite open coverings of E_α, so the last equation can be written in terms of the $\mathscr{U}$ coordinates. If $\mathscr{U}$ is a finite open covering of E_α and $\mathscr{U}' = \pi_\alpha^{-1}(\mathscr{U})$ then the appropriate coordinates $\phi_{\mathscr{U}'}$ and $\theta_{\alpha\mathscr{U}}$ of ϕ and θ_α must satisfy the equation

$$\pi_{\alpha\mathscr{U}}(\phi_{\mathscr{U}'}) = \theta_{\alpha\mathscr{U}}$$

where

$$\pi_{\alpha\mathscr{U}}: H_p(E, F; \mathscr{U}') \to H_p(E_\alpha, F_\alpha; \mathscr{U})$$

is induced by a simplicial map of the nerve of $\mathscr{U}'$ into the nerve of $\mathscr{U}$ to correspond, as in Section 5-3, to the continuous map π_α.

Also note that, for $\beta > \alpha$, the compatibility conditions $\theta_\alpha = \pi_{\alpha\beta*}(\theta_\beta)$ can be expressed in terms of the coordinates of the elements θ_α and θ_β of $\check{H}_p(E_\alpha, F_\alpha)$ and $\check{H}_p(E_\beta, F_\beta)$, considered as inverse limits; namely if $\mathscr{U}'' = \pi_{\alpha\beta}^{-1}(\mathscr{U})$ and if

$$\pi_{\alpha\beta\mathscr{U}}: \check{H}_p(E_\beta, F_\beta; \mathscr{U}'') \to \check{H}_p(E_\alpha, F_\alpha; \mathscr{U})$$

is induced by a simplicial map corresponding to $\pi_{\alpha\beta}$, as in Section 5-3, then the compatibility conditions on the θ_α are expressed as

$$\theta_{\alpha\mathscr{U}} = \pi_{\alpha\beta\mathscr{U}}(\theta_{\beta\mathscr{U}''})$$

Also, the relations $\pi_\alpha = \pi_{\alpha\beta}\,\pi_\beta$ for $\beta > \alpha$ imply that

$$\pi_{\alpha\mathscr{U}} = \pi_{\alpha\beta\mathscr{U}}\,\pi_{\beta\mathscr{U}''}$$

(cf. Theorem 5-4).

From this last relation it follows that, if an inverse image could be found for $\theta_{\beta\mathscr{U}''}$ under $\pi_{\beta\mathscr{U}''}$, an inverse image would automatically be found for $\theta_{\alpha\mathscr{U}}$ under $\pi_{\alpha\mathscr{U}}$. The essential part of the proof will be to show that β can be chosen so that $\theta_{\beta\mathscr{U}''}$ has an inverse image under $\pi_{\beta\mathscr{U}''}$. In fact, it will be shown that, if $\mathscr{U}$ is taken from a suitable class of coverings, then $\pi_{\beta\mathscr{U}''}$ is an isomorphism onto for sufficiently large β. The corresponding simplicial map $\pi^1_{\beta\mathscr{U}''}$ actually will be seen to be a homeomorphism.

An inverse image $\phi_{\mathscr{U}'}$ of $\theta_{\alpha\mathscr{U}}$ under $\pi_{\alpha\mathscr{U}}$ having been obtained in this way, it will have to be shown to be the $\mathscr{U}'$ coordinate of some element in $\check{H}_p(E, F)$. This will be a two-fold operation, to check first that the $\phi_{\mathscr{U}'}$ satisfy the compatibility conditions for coordinates of an element of the inverse limit and second that the coverings so obtained are cofinal in the set of all finite open coverings of E.

When the details of the above have been filled in, showing that π is onto, the proof that its kernel is zero will turn out to be quite an easy matter.

The details of the proof will be arranged now in a series of lemmas. The first three describe the type of coverings to be used.

Lemma 6-12. *Let X be a compact space, C a subset of X, and $\mathscr{U} = \{\mathscr{U}_i\}$, $i = 1, 2, \ldots, n$ a finite open covering of X. Then there is a refinement $\mathscr{V} = \{V_i\}$ of $\mathscr{U}$ such that $\overline{V}_i \subset U_i$ for each i, and a collection of the V_i has a nonempty intersection with C if and only if the same is true of the corresponding U_i.*

Proof. Assume that $V_1, V_2, \ldots, V_r$ have already been constructed so that

(1) $\overline{V}_i \subset U_i, i = 1, 2, \ldots, r$,

(3) $V_1, V_2, \ldots, V_r, U_{r+1}, \ldots, U_n$ is an open covering of X,

(3) a collection of the sets $V_1, V_2, \ldots, V_r, U_{r+1}, \ldots, U_n$ has nonempty intersection with C if and only if the correspondingly numbered U_i have this property.

Write $F = \complement (V_1 \cup V_2 \cup \cdots \cup V_r \cup U_{r+2} \cup \cdots \cup U_n)$. This is a closed set and is contained in U_{r+1}, since the sets listed in (2) form a covering of X. For each collection of sets in (2) having a nonempty intersection with $C \cap U_{r+1}$, take a point in this intersection and adjoin this point to F. Call the resulting set F'. X is compact, hence normal, so there is an open set V_{r+1} such that $F' \subset V_{r+1} \subset \overline{V}_{r+1} \subset U_{r+1}$. This means, in particular, that $V_{r+1} \supset F$, so the sets $V_1, V_2, \ldots, V_{r+1}, U_{r+2}, \ldots, U_n$ form a covering of X. Since $V_{r+1} \supset F'$, it follows that, whenever a collection of the sets listed under (2) in the statement of the lemma has a nonempty intersection with C, the same is true with U_{r+1} replaced by V_{r+1}. Thus properties (1), (2), and (3) hold with r replaced by $r + 1$, so the required result follows by induction. ∎

Lemma 6-13. *Let X be a compact space and C a subset. Any finite open covering $\mathscr{U}$ of X has a refinement $\mathscr{V}$ with the property that a collection of the sets of $\mathscr{V}$ has a nonempty intersection with C if and only if their closures have this property.*

Proof. The covering $\mathscr{V} = \{V_i\}$ of the last lemma has this property. If $\overline{V}_i \cap \overline{V}_j \cap \cdots \cap \overline{V}_k \cap C \neq \varnothing$ then $U_i \cap U_j \cap \cdots \cap U_k \cap C \neq \varnothing$ and so also $V_i \cap V_j \cap \cdots \cap V_k \cap C \neq \varnothing$. The converse is obvious. ∎

Another way of stating this lemma is to say that the coverings with the property of $\mathscr{V}$ are cofinal in the set of all finite open coverings of X.

Now consider the situation involved in the continuity theorem. E is the inverse limit of a system of compact spaces $\{E_\alpha, \pi_{\alpha\beta}, A\}$. For some α, take a finite open covering $\mathscr{U}$ of E_α and let $\mathscr{U}' = \pi_\alpha^{-1}(U)$. It will be shown that the collection of coverings of E obtained in this way for various α and $\mathscr{U}$ is cofinal in the set of all finite open coverings of E.

Lemma 6-14. *Let $\mathscr{V}$ be a finite open covering of E. Then for some γ, there is a refinement of $\mathscr{V}$ of the form $\mathscr{U}' = \pi_\gamma^{-1}(\mathscr{U})$, where $\mathscr{U}$ is a finite open covering of E_γ. In addition, the γ can always be taken so that $\gamma > \alpha$ for some preassigned α.*

Proof. It has already been seen (cf. remark following Lemma 6-11) that sets of the type $\pi_\alpha^{-1}(U_\alpha)$, where U_α is open in E_α, form a basis for the topology of E. Hence $\mathscr{V}$ may be refined by a covering $\mathscr{V}'$ consisting of sets of this type, and since E is compact, $\mathscr{V}'$ can be taken to be finite. Take an index greater than all the α appearing among the $\pi_\alpha^{-1}(U_\alpha)$ of $\mathscr{V}'$. Incidentally, note that, at this point, γ can also be taken to exceed a preassigned α as well. Then, since $\pi_\alpha = \pi_{\alpha\gamma}\pi_\gamma$, it follows that

$$\pi_\alpha^{-1}(U_\alpha) = \pi_\gamma^{-1}\pi_{\alpha\gamma}^{-1}(U_\alpha)$$

Hence each set of $\mathscr{V}'$ is of the form $\pi^{-1}(U_i)$ for some open set U_i of E_γ, γ now being fixed. The open sets U_i so obtained, with i running from 1 to n, for instance, cover $\pi_\gamma(E)$. Let $U_0 = \complement\, \pi_\gamma(E)$. This is an open set since $\pi_\gamma(E)$ is compact in the Hausdorff space E_γ and, therefore, closed. Then $\mathscr{U} = \{U_0, U_1, \ldots, U_n\}$ is a covering of E_γ and $\mathscr{V}' = \pi_\gamma^{-1}(\mathscr{U})$. The lemma is thus proved. ∎

Now let $\{(E_\alpha, F_\alpha), \pi_{\alpha\beta}, A\}$ be an inverse system of pairs of compact spaces and let (E, F) be the corresponding inverse limit pair. For a given covering $\mathscr{U}$ of E_α, the relation between the nerves of $\pi_\alpha^{-1}(\mathscr{U})$ and $\pi_{\alpha\beta}^{-1}(\mathscr{U})$ for $\beta > \alpha$ should be examined.

Lemma 6-15. *Let $\mathscr{U}$ be a finite open covering of E_α such that a collection of the U_i in $\mathscr{U}$ has a nonempty intersection with $\pi_\alpha(E)$ or with $\pi_\alpha(F)$ if and only if their closures have this property. Then there is a $\beta > \alpha$ such that, if $\mathscr{U}' = \pi_\alpha^{-1}(\mathscr{U}) = \pi_\beta^{-1}\ \pi_{\alpha\beta}^{-1}(\mathscr{U})$ and $\mathscr{U}'' = \pi_{\alpha\beta}^{-1}(\mathscr{U})$, then the simplicial map $\pi_{\beta\mathscr{U}''}^1$ that is induced by π_β, as in Section 5-3, is a homeomorphism of the pair $(K_{\mathscr{U}'}, K'_{\mathscr{U}'})$ onto the pair $(K_{\mathscr{U}''}, K'_{\mathscr{U}''})$.*

Here $K_{\mathscr{U}'}$ and $K_{\mathscr{U}''}$ are, as usual, the nerves of $\mathscr{U}'$ and $\mathscr{U}''$, while $K'_{\mathscr{U}'}$ and $K'_{\mathscr{U}''}$ are the subcomplexes associated with F and F_β, respectively.

Proof. The simplicial map $\pi_{\beta\mathscr{U}''}^1$ is generated by the following natural assignment of vertexes. If U is a set in the covering $\mathscr{U}$ and $\pi_\alpha^{-1}(U) \neq \varnothing$, then the vertex $\pi_\alpha^{-1}(U)$ of $K_{\mathscr{U}'}$ is mapped by $\pi_{\beta\mathscr{U}''}^1$ on the vertex $\pi_{\alpha\beta}^{-1}(U)$ of $K_{\mathscr{U}''}$. It is a general property of such vertex assignments (making it possible to generate a simplicial map in this way) that, whenever $U_0', U_1', \ldots, U_r'$ are vertexes of a simplex in $K_{\mathscr{U}'}$, their images, defined as before, are vertexes of a simplex in $K_{\mathscr{U}''}$. The essential step in this lemma is getting a converse statement; that is, if β is large enough, then whenever $U_0', U_1', \ldots, U_r' = \pi_\alpha^{-1}(U_0), \ldots, \pi_\alpha^{-1}(U_r)$ are not vertexes of a simplex in $K_{\mathscr{U}''}$. Now to say that the $\pi_\alpha^{-1}(U_i)$, $i = 0, 1, 2, \ldots, r$, are not vertexes of a simplex in $K_{\mathscr{U}'}$ is equivalent to saying that

$$\pi_\alpha^{-1}(U_0) \cap \pi_\alpha^{-1}(U_1) \cap \cdots \cap \pi_\alpha^{-1}(U_r) = \varnothing$$

that is,

$$U_0 \cap U_1 \cap \cdots \cap U_r \cap \pi_\alpha(E) = \varnothing$$

By the assumption on the covering $\mathscr{U}$, this is equivalent to

$$\bar{U}_0 \cap \bar{U}_1 \cap \cdots \cap \bar{U}_r \cap \pi_\alpha(E) = \varnothing$$

and by Lemma 6-11 this is equivalent to

$$\bar{U}_0 \cap \bar{U}_1 \cap \cdots \cap \bar{U}_r \cap \bigcap_{\beta > \alpha} \pi_{\alpha\beta}(E_\beta) = \varnothing \tag{6-9}$$

But $\bar{U}_0 \cap \bar{U}_1 \cap \cdots \cap \bar{U}_r$ is closed and, therefore, compact, and the sets $\pi_{\alpha\beta}(E)$ are all closed. Thus, if the intersection (6-9) is empty, the same must be true for some finite collection of the $\pi_{\alpha\beta}(E_\beta)$ (using here the definition of compactness by the finite intersection property of closed sets). So there is a finite set of indexes $\beta_1, \beta_2, \ldots, \beta_n$ such that

$$\bar{U}_0 \cap \bar{U}_1 \cap \cdots \cap \bar{U}_r \cap \pi_{\alpha\beta_1}(E_{\beta_1}) \cap \cdots \cap \pi_{\alpha\beta_n}(E_{\beta_n}) = \varnothing$$

Let β be such that $\beta > \beta_i$ for each i. Then

$$\pi_{\alpha\beta}(E_\beta) = \pi_{\alpha\beta_i}\pi_{\beta_i\beta}(E_\beta) \subset \pi_{\alpha\beta_i}(E_{\beta_i})$$

for each i, so

$$\bar{U}_0 \cap \bar{U}_1 \cap \cdots \cap \bar{U}_r \cap \pi_{\alpha\beta}(E_\beta) = \varnothing$$

and this implies that

$$U_0 \cap U_1 \cap \cdots \cap U_r \cap \pi_{\alpha\beta}(E_\beta) = \varnothing$$

But this means that

$$\pi_{\alpha\beta}^{-1}(U_0) \cap \pi_{\alpha\beta}^{-1}(U_1) \cap \ldots \cap \pi_{\alpha\beta}^{-1}(U_r) = \varnothing$$

so the vertexes $\pi_{\alpha\beta}^{1-}(U_i)$, $i = 0, 1, 2, \cdots, r$, of $K_{\mathcal{U}''}$, for this choice of β, are not vertexes of a simplex.

Note that the same argument, involving one vertex only, shows that if $\pi_\alpha^{-1}(U)$ is empty, for U in $\mathcal{U}$, then β can be chosen so that $\pi_{\alpha\beta}^{-1}(U)$ is empty.

Since there are only finitely many possible combinations of vertexes of $K_{\mathcal{U}'}$, it follows that β can be chosen to satisfy all the following conditions:

(1) whenever $\pi_\alpha^{-1}(U)$ is not a vertex of $K_{\mathcal{U}'}$, then $\pi_{\alpha\beta}^{-1}((U)$ is not a vertex of $K_{\mathcal{U}''}$,

(2) whenever $\pi_\alpha^{-1}(U_0)$, $\pi_\alpha^{-1}(U_1), \ldots, \pi_\alpha^{-1}(U_r)$ are not vertexes of any simplex of $K_{\mathcal{U}'}$, then $\pi_{\alpha\beta}^{-1}(U_0), \ldots, \pi_{\alpha\beta}^{-1}(U_r)$ are not vertexes of a simplex of $K_{\mathcal{U}''}$.

Now consider the simplicial map $\pi_{\beta\mathcal{U}''}^1$ corresponding, as in Section 5-3 to π_β. This is a map of $K_{\mathcal{U}'}$ into $K_{\mathcal{U}''}$, and in particular, it maps the vertex $\pi_{\alpha\beta}^{-1}(U)$ of the first complex on the vertex $\pi_{\alpha\beta}^{-1}(U)$ of the second, where U is a set in $\mathcal{U}$. If U is any set of $\mathcal{U}$ such that $\pi_{\alpha\beta}^{-1}(U)$ is not empty and so represents a vertex of $K_{\mathcal{U}''}$, then condition (1) says that $\pi_\alpha^{-1}(U)$ is not empty, so it represents a vertex of $K_{\mathcal{U}'}$. It follows that the vertex $\pi_{\alpha\beta}^{-1}(U)$ appears in the image of the simplicial map $\pi_{\beta\mathcal{U}''}^1$. Thus $\pi_{\beta\mathcal{U}''}^1$ is onto so far as the mapping of the vertexes is concerned. It is also one-to-one, since, clearly, the only vertex of $K_{\mathcal{U}'}$ mapping onto $\pi_{\alpha\beta}^{-1}(U)$ under $\pi_{\beta\mathcal{U}''}^1$ is $\pi_\beta^{-1}\pi_{\alpha\beta}^{-1}(Y) = \pi_\alpha^{-1}(U)$. Next, condition (2) implies that vertexes of $K_{\mathcal{U}'}$ are vertexes of a simplex if and only if their images under $\pi_{\beta\mathcal{U}''}^1$ are vertexes of a simplex in $K_{\mathcal{U}''}$. It follows that $\pi_{\beta\mathcal{U}''}^1$ is a simplicial homeomorphism between $K_{\mathcal{U}'}$ and $K_{\mathcal{U}''}$, as asserted.

To complete the proof it must be shown that $\pi_{\beta\mathcal{U}''}^1$ maps the subcomplex $K'_{\mathcal{U}'}$ of $K_{\mathcal{U}'}$ onto $K'_{\mathcal{U}''}$ in $K_{\mathcal{U}''}$; it is, of course, certainly mapped *into* $K'_{\mathcal{U}''}$ by the general definition of the induced simplicial map given in Section 5-3. To check this it is sufficient to show that, U being a set in $\mathcal{U}$, if $\pi_{\alpha\beta}^{-1}(U)$ is a vertex of $K'_{\mathcal{U}''}$, then $\pi_\alpha^{-1}(U)$ is a vertex of $K'_{\mathcal{U}'}$. Now if $\pi_\alpha^{-1}(U)$ is not a vertex of $K'_{\mathcal{U}'}$, the set $\pi_\alpha^{-1}(U)$ satisfies

$$\pi_\alpha^{-1}(U) \cap F = \varnothing$$

that is,

$$U \cap \pi_\alpha(F) = \varnothing$$

By hypothesis, this implies

$$\overline{U} \cap \pi_\alpha(F) = \varnothing$$

which again is equivalent to

$$\bar{U} \cap \bigcap_{\gamma > \alpha} \pi_{\alpha\gamma}(F_\gamma) = \varnothing \tag{6-10}$$

$\bar{U}$ is closed in the compact space E_α, so it is compact. Hence the last condition means that

$$\bar{U} \cap \pi_{\alpha\gamma_1}(F_{\gamma_1}) \cap \pi_{\alpha\gamma_2}(F_{\gamma_2}) \quad \cdots \cap \pi_{\alpha\gamma_k}(F_{\gamma_k}) = \varnothing \tag{6-11}$$

for a finite collection of indexes $\gamma_1, \gamma_2, \ldots, \gamma_k$. Choose β so that $\beta > \gamma_i$ for each i. Then

$$\pi_{\alpha\beta}(F_\beta) = \pi_{\alpha\gamma_i} \pi_{\gamma_i\beta}(F_\beta) \subset \pi_{\alpha\gamma_i}(F_{\gamma_i})$$

for each γ_i, so (6-10) implies that

$$\bar{U} \cap \pi_{\alpha\beta}(F_\beta) = \varnothing$$

which in turn implies that $U \cap \pi_{\alpha\beta}(F_\beta) = \varnothing$, or in other words,

$$\pi_{\alpha\beta}^{-1}(U) \cap F_\beta = \varnothing$$

Hence $\pi_{\alpha\beta}^{-1}(U)$ is not a vertex of $K'_{\mathscr{U}''}$ under the assumption that $\pi_\alpha^{-1}(U)$ is not a vertex of $K'_{\mathscr{U}'}$. Turn this the other way round and it follows that $\pi_\alpha^{-1}(U)$ is a vertex of $K'_{\mathscr{U}'}$ whenever $\pi_{\alpha\beta}^{-1}(U)$ is a vertex of $K'_{\mathscr{U}''}$, so long as β is greater than all the γ_i appearing in (6-11).

Thus, if β satisfies this condition in addition to the conditions (1) and (2), $\pi_{\beta\mathscr{U}''}$ is a homeomorphism of the pairs. ∎

The continuity theorem will now be proved.

Theorem 6-16. *For an inverse system of compact pairs (E_α, F_α), the homomorphism π: $\check{H}_p[\lim (E, F)] \to \lim \check{H}_p(E_\alpha, F_\alpha)$ is an isomorphism onto.*

Proof. To prove the onto part, take $\theta \in \lim \check{H}_p(E_\alpha, F_\alpha)$ and try to find $\phi \in \check{H}_p(E, F)$ such that $\pi_{\gamma*}(\phi) = \theta_\gamma$ (the γ coordinate of θ) for each γ in A; that is, the coordinate of this element ϕ in $\check{H}_p(E, F)$ that corresponds to any finite open covering of E must be found. Let $\mathscr{V}$ be a given finite open covering of E. By Lemma 6-14, there is a refinement $\mathscr{U}'$ of $\mathscr{V}$ of the form $\pi_\alpha^{-1}(\mathscr{U})$ for some α and some finite open covering $\mathscr{U}$ of E_α. Clearly, this statement remains true if $\mathscr{U}$ is replaced by some refinement, so, if Lemma 6-13 is applied it can be assumed that $\mathscr{U}$ has the property that, if any collection of its sets has an empty intersection with $\pi_\alpha(E)$ or with $\pi_\alpha(F)$, then the same is true of the corresponding closures. Also, (cf. Lemma 6-14) α can be assumed to be greater than a preassigned index.

Apply Lemma 6-15, starting with the covering $\mathscr{U}$ of E_α and writing $\mathscr{U}' = \pi_\alpha^{-1}(\mathscr{U})$ and $\mathscr{U}'' = \pi_{\alpha\beta}^{-1}(\mathscr{U})$. Then β can be chosen so that the corresponding simplicial map $\pi_{\beta}{}^{1}{}_{\mathscr{U}''}$ is a homeomorphism between the pairs $(K_{\mathscr{U}'}, K'_{\mathscr{U}'})$ and $(K_{\mathscr{U}''}, K'_{\mathscr{U}''})$ (by Lemma 6-15). Hence the induced homomorphism

$$\pi_{\beta\mathscr{U}''}: H_p(E, F; \mathscr{U}') \to H_p(E_\beta, F_\beta; \mathscr{U}'')$$

is an isomorphism onto.

Return now to $\theta \in \lim \check{H}_p(E_\alpha, F_\alpha)$. Take its β coordinate $\theta_\beta \in \check{H}_p(E_\beta, F_\beta)$ and the representative $\theta_{\beta\mathscr{U}''}$ in $H_p(E_\beta, F_\beta; \mathscr{U}'')$ corresponds to the covering $\mathscr{U}''$. Define

$$\phi_{\mathscr{U}'} = \pi_{\beta\mathscr{U}''}^{-1}(\theta_{\beta\mathscr{U}''})$$

Then if $f_{\mathscr{V}\mathscr{U}'}$ is the projection

$$f_{\mathscr{V}\mathscr{U}'}: H_p(E, F; \mathscr{U}') \to H_p(E, F; \mathscr{V})$$

associated with the pair $\mathscr{V} < \mathscr{U}'$ of coverings, write

$$\phi_{\mathscr{V}} = f_{\mathscr{V}\mathscr{U}'}(\phi_{\mathscr{U}'})$$

It must now be shown that the $\phi_{\mathscr{V}}$ so obtained are the coordinates of an element ϕ of $\check{H}_p(E, F)$ such that $\pi_{\gamma *}(\phi) = \theta_\gamma$ for each γ. In the first place, the independence of $\phi_{\mathscr{V}}$ of the various choices made must be established.

To formulate this more precisely, $\phi_{\mathscr{V}}$ is defined as

$$\phi_{\mathscr{V}} = f_{\mathscr{V}\mathscr{U}'}\pi_{\beta\mathscr{U}''}^{-1}(\theta_{\beta\mathscr{U}''})$$

where $\mathscr{U}' = \pi_\alpha^{-1}(\mathscr{U})$ is a refinement of $\mathscr{V}$ and β is chosen so that, if $U'' = \pi_{\alpha\beta}^{-1}(U)$, $\pi_{\beta\mathscr{U}''}$ is an isomorphism onto. The elements that can be chosen here are α, $\mathscr{U}$, β, and it will now be checked that changes in these choices leave $\phi_{\mathscr{V}}$ unchanged.

Let β' be chosen instead of β, α and $\mathscr{U}$ being as before. There is a γ in A such that $\gamma > \beta$ and $\gamma > \beta'$, so it is sufficient to prove that the above construction gives the same $\phi_{\mathscr{V}}$ if β is replaced by γ, with $\gamma > \beta$. Write $\mathscr{U}''' = \pi_{\alpha\gamma}^{-1}(\mathscr{U})$. The new $\phi_{\mathscr{V}}$, constructed with γ instead of β, is $f_{\mathscr{V}\mathscr{U}'}\pi_{\alpha\gamma}^{-1}(\theta_{\gamma\mathscr{U}'''}^{-1})$. Now $\pi_\beta = \pi_{\beta\gamma}\pi_\gamma$, and, so for the induced maps on the homology groups of the appropriate nerves

$$\pi_{\beta\mathcal{U}''} = \pi_{\beta\gamma\mathcal{U}''}\,\pi_{\gamma\mathcal{U}'''}$$

Here $\pi_{\beta\mathcal{U}'''}$, $\pi_{\gamma\mathcal{U}'''}$ are isomorphisms onto because of the choice of β and γ, so $\pi_{\gamma\beta\mathcal{U}''}$ is also an isomorphism onto, and the last equation can be inverted to give

$$\pi^{-1}_{\beta\mathcal{U}''}\,\pi_{\beta\gamma\mathcal{U}''} = \pi^{-1}_{\gamma\mathcal{U}'''}$$

Then the new $\phi_{\mathcal{V}}$ (using γ) can be written as

$$f_{\mathcal{V}\mathcal{U}'}\,\pi^{-1}_{\beta\mathcal{U}''}\,\pi^{-1}_{\beta\gamma\mathcal{U}''}(\theta_{\gamma\mathcal{U}'''}) = f_{\mathcal{V}\mathcal{U}'}\,\pi^{-1}_{\beta\mathcal{U}''}(\theta_{\beta\mathcal{U}''})$$

using at this last step the relation $\theta_{\beta\mathcal{U}''} = \pi_{\beta\gamma\mathcal{U}''}(\theta_{\gamma\mathcal{U}'''})$, expressed in terms of coordinates. This is the same as the original $\phi_{\mathcal{V}}$, as was to be shown.

Next consider the effect of changing the choice of covering of E_α. Again, it is clearly sufficient to consider the effect of replacing $\mathcal{U}$ by $\mathcal{W}$ where $\mathcal{W} > \mathcal{U}$. Write

$$\mathcal{W}' = \pi_\alpha^{-1}(\mathcal{W}),\ \mathcal{W}'' = \pi_{\alpha\beta}^{-1}(\mathcal{W})$$

and choose β so that both $\pi_{\beta\mathcal{U}''}$ and $\pi_{\beta\mathcal{W}''}$ are isomorphisms onto. Then, using the above construction,

$$\phi_{\mathcal{U}'} = \pi^{-1}_{\beta\mathcal{U}''}(\theta_{\beta\mathcal{U}''}),\ \phi_{\mathcal{W}'} = \pi^{-1}_{\beta\mathcal{W}''}(\theta_{\beta\mathcal{W}''})$$

By Theorem 5-6 the following diagram is commutative:

$$\begin{array}{ccc}
H_p(E, F; \mathcal{W}') & \xrightarrow{\pi_{\beta\mathcal{W}''}} & H_p(E_\beta, F_\beta; \mathcal{W}'') \\
\downarrow{\scriptstyle f_{\mathcal{U}'\mathcal{W}'}} & & \downarrow{\scriptstyle f_{\mathcal{U}''\mathcal{W}''}} \\
H_p(E, F; \mathcal{U}') & \xrightarrow{\pi_{\beta\mathcal{U}''}} & H_p(E_\beta, F_\beta; \mathcal{U}'')
\end{array}$$

Hence

$$\begin{aligned}
f_{\mathcal{V}\mathcal{W}'}\,\pi^{-1}_{\beta\mathcal{W}''}(\theta_{\beta\mathcal{W}}) &= f_{\mathcal{V}\mathcal{U}'}\,f_{\mathcal{U}'\mathcal{W}'}\,\pi^{-1}_{\beta\mathcal{W}''}(\theta_{\beta\mathcal{W}''}) \\
&= f_{\mathcal{V}\mathcal{U}'}\,\pi^{-1}_{\beta\mathcal{U}''}\,f_{\mathcal{U}''\mathcal{W}''}(\theta_{\beta\mathcal{W}''}) \\
&= f_{\mathcal{V}\mathcal{U}'}\,\pi^{-1}_{\beta\mathcal{U}''}(\theta_{\beta\mathcal{U}''})
\end{aligned}$$

In other words, the $\phi_{\mathcal{V}}$ constructed with $\mathcal{W}$ is the same as that constructed with $\mathcal{U}$.

Finally, the effect of changing α must be examined, and once again, it is enough to consider the replacement of α by γ with $\gamma > \alpha$. Some covering must be taken for E_γ; it has just been shown that the construction is independent of this choice. Take the covering $\mathscr{W} = \pi_{\alpha\gamma}^{-1}(\mathscr{U})$. This satisfies the condition $\mathscr{U}' = \pi_\gamma^{-1}(\mathscr{W}) > \mathscr{V}$, the only condition needed. β must now be chosen so that $\beta > \gamma$. Clearly $\mathscr{U}'' = \pi_{\gamma\beta}^{-1}(\mathscr{W})$. Then the new $\phi_{\mathscr{V}}$ constructed using γ, $\mathscr{W}$, β instead of α, $\mathscr{U}$, β is $f_{\mathscr{V}\mathscr{U}'}\pi_{\beta\mathscr{U}''}^{-1}(\theta_{\beta\mathscr{U}''})$, and this is the same as before.

It has now been shown that $\phi_{\mathscr{V}}$ depends only on the covering $\mathscr{V}$ and not on the intermediate choices of α, $\mathscr{U}$, β.

Let $\mathscr{V}$, $\mathscr{W}$ be two finite open coverings of E with $\mathscr{W} > \mathscr{V}$. Choose the index α and the covering $\mathscr{U}$ of E_α so that $\mathscr{U}' = \pi_\alpha^{-1}(\mathscr{U}) > \mathscr{W}$. The construction gives

$$\phi_{\mathscr{V}} = f_{\mathscr{V}\mathscr{U}'}\pi_{\beta\mathscr{U}''}^{-1}(\theta_{\beta\mathscr{U}''})$$

$$\phi_{\mathscr{W}} = f_{\mathscr{W}\mathscr{U}'}\pi_{\beta\mathscr{U}''}^{-1}(\theta_{\beta\mathscr{U}''})$$

and it is a matter of simple computation to see that $\phi_{\mathscr{V}} = f_{\mathscr{V}\mathscr{W}}\phi_{\mathscr{W}}$; that is, the $\phi_{\mathscr{V}}$ constructed in this way are the coordinates of an element ϕ of $\check{H}_p(E, F)$.

Now it will be shown that $\pi_{\alpha *}(\phi) = \theta_\alpha$, for each index α. Given an index α, it is actually sufficient to prove that $\pi_{\gamma *}(\phi) = \theta_\gamma$ for some $\gamma > \alpha$, and the required relation follows by applying $\pi_{\alpha\gamma}$. But if $\mathscr{U}$ is any finite open covering of E_γ, this is equivalent to checking that $\pi_{\gamma\mathscr{U}}(\phi_{\mathscr{U}'}) = \theta_{\gamma\mathscr{U}}$, where $\mathscr{U}' = \pi_\gamma^{-1}(\mathscr{U})$. In fact, it is sufficient to check this with the covering $\mathscr{U}$ picked from a family of coverings of E_γ that is cofinal in the set of all finite open coverings of E_γ. Take, for example, the cofinal family satisfying the conclusion if Lemma 6-15. In this case,

$$\begin{aligned} \pi_{\gamma\mathscr{U}}(\phi_{\mathscr{U}'}) &= \pi_{\gamma\mathscr{U}}\pi_{\beta\mathscr{U}''}^{-1}(\theta_{\beta\mathscr{U}''}) \qquad \text{for suitable } \beta > \gamma \\ &= \pi_{\gamma\beta\mathscr{U}}(\theta_{\beta\mathscr{U}''}) \\ &= \theta_{\gamma\mathscr{U}} \end{aligned}$$

as was to be shown.

Thus, for given α, $\phi \in \check{H}_p(E, F)$ has been found such that $\pi_{\alpha *}(\phi) = \theta_\alpha$, and this completes the proof that π is onto.

It must now be shown that the kernel of π is zero. Let ϕ be an element of the kernel, namely, an element of $\check{H}_p(E, F)$ such that $\pi(\phi) = 0$. Then $\pi_{\alpha *}(\phi) = 0$ for each α in A. The object is to show that all the coordinates of ϕ in the inverse limit $\check{H}_p(E, F)$ are zero, that is, that $\phi_{\mathscr{V}} = 0$ for each finite open covering $\mathscr{V}$ of E. Lemma 6-14 implies that, given $\mathscr{V}$, there is an index α and a covering $\mathscr{U}$ of E_α such that $\mathscr{U}' = \pi_\alpha^{-1}(\mathscr{U}) > \mathscr{V}$. Lemma 6-15 says that there is an index β such that if $\mathscr{U}'' = \pi_{\alpha\beta}^{-1}(\mathscr{U})$, then the homomorphism

$$\pi_{\beta\mathscr{U}''} \colon H_p(E, F; \mathscr{U}') \to H_p(E_\beta, F_\beta; \mathscr{U}'')$$

is an isomorphism onto. The hypothesis on ϕ is that $\pi_{\beta*}(\phi) = 0$. This means that, in particular, the $\mathscr{U}'$ coordinate of $\pi_{\beta*}(\phi)$ in the inverse limit $H(E, F; \mathscr{U}')$ is zero; in other words $\pi_{\beta\mathscr{U}''}(\phi_{\mathscr{U}'}) = 0$. Since $\pi_{\beta\mathscr{U}''}$ is an isomorphism, $\phi_{\mathscr{U}'} = 0$, so for the given covering $\mathscr{V}$, the $\mathscr{V}$ coordinate of ϕ is $\phi_{\mathscr{V}} = f_{\mathscr{V}\mathscr{U}'} \phi_{\mathscr{U}'} = 0$. Thus the kernel of π is zero, and the proof of the continuity theorem is complete. ∎

Example

6-2. The compactness condition cannot be dropped from this theorem. Let E be the open segment $-1 < x < 1$ and F_n be the union of the segments $-1 < x \leqq 1/n - 1$ and $1 - 1/n \leqq x < 1$. Then $\check{H}_1(E, F_n)$ is isomorphic to the coefficient group $\mathscr{G}$. This can be seen directly, since simple coverings of E (cf. Lemma 6-2) are cofinal in the set of all finite open coverings of E and the nerve of such a covering is a closed segment, while the subcomplex associated with F is the union of two subsegments at the ends. In another argument, two open segments can be removed from the ends of E with closures contained in F_n. Then, by the excision theorem, $\check{H}_1(E, F_n)$ is the same as the one-dimensional homology of a closed segment modulo subsegments at its ends. This is a simplicial pair. The required result follows.

The pairs (E, F_n), indexed by the natural numbers and having inclusion maps $\pi_{mn} \colon (E, F_n) \to (E, F_m)$ for $n > m$, form an inverse system. The groups $\check{H}_1(E, F_n)$ are all equal to $\mathscr{G}$, and the induced homomorphisms π_{mn*} are all isomorphisms, so

$$\lim \check{H}_1(E, F_n) \cong \mathscr{G}$$

But $\lim (E, F_n) = (E, \varnothing)$ (cf. Exercise 6-1) and $\check{H}_1(E) = 0$; again this can be seen by using simple coverings. Hence

$$\lim \check{H}_1(E, F_n) \neq \check{H}_1(\lim (E, F_n))$$

6-6. COMPARISON OF SINGULAR AND ČECH HOMOLOGY THEORIES

While the singular and Čech homology theories share a number of properties, there are two major differences: the singular theory satisfies the exactness property, which appears only in weak form in the Čech theory, and on the other hand, the Čech theory has the continuity property, which does not appear at all in the singular theory. Of course, these differences might be more apparent than real, due to accidents in the method of proof. In order to show that the two homology theories really are different, a space must be exhibited whose singular and Čech groups are different. It already has been suggested in Example 5-14 how this could be done, and now the indications given there can be proved.

Example

6-3. Let E be defined as in Example 5-14; that is, E is the subspace of the plane consisting of points $(0, y)$ and $(2/\pi, y)$ for all y such that $0 \geqq y \geqq -2$, all points $(x, -2)$ with $0 \leqq x \leqq 2/\pi$ and the part of the graph of $y = \sin(1/x)$ with $0 < x \leqq 2/\pi$. Let E_i be the union of E with the rectangle of points (x, y) such that $0 \leqq x \leqq 1/i$, $-2 \leqq y \leqq 1$. Then it is clear that $E = \bigcap E_i$. But the E_i are a family of spaces indexed by the natural numbers, and if the maps $\pi_{ij} (j > i)$ are inclusions, then $\{E_i, \pi_{ij}\}$ is an inverse system whose limit, by Exercise 6-1, is $\bigcap E_i = E$. Also, the E_i are compact, so, by Theorem 6-16,

$$\check{H}_1(E) = \lim \check{H}_1(E_i)$$

Here E_i is triangulable, so $\check{H}_1(E_i) = H_1(E_i) = \mathscr{G}$, the coefficient group, and it is easy to see that the π_{ij} induce isomorphisms onto. Hence (Exercise 5-9) $\lim \check{H}_1(E_i) = \mathscr{G}$; that is, $\check{H}_1(E) = \mathscr{G}$.

In the singular theory, it has already been seen that $H_1(E) = 0$. So, as was to be shown, the singular and Čech theories are different.

Note that this example actually shows that the singular theory does not satisfy the continuity property. If it did $H_1(E)$ would be $\lim H_1(E_i) = \mathscr{G}$, whereas $H_1(E) = 0$.

6-7. *RELATIVE HOMEOMORPHISMS*

The continuity theorem can be combined with the excision theorem to give a rather strong form of excision in Čech homology theory for compact pairs of spaces. This will be formulated in terms of a special kind of continuous map, which will now be defined.

Definition 6-6. Let $f: (E, F) \to (E', F')$ be a continuous map whose restriction to $E - F$ is a homeomorphism of $E - F$ onto $E' - F'$. Then f is called *a relative homeomorphism.*

Theorem 6-17. *Let $f: (E, F) \to (E', F')$ be a relative homeomorphism and suppose that E, F, E', F' are all compact. Then f induces isomorphisms onto*

$$f_*: \check{H}_p(E, F) \to \check{H}_p(E', F')$$

for all p.

Proof. In effect, what will be done here is to excise F and F', using the fact that f is a homeomorphism on the remainder. Of course, this cannot be done directly as the conditions of the excision theorem may not be satisfied. So the idea is to find families of pairs to which the excision theorem can be applied, which have as limits the pairs (E, F) and (E', F') and, at the same time chosen so that f_* is an isomorphism between the Čech homology groups of corresponding pairs.

To do this, pick first a closed neighborhood U of F', that is, a closed set containing F' in its interior. Then define sets A_U, B_U, C_U, D_U as follows. $A_U = U$. B_U is an open set such that $F' \subset B_U \subset \bar{B}_U \subset \mathring{A}_U$. A set B_U can always be found with this property since E', being compact, is normal and F' is a closed subset. Next, define $C_U = f^{-1}(A_U)$ and $D_U = f^{-1}(B_U)$. Note that, since f is continuous, C_U is a closed set and D_U is open with the property $F \subset D_U \subset \bar{D}_U \subset \mathring{C}_U$. Now the set of all closed neighborhoods of F' is a directed set, where the order relation $U > V$ means $U \subset V$. So the collection of A_U is an inverse system, if the maps π_{VU} are inclusion. By Exercise 6-1, $\lim A_U = \bigcap A_U = F'$. Thus the collection of pairs (E, A_U) is an inverse system with $\lim (E', A_U) = (E', F')$.

On the other hand, the C_U also form an inverse system indexed by the closed neighborhoods of F' and $\lim C_U = \bigcap C_U$. Clearly, this contains F. But if $p \notin F, f(p) \notin F'$, so there is a closed neighborhood of F' which does not contain $f(p)$. It follows at once that $p \in \bigcap C$ so $\lim C_U = \bigcap C_U = F$. So, of course, the pairs (E, C_U) form an inverse system with $\lim (E, C_U) = (E, F)$.

Now, for each U, there is a commutative diagram

$$\begin{array}{ccc} (E - D_U, C_U - D_U) & \rightarrow & (E, C_U) \\ {\scriptstyle f_U}\downarrow & & \downarrow{\scriptstyle f_U} \\ (E' - B_U, A_U - B_U) & \rightarrow & (E', A_U) \end{array}$$

where the horizontal maps are inclusions. Also, these inclusions satisfy the conditions for the excision theorem (Theorem 6-8), so they induce isomorphisms of the Čech homology groups. The maps f_U are the appropriate restrictions of f. The one on the left is a homeomorphism and so induces an isomorphism of the Čech homology groups. Put all this together and it follows that the map f_U on the right induces an isomorphism onto

$$f_{U*}: \check{H}_p(E, C_U) \rightarrow \check{H}_p(E', A_U)$$

If $V > U$ (that is, $V \subset U$), there is a commutative diagram

$$\begin{array}{ccc} (E, C_U) & \xleftarrow{\pi_{UV}} & (E, C_V) \\ {\scriptstyle f_U}\downarrow & & {\scriptstyle f_V}\downarrow \\ (E', A_U) & \xleftarrow{\pi_{UV}} & (E', A_U) \end{array}$$

the π_{UV} being inclusion maps and f_U, f_V being induced by f, as before. Hence there is a commutative diagram

$$\begin{array}{ccc} \check{H}_p(E, C_U) & \xleftarrow{\pi_{UV*}} & \check{H}_p(E, C_V) \\ \downarrow f_{U*} & & \downarrow f_{V*} \\ \check{H}_p(E', A_U) & \xleftarrow{\pi_{UV*}} & \check{H}_p(E', A_V) \end{array}$$

That is, the f_{U*} form an inverse system of homomorphisms of the inverse system $\{\check{H}_p(E, C_U), \pi_{UV*}\}$ into $\{\check{H}_p(E', A_U), \pi_{UV*}\}$, the index set in both cases being the set of closed neighborhoods of F'. But the f_{U*} are isomorphisms onto, so $f_* = \lim f_{U*}$ is an isomorphism onto (Exercise 5-6). By the continuity (Theorem 6-16)

$$\lim \check{H}_p(E, C_U) = \check{H}_p(E, F)$$
$$\lim \check{H}_p(E', A_U) = \check{H}_p(E', F')$$

so

$$f_*: \check{H}_p(E, F) \to \check{H}_p(E', F')$$

is an isomorphism onto, as required. ∎

Example

6-4. To illustrate the preceding theorem, consider again the space E of Example 6-3, and let G be the set in the plane with the curve E as its boundary. Also let G' be the rectangle of points (x, y) such that $0 \leqq x \leqq 1/\pi$. $-2 \leqq y \leqq 1$, and let E' be the union of the boundary of G' and the part of G' above the curve $y = \sin(1/x)$. Then the relative homeomorphism theorem implies that

$$\check{H}_p(G, E) \cong \check{H}_p(G', E')$$

the isomorphism being induced by inclusion. By the homotopy theorem (Theorem 6-7), $\check{H}_p(G', E')$ is the same as the p-dimensional Čech homology group of G' modulo its boundary. Thus the Čech homology groups of the pair (G, E) are the same as those of a 2 cell modulo its boundary.

6-8. THE UNIQUENESS THEOREM FOR ČECH HOMOLOGY

The motivation for Čech homology was the somewhat vague notion of approximating a space by simplicial complexes, namely, the nerves of finite open coverings. In the case of compact pairs of spaces, this notion can be given a more precise meaning. It can be shown that a compact pair can be expressed as the inverse limit of a system of simplicial pairs. This has important consequences: if a homology theory agrees with the simplicial theory on simplicial pairs and if it satisfies the continuity condition, then it must coincide with the Čech theory on compact pairs of spaces.

The idea of the limit theorem for compact pairs is as follows. First, note that, if a compact space E is contained in some Euclidean space, then it admits a finite covering by arbitrarily small rectangular neighborhoods. The union of these neighborhoods is triangulable, so E is contained in a simplicial complex. On the other hand, this complex can be assumed to lie in an arbitrarily small neighborhood of E. Thus E is the intersection of complexes obtained in this way; that is, it is their inverse limit. This is a special case but it suggests the general procedure, for it happens that a compact space can be embedded in an infinite product of line intervals and there are (in a suitable sense) good approximations of these embedding in finite subproducts. These are contained in Euclidean spaces.

This plan will now be carried out in detail.

Lemma 6-18. *Let E be a compact space contained in Euclidean n space R^n and let U be a neighborhood of E. Then there is a simplicial complex K such that*

$$E \subset \mathring{K} \subset K \subset U$$

Proof. A rectangular set in R^n is a product of n line intervals, one contained in each copy of R. Open rectangular sets form a basis for the topology of R^n. Each point of E is contained in a rectangular open set whose closure is contained in U. These rectangular open sets form an open covering of E which can be reduced to a finite subcovering. Thus there exists a finite union K of closed rectangular sets such that $E \subset \mathring{K} \subset K \subset U$. It remains to be shown that K is triangulable. Certainly, K is contained in some cube I^n, where I is a closed line interval, so it is sufficient to triangulate I^n so that K is a subcomplex. For each rectangular set constituting K, each face is contained in a hyperplane, and the set of all these hyperplanes divides I^n into rectangular subsets. Then, if I^n can be triangulated so that the union of all the dividing hyperplanes becomes a subcomplex, K, in particular, will be triangulated. Such a triangulation of I^n can be constructed inductively. The result is trivial if $n = 1$. Now suppose that a rectangular subdivision of I^{n-1} can be triangulated so that each rectangular subset is a union of simplexes. Then a triangulation of $I^n = I^{n-1} \times I$ in a suitable number of layers (cf. proof of Lemma 6-3) gives the triangulation that is wanted here. This completes the proof of the lemma. ∎

The following theorem of general topology, the Tietze embedding theorem, is needed (cf. [66], pp. 115-118).

Lemma 6-19. *Given a compact space E, there is a product I^ω of copies of the unit interval I such that E can be identified with a subspace of I^ω.*

From now on it will be assumed that the given compact space E is embedded in the infinite cube I^ω, and an inverse system of complexes will be constructed of which E will be the inverse limit.

Let J be a finite subproduct of I^ω; that is, it is the product of a finite number of the factors making up I^ω. J is a finite cube. There is a well defined projection of I^ω onto J which is obtained by setting all coordinates of a point of I^ω equal to 0 except those corresponding to factors of J. Coordinates corresponding to the factors of J are left unchanged in this projection. The projection is a continuous map, so the projection of E into J will be a compact set in J.

Now let A be the set of all pairs (J, K), where J is a finite subproduct of I^ω and K is a simplicial complex in J that contains the projection of E into J in its interior.

Lemma 6-20. *A can be partially ordered as a directed set.*

Proof. Let $\alpha = (J, K)$ and $\alpha' = (J', K')$. Then define $\alpha' > \alpha$ to mean

(1) J is a subproduct of J' (that is, the product of some of the factors of J')

(2) the projection of K' into J is contained in the interior of K.

In (2), projection again means setting equal to 0 all coordinates except those corresponding to factors of the product J. It is clear that if $\alpha'' > \alpha'$ and $\alpha' > \alpha$, then $\alpha'' > \alpha$, so this defines a partial ordering.

Now to show that A is directed by this ordering, let $\alpha_1 = (J_1, K_1)$ and $\alpha_2 = (J_2, K_2)$ be in A. It is necessary to find α in A such that $\alpha > \alpha_1$ and $\alpha > \alpha_2$.

J_1 and J_2 are both products of finitely many factors of I. Let J be the product of all factors appearing in either J_1 or J_2. Let p_1 and p_2 be the projections of J on J_1 and J_2, respectively. The projection of E into J_1 is contained in $\mathring{K}_1$, so the projection of E into J is contained in $p_1^{-1}(\mathring{K}_1)$. Similarly, the projection of E into J is contained in $p_2^{-1}(\mathring{K}_2)$. Hence the projection of E into J is contained in $U = p_1^{-1}(\mathring{K}_1) \cap p_2^{-1}(\mathring{K}_2)$. Lemma 6-18 then implies that there is a simplicial complex K in J such that the projection of E into J is contained in $\mathring{K}$ and $K \subset U$. Hence $(J, K) = \alpha$ is in A. Also, J has J_1 as a subproduct and $K \subset p_1^{-1}(\mathring{K}_1)$; that is, K projects into the interior of K_1. Hence $\alpha > \alpha_1$. Similarly, $\alpha > \alpha_2$, so A is directed, as was to be shown. ∎

The complexes of the inverse system to be constructed are the K appearing in pairs (J, K) of A, and the mappings of the inverse system will be projections. More precisely, for each $\alpha = (J, K)$ in A, write $K_\alpha = K$. If $\alpha = (J, K)$ and $\beta = (H, L)$ are in A and $\beta > \alpha$, $\pi_{\alpha\beta}: K_\beta \to K_\alpha$ is defined as the restriction to $K_\beta = L$ of the projection of H onto J. The properties of projection, along with the meaning of order in A, imply that $\pi_{\alpha\beta}\pi_{\beta\gamma} = \pi_{\alpha\gamma}$ when $\gamma > \beta > \alpha$, and $\pi_{\alpha\alpha}$ is the identity. Hence $\{K_\alpha, \pi_{\alpha\beta}, A\}$ is an inverse system.

The main theorem can now be stated.

Theorem 6-21. *E is homeomorphic to* $\lim \{K_\alpha, \pi_{\alpha\beta}, A\}$.

Proof. For each α, there is automatically given a map $f_\alpha: E \to K_\alpha$. In other words, K_α is a complex in a finite subproduct J of I^ω, and part of the definition of A is that E should project into the interior of K_α. Then f_α is defined as the projection map of I^ω onto J, restricted to E.

Thus, for x in E, a point $f(x)$ with coordinates $\{f_\alpha(x)\}$ is obtained in ΠK_α. The object of this proof is to show that $f(x)$ is actually in $\lim K_\alpha$ and, then, that f is a homeomorphism between E and $\lim K_\alpha$.

The first point here is immediate, since the projection maps satisfy a transitivity condition. Thus if $\alpha \subset \beta$ with $\alpha = (J, K)$ and $\beta = (J', K')$, then J is a factor of J' and the projection of I^ω to J', followed by projection to J, is the same as projection of I onto J. This implies that $\pi_{\alpha\beta} f_\beta(x) = f_\alpha(x)$, so $f(x)$ is in $\lim K_\alpha$.

Certaintly f is continuous as a map into ΠK_α, since each coordinate $f_\alpha(x)$ is a continuous function of x. Hence, it is continuous as a map into $\lim K_\alpha$ with the subspace topology.

f is one-to-one. For suppose $f(x) = f(y)$; that is, $f_\alpha(x) = f_\alpha(y)$ for all α. This means that all the coordinates of x in the product I^ω which correspond to the factors in a subproduct J are the same as the corresponding coordinates of y. This is true for all finite subproducts J, so all the coordinates of x are the same as the corresponding coordinates of y; that is, $x = y$.

Thus f is a one-to-one continuous map of the compact space E into the compact space $\lim K_\alpha$ and so f is a homeomorphism into $\lim K_\alpha$. It remains to be shown that f is onto. So take y in $\lim K_\alpha$. y has coordinates $\{y_\alpha\}$, for example, in ΠK_α. To show that there is an x such that $y = f(x)$, coordinates of x in I^ω must be found. So pick one factor I_0 of the product I^ω. Let $\alpha = (J, K)$ with J containing the factor I_0. Define the coordinate x_0 of x corresponding to I_0 as the coordinate of y in J corresponding to the factor I_0. Suppose now that $\beta > \alpha$, with $\beta = (J', K')$. Then the coordinate of y_β corresponding to I_0, which is a factor of J' is the same as the coordinate of y_α corresponding to I_0 because of the condition $\pi_{\alpha\beta} y_\beta = y_\alpha$. Hence it follows that the definition of x_0 is independent of the α chosen. If all the coordinates are defined in this way, a point x in I^ω is obtained. Now this point x must be in E; if not, there is a neighborhood V of x that does not meet E and that can be assumed to belong to the basis of the topology of

I^{ω}. In other words, it can be assumed to be of the form $V_0 \times I^{\omega\prime}$, where V_0 is a rectangular open set in a finite subproduct J while $I^{\omega\prime}$ is the complementary factor of J in the product I^{ω}. Then the projection of E into J does not meet V_0 and so does not contain the projection of x into J. Thus the complement of the projection of x in J is an open set containing the projection of E, so Lemma 6-18 says that there is a simplicial complex K in J containing the projection of E in its interior, but not containing the projection of x. Write $\alpha = (J, K)$. Then the coordinates of x corresponding to factors of J are, by definition, the coordinates of y_α in J, that is, the coordinates of a point of $K_\alpha = K$. Thus K would contain the projection of x into J. This is a contradiction, so x is in E as was to be shown. Also, with any $\alpha = (J, K), f_\alpha(x)$ has the same J coordinates as x; that is, $f_\alpha(x) = y_\alpha$. So $f(x) = y$. f is thus a homeomorphism onto $\lim K_\alpha$, and the proof of the theorem is complete. ∎

As indicated earlier, the main application of this theorem is to show that, in a sense, the continuity property uniquely defines the Čech homology theory. Suppose that some construction attaches a system of groups $\overline{H}_p(E)$ to each compact space for each integer $p = 0$, and to each continuous map $f\colon E \to F$ a homomorphism $\bar{f}\colon \overline{H}_p(E) \to \overline{H}_p(F)$ satisfying the conditions $\overline{fg} = \bar{f}\bar{g}$ and $\overline{\text{(identity)}} = \text{identity}$. Also suppose that, when E is a simplicial complex, $\overline{H}_p(E) = \mathscr{H}_p(E)$ and so is also equal to $H_p(E)$ and $\check{H}_p(E)$, and that for such spaces $\bar{f} = f_*$. Suppose also that the continuity property is satisfied; that is, if $\{E_\alpha, \pi_{\alpha\beta}, A\}$ is an inverse system of spaces then

$$\overline{H}_p(\lim E_\alpha) = \lim \overline{H}_p(E_\alpha)$$

Theorem 6-22. *Under the preceding conditions $\overline{H}_p(E) = \check{H}_p(E)$ for any compact space E.*

Proof. Let E be a compact space expressed, as in Theorem 6-21, as $\lim K$ where $\{K_\alpha, \pi_{\alpha\beta}, A\}$ is an inverse system of simplicial complexes. The conditions of the theorem indicate that $\{\overline{H}_p(K_\alpha), \bar{\pi}_{\alpha\beta}, A\}$ is an inverse system coinciding with $\{\check{H}_p(K_\alpha), \pi_{\alpha\beta *}, A\}$, so the limits of these systems are the same. The first limit is $\overline{H}_p(E)$, by the assumed continuity property, and the second is $\check{H}_p(E)$, by Theorem 6-16. This completes the proof. ∎

Exercises. 6-2. Theorems 6-21 and 6-22 are formulated for single spaces. Reformulate and prove them for pairs of spaces.

6-3. This exercise gives an illustration of the use of the continuity theorem for pairs of spaces and, at the same time, shows the failure of exactness for the Čech homology theory.

The projective plane P^2 can be obtained from a disk by identifying diametrically opposite points on the circumference. If, in particular, the disk is taken as the unit disk in the complex plane, z is identified with $-z$ for points on the circumference. Let C be the image of this circumference in p^2 (note that C is a circle). Verify that the map f of the disk on itself which is defined by $f(z) = z^3$ induces a map of the pair (p^2, C) into itself, also denoted by f.

Define an inverse system $\{(E_\alpha, F_\alpha), \pi_{\alpha\beta}, A\}$ of pairs of spaces as follows: A is the set of positive integers, each (E_α, F_α) is the pair (P^2, C), and $\pi_{\alpha, \alpha+1}$ is defined as f, while the remaining $\pi_{\alpha\beta}$ are defined by the transitivity requirement. Use the continuity theorem to compute the Čech homology groups of $(E, F) = \lim (E_\alpha, F_\alpha)$, and show that the homology sequence of this pair is not exact.

7

Čech Cohomology Theory

The object of this chapter is to construct a cohomology theory corresponding to Čech homology theory. Again, nothing new will be learned for simplicial complexes, but for general spaces a continuity theorem will be proved analogous to that for Čech homology theory.

7-1. INTRODUCTION

A procedure similar to that used in Chapter 5 will be used to construct the Čech cohomology groups of a pair of spaces. Specifically, limits of cohomology groups of nerves of finite open coverings will be taken. Thus, in the notation of Section 5-3, a family of groups $\mathscr{H}^p(K_{\mathscr{U}}, K'_{\mathscr{U}})$ will be obtained, and there will be homomorphisms

$$f^{\mathscr{V}\mathscr{U}}: \mathscr{H}^p(K_{\mathscr{U}}, K'_{\mathscr{U}}) \to \mathscr{H}^p(K_{\mathscr{V}}, K'_{\mathscr{V}})$$

induced by the simplicial map

$$f^1_{\mathscr{U}\mathscr{V}}: (K_{\mathscr{V}}, K'_{\mathscr{V}}) \to (K_{\mathscr{U}}, K'_{\mathscr{U}})$$

when $\mathscr{V} > \mathscr{U}$. Note that now the homomorphisms go in the opposite direction to that of the $f_{\mathscr{U}\mathscr{V}}$ in Section 5-3. Thus a different kind of limit process needs to be defined here.

It can already be seen from this that there is an essential difference between Čech theory and singular or simplicial theory. In the latter cases, the homology groups are obtained from a chain complex and the cohomology groups from the dual cochain complex. So there is a fairly simple relation between homology and cohomology. In the Čech theory, however, the relation between homology and cohomology is complicated by the intervention of limiting processes.

At this point, a digression must be made to describe the notion of direct limit, the limit concept that will be used in defining the Čech cohomology groups.

7-2. DIRECT LIMITS

Definition 7-1. Let A be a directed set. *A direct system of groups* $\{G_\alpha, \pi^{\alpha\beta}, A\}$ is a family of groups G indexed by A, and for each $\alpha > \beta$, a homomorphism

$$\pi^{\alpha\beta}: G_\beta \to G_\alpha$$

satisfying the conditions

$$\pi^{\alpha\beta}\pi^{\beta\gamma} = \pi^{\alpha\gamma} \qquad (\alpha > \beta > \gamma)$$
$$\pi^{\alpha\alpha} = \text{identity}$$

Note that this is similar to an inverse system, except that the homomorphisms go forward, in the sense of the ordering of A, instead of backwards.

Now let ΣG_α be the direct sum of the G_α; its elements can be written uniquely as finite sums $x_{\alpha_1} + x_{\alpha_2} = \cdots + x_{\alpha_n}$ with $x_{\alpha_i} \in G_{\alpha_i}$ for each i. Addition of elements of the direct sum is carried out termwise in the separate G_α. Let R be the subgroup of ΣG_α generated by all elements of the form $x_\alpha - \pi^{\beta\alpha}x_\alpha$ $(\beta > \alpha)$.

Definition 7-2. The *direct limit of the direct system* $\{G_\alpha, \pi^{\alpha\beta}, A\}$, written as $\lim \{G_\alpha, \pi^{\alpha\beta}, A\}$ or $\lim_A G_\alpha$ or simply $\lim G_\alpha$, is defined to be the quotient group $(\Sigma G_\alpha)/R$.

The same notation is used here as that for inverse limits, but the context always shows which is intended. Note that the direct limit process has the effect of identifying $x_\alpha \in G_\alpha$ with its image $\pi^{\beta\alpha}x_\alpha$ in any later group G_β.

Exercises. 7-1. Let $x \in \lim G_\alpha$, where $\{G_\alpha, \pi^{\alpha\beta}, A\}$ is a direct system. Show that x has a representative in $\Sigma\, G_\alpha$ which is actually an element x_β of some G_β.

7-2. In the notation of Definition 7-2, suppose that $x_\alpha \in \Sigma\, G_\alpha$ and that $x_\alpha \in R$. Show that $\pi^{\beta\alpha} x_\alpha = 0$ for some $\beta > \alpha$.

(*Hint*: Start by writing

$$x_\alpha = \Sigma(x_{\alpha i} - \pi^{\beta_i \alpha_i} x_{\alpha_i})$$

Show that

$$x_\beta - \pi^{\gamma\beta} x_\beta + y_\beta - \pi^{\delta\beta} y_\beta = (x_\beta + y_\beta) - \pi^{\lambda\beta}(x_\beta + y_\beta) - (\pi^{\gamma\beta} x_\beta - \pi^{\lambda\beta} x_\beta) - (\pi^{\delta\beta} y_\beta - \pi^{\lambda\beta} y_\beta)$$

Hence show that all the α_i can be assumed to be different. Then assume that $\alpha_1 < \alpha_2 < \cdots$; note that, in $\Sigma\, G_\beta$, x_α must be equal to x_{α_1} and

$$\pi^{\beta_1 \alpha_1} x_{\alpha_1} = \sum_{i \geqq 2} (x_{\alpha_i} - \pi^{\beta_i \alpha_i} x_{\alpha_i})$$

and proceed inductively.)

7-3. Let G be a family of groups indexed by a set $\mathscr{I}$ and let A be the set of all finite subsets of $\mathscr{I}$, ordered by inclusion; that is, $\alpha < \beta$ means $\alpha \subset \beta$. Let G_α be the direct sum of the G_i for i in the set α. Thus G_α is a finite direct sum. If $\alpha < \beta$, there is a homomorphism $\pi^{\beta\alpha}: G_\alpha \to G_\beta$ which maps an element $x_i + x_j + \cdots + x_k$ of G_α on the same element of G_β, the contribution of the summands in G_β being 0 for indexes i in β but not in α. Verify that $\{G_\alpha, \pi^{\beta\alpha}, A\}$ is a direct system. Construct a map

$$\phi: \lim G_\alpha \to \sum_{i \in \mathscr{I}} G_i$$

as follows. An element x of $\lim G_\alpha$ can be represented by an element of $\Sigma\, G_\alpha$ and, in particular, by an element x_α in some G_α. But G_α is the direct sum of a finite number of the G_i and so can be identified with a subgroup of $\Sigma\, G_i$. Thus, x can be identified with an element of $\Sigma\, G_i$; call this element $\phi(x)$. Prove that ϕ is an isomorphism of $\lim G_\alpha$ onto $\Sigma\, G_i$.

Note that this result makes an infinite direct sum appear as the direct limit of finite direct sums (cf. Example 5-11).

Some properties of homomorphisms associated with direct limits will be needed. Note that these are dual to the corresponding properties of inverse limits.

First, let A' be a subset of the directed set A, and suppose that A' is also directed (by the order relation in A). Let $\{G_\alpha, \pi^{\alpha\beta}, A\}$ be a direct system. Then, automatically, $\{G_\alpha, \pi^{\alpha\beta}, A'\}$ is a direct system, and the two direct limits $\lim_A G_\alpha$ and $\lim_{A'} G_\alpha$ can be constructed. To compare these limits, note first that $\sum_{A'} G_\alpha$ can be identified with a subgroup of $\sum_A G_\alpha$. Let R be as in Definition 7-2, and let R' be the corresponding group for the index set A', that is, the subgroup of $\sum_{A'} G_\alpha$ generated by elements $x_\alpha - \pi^{\beta\alpha} x_\alpha$ with α and β in A'. R' is clearly a subgroup of R. Thus the inclusion homomorphism $\sum_{A'} G_\alpha \to \sum_A G_\alpha$ induces a homomorphism of $\sum_{A'} G_\alpha / R'$ into $\sum_A G_\alpha / R$. This is denoted by

$$\pi^{A'}: \lim_{A'} G_\alpha \to \lim_A G_\alpha$$

In particular, if A' reduces to one element α, $\pi^{A'}$ becomes a homomorphism

$$\pi^\alpha: G_\alpha \to \lim_A G_\alpha$$

Exercise. 7-4. Prove that if $\alpha > \beta$, then $\pi^\alpha \pi^{\alpha\beta} = \pi^\beta$.

The most interesting case of the homomorphism $\pi^{A'}$ occurs when A' is cofinal in A (cf. Theorem 5-10).

Theorem 7-1. *If A' is cofinal in A, then $\pi^{A'}: \lim_{A'} G_\alpha \to \lim_A G_\alpha$ is an isomorphism onto.*

Proof. Take x in $\lim_A G_\alpha$ and represent it by $x_\alpha \in G_\alpha$ for some α in A; this can be done by Exercise 7-1. Since A' is cofinal in A, there is a β in A' such that $\beta > \alpha$. Then x can also be represented by $x_\beta = \pi^{\beta\alpha} x_\alpha$. Now x_β is in $\sum_{A'} G_\alpha$, so it represents an element x' of $\lim_{A'} G_\alpha$. The definition of $\pi^{A'}$ then implies that $x = \pi^{A'}(x')$. Hence $\pi^{A'}$ is onto.

Next, take x' in the kernel of $\pi^{A'}$. Represent x' by $x_\alpha \in G_\alpha$ for some α in A'. The relation $\pi^{A'}(x') = 0$ implies that x_α is in R, so by Exercise 7-2, $\pi^{\beta\alpha} x_\alpha = 0$ for some β in A, $\beta > \alpha$. Take γ in A' such that $\gamma > \beta$, using the cofinality of A' in A. Then $\pi^{\gamma\alpha} x_\alpha = \pi^{\gamma\beta} \pi^{\beta\alpha} x_\alpha = 0$ and so $x_\alpha = x_\alpha - \pi^{\gamma\alpha} x_\alpha$ with α and γ in A'; that is, x_α is in the relation group R'. It follows that $x' = 0$, so the kernel of $\pi^{A'}$ is 0, and the proof of the theorem is completed. ∎

Limits of certain families of homomorphisms can be constructed (cf. Definition 5-15). Let $\{G_\alpha, \pi^{\alpha\beta}, A\}$ and $\{H_\alpha, \kappa^{\alpha\beta}, A'\}$ be two direct systems and let ϕ be an order-preserving map of A into A'. For convenience of notation, write $\phi(\alpha) = \alpha'$. Suppose that a family of homomorphisms f_α is given such that the following diagrams are commutative for all $\beta > \alpha$.

$$\begin{array}{ccc} G_\alpha & \xrightarrow{f_\alpha} & H_{\alpha'} \\ {\scriptstyle \pi^{\beta\alpha}} \downarrow & & \downarrow {\scriptstyle \kappa^{\beta'\alpha'}} \\ G_\beta & \xrightarrow{f_\beta} & H_{\beta'} \end{array}$$

Definition 7-3. Under the conditions just stated, the f_α are said to form a *direct system of homomorphisms*.

It will now be shown that the f_α induce a homomorphism f of $\lim G_\alpha$ into $\lim H_\alpha$. First, note that the family of f_α induces a homomorphism

$$\Sigma f_\alpha : \Sigma G_\alpha \to \Sigma H_\alpha$$

by mapping an element Σx_α on the element $\Sigma f_\alpha(x_\alpha)$. Now in this mapping, $x_\alpha - \pi^{\beta\alpha}x_\alpha$ is mapped on $f_\alpha(x_\alpha) - f_\beta \pi^{\beta\alpha}x_\alpha = f_\alpha(x_\alpha) - \kappa^{\beta'\alpha'}f_\alpha(x_\alpha)$. Thus elements of the relation group R in ΣG_α are carried into the relation group R' corresponding to the direct system $\{H_\alpha, \kappa^{\alpha\beta}, A'\}$ and Σf_α induces a homomorphism of the quotient groups

$$f : \lim_A G_\alpha \to \lim_{A'} H_\alpha$$

Definition 7-4. This homomorphism f is called *the direct limit of the family* f_α and is denoted by $\lim f_\alpha$. Note that f is determined in part by the map ϕ, but as this is usually clear from the context, it is unnecessary to include it in the notation.

Exercises. 7-5. Let $\{G_\alpha, \pi^{\alpha\beta}, A\}$, $\{H_\alpha, \kappa^{\alpha\beta}, A'\}$, and $\{L_\alpha, \lambda^{\alpha\beta}, A''\}$ be direct systems, and let $\phi: A \to A'$ and $\psi: A' \to A''$ be order-preserving maps. As before, write $\alpha' = \phi(\alpha)$. Let the f_α be a direct system of homomorphisms from the G_α to the H_α and let the g_α be a direct system of homomorphisms from the H_α to the L_α, corresponding to the maps ϕ and ψ, respectively. Check that the compositions $g_{\alpha'} f_\alpha$ form a direct system of homomorphisms from the G_α to the L_α, and prove that

$$\lim g_{\alpha'} f_\alpha = (\lim_{A'} g_\alpha)(\lim_A f_\alpha)$$

7-6. In the notation of the last exercise, suppose that, for each α, the image of f_α is equal to the kernel of $g_{\alpha'}$. Prove that the image of $\lim f_\alpha$ is equal to the kernel of $\lim g_A$.

This exercise is very important, for its shows the essential difference between the direct limit and the inverse limit. It can be formulated briefly by saying that the direct limit of exact sequences is exact (cf. Section 6-4 for the weaker result in the case of inverse limits).

7-7. In Definition 7-3, suppose that all the f_α are isomorphisms onto. Prove that $\lim f_\alpha$ is an isomorphism onto.

7-8. Show that Exercise 5-8, suitably formulated, holds for direct systems.

7-3. THE DEFINITION OF ČECH COHOMOLOGY GROUPS

Let (E, F) be a pair of topological spaces and assume that a coefficient group $\mathcal{G}$ has been fixed once and for all. For each finite open covering $\mathcal{U}$ of E, construct the simplicial pair $(K_\mathcal{U}, K'_\mathcal{U})$ as in Section 5-3, $K_\mathcal{U}$ being the nerve of $\mathcal{U}$ and $K'_\mathcal{U}$ the subcomplex associated with F. If $\mathcal{V}$ is a refinement of

$\mathscr{U}$, there is a simplicial map

$$f^1_{\mathscr{U}\mathscr{V}} : (K_{\mathscr{V}}, K'_{\mathscr{V}}) \to (K_{\mathscr{U}}, K'_{\mathscr{U}})$$

constructed as in Section 5-3.

Let

$$f^{\mathscr{V}\mathscr{U}} : \mathscr{H}^p(K_{\mathscr{U}}, K'_{\mathscr{U}}) \to \mathscr{H}^p(K_{\mathscr{V}}, K'_{\mathscr{V}})$$

be the induced homomorphism on simplicial cohomology groups, and write

$$H^p(E, F; \mathscr{U}) = \mathscr{H}^p(K_{\mathscr{U}}, K'_{\mathscr{U}})$$

Lemma 7-2. *$\{H^p(E, F; \mathscr{U}), f^{\mathscr{V}\mathscr{U}}\}$, the indexing set being the set of all finite open coverings of E, is a direct system.*

Proof. The transitivity condition on the $f^{\mathscr{V}\mathscr{U}}$ follows at once, as in Theorem 5-5. Also, $f^1_{\mathscr{U}\mathscr{U}}$ is the identity (cf. Theorem 5-5), so $f^{\mathscr{U}\mathscr{U}}$ is the identity. This completes the proof. ∎

Definition 7-5. Lim $\{H^p(E, F; \mathscr{U}), f^{\mathscr{V}\mathscr{U}}\}$ is written as $\check{H}^p(E, F)$ and is called *the* pth *Čech cohomology group of* (E, F). If F is the empty set, then $\check{H}^p(E, F)$ is written as $\check{H}^p(E)$.

7-4. INDUCED HOMOMORPHISMS

Let $f: (E, A) \to (F, B)$ be a continuous map. If $\mathscr{U}$ is a finite open covering of F and $\mathscr{U}' = f^{-1}(\mathscr{U})$, then a simplicial map

$$f_{\mathscr{U}}^1 : (K_{\mathscr{U}'}, K'_{\mathscr{U}'}) \to (K_{\mathscr{U}}, K'_{\mathscr{U}})$$

was constructed in Section 5-3. This induces a homomorphism

$$f^{\mathscr{U}} : H^p(F, B; \mathscr{U}) \to H^p(E, A; \mathscr{U}')$$

for each p. The commutativity relation (5-2) of Section 5-3 implies the commutativity of the following diagram:

$$\begin{array}{ccc} H^p(F, B; \mathscr{U}) & \xrightarrow{f^{\mathscr{U}}} & H^p(E, A; \mathscr{U}') \\ {\scriptstyle f^{\mathscr{V}\mathscr{U}}}\downarrow & & \downarrow{\scriptstyle f^{\mathscr{V}'\mathscr{U}'}} \\ H^p(F, B; \mathscr{V}) & \xrightarrow{f^{\mathscr{V}}} & H^p(E, A; \mathscr{V}') \end{array}$$

whenever $\mathscr{V} > \mathscr{U}$. Also note that the assignment of $\mathscr{U}'$ to $\mathscr{U}$ is an order-preserving map of the finite open coverings of F into the finite open coverings of E. Thus, the $f^{\mathscr{U}}$ form a direct system of homomorphisms (Definition 7-3), so their direct limit can be constructed (Definition 7-4).

Definition 7-6. $\operatorname{Lim} f^{\mathscr{U}}$ is denoted by

$$f^*: \check{H}^p(F, B) \to \check{H}^p(E, A)$$

It is called the *homorphism on Čech cohomology groups induced by f.*

Theorem 7-3. *Let* $f: (E, A) \to (F, B)$ *and* $g: (F, B) \to (G, C)$ be continuous *maps. Then* $(gf)^* = f^*g^*$. *Also, the homomorphism induced by the identity is the identity.*

Proof. The first part follows from the proof of Theorem 5-4, along with Exercise 7-5. The second part is trivial. ∎

7-5. THE EXCISION AND HOMOTOPY THEOREMS

The homotopy theorem will be proved for Čech cohomology theory, as for Čech homology, only for compact pairs of spaces, that is pairs (E, F) with E and F both compact. It is to be shown that if f and g are homotopic maps of the compact pair (E, A) in to the compact pair (F, B) then the induced homomorphisms f^* and g^* on the Čech cohomology groups are the same. As in Section 6-2, it is clearly sufficient to prove the following theorem.

Theorem 7-4. *Let* (E, F) *be a compact pair of spaces and let the maps* f *and* g *of* (E, F) *into* $(E \times I, F \times I)$ *be defined by* $f(x) = (x, 0)$ *and* $g(x) = (x, 1)$. *Then* $f^* = g^*$.

Proof. Let $\mathscr{U} \times \mathscr{I}$ be a product covering of $E \times I$ (cf. Section 6-2) where $\mathscr{I}$ is a simple covering of I. Then $f^{-1}(\mathscr{U} \times \mathscr{I}) = g^{-1}(\mathscr{U} \times \mathscr{I}) = \mathscr{U}$ and the maps f and g induce simplicial maps $g^1_{\mathscr{U} \times \mathscr{I}}$ and $g^1_{\mathscr{U} \times \mathscr{I}}$ of $(K_{\mathscr{U}'}, K'_{\mathscr{U}})$ into $(K_{\mathscr{U} \times \mathscr{I}}, K'_{\mathscr{U} \times \mathscr{I}})$ which were proved in Section 6-2 to be homotopic. Hence the two induced homomorphisms

$$f^{\mathscr{U} \times \mathscr{I}}, g^{\mathscr{U} \times \mathscr{I}}: H^p(E \times I, F \times I; \mathscr{U} \times \mathscr{I}) \to H^p(E, F; \mathscr{U})$$

coincide. But coverings of the form $\mathscr{U} \times \mathscr{I}$ are cofinal in the set of all finite open coverings of $E \times I$ (*cf.* Lemma 6-3), and so the required result follows by taking direct limits. ∎

The excision theorem for Čech cohomology depends on that for Čech homology (Theorem 6-8) and holds under the same conditions.

Theorem 7-5. *Let (E, F) be a pair of spaces and let U be an open set in E such that $\overline{U} \subset \mathring{F}$. Then the inclusion map*

$$i: (E - U, F - U) \to (E, F)$$

induces isomorphisms onto

$$i^*: \check{H}^p(E, F) \to \check{H}^p(E - U, F - U)$$

for each p.

Proof. Let $\mathscr{U}$ be a finite open covering of E and let $\mathscr{U}'$ be its trace on $E - U$ (that is to say its inverse image under inclusion). Also assume that $\mathscr{U}$ is a refinement of the covering formed by the two sets $\mathring{F}$ and $\complement \overline{U}$. It was shown in Section 6-3 that such coverings are confinal in the set of all finite open coverings of E.

It was also shown in Section 6-3 that the simplicial map

$$i_{\mathscr{U}}{}^1: (K_{\mathscr{U}'}, K'_{\mathscr{U}'}) \to (K_{\mathscr{U}}, K'_{\mathscr{U}})$$

induces isomorphisms onto of the homology groups, assumed defined over the integers as coefficient group. Hence, by Theorem 1-6, $i^1_{\mathscr{U}}$ has an algebraic homotopy inverse, and so for any coefficient group

$$i^{\mathscr{U}}: H^p(E, F; \mathscr{U}) \to H^p(E - U, F - U; U')$$

is an isomorphism onto. If direct limits are taken and Exercise 7-7 is used the required result follows. ∎

7-6. THE EXACTNESS THEOREM

At this point, it is important to note that there is an essential difference between the description of cohomology theory and that of the Čech homology theory, due to the fact that here direct limits are being used, so that exactness is preserved in the limiting process (cf. Exercise 7-6). Remember that in the Čech homology theory the exactness was lost in the process of taking the inverse limits.

As in Section 6-4, let (E, F) be a pair of spaces, both compact. Let $\mathscr{U}$ be a finite open covering of E. There is an exact cohomology sequence

(Exercise 3-6 applied to simplicial cohomology)

$$\mathscr{H}^p(K'_{\mathscr{U}}) \xleftarrow{i'^{\mathscr{U}}} \mathscr{H}^p(K_{\mathscr{U}}) \xleftarrow{j^{\mathscr{U}}} \mathscr{H}^p(K_{\mathscr{U}}, K'_{\mathscr{U}}) \xleftarrow{\partial'^{\mathscr{U}}} \mathscr{H}^{p-1}(K'_{\mathscr{U}})$$

$j^{\mathscr{U}}$ is associated with the inclusion map $E \to (E, F)$, so when the direct limit is taken over finite open coverings of E, the limit of $j^{\mathscr{U}}$ is

$$j^*: \check{H}^p(E, F) \to \check{H}^p(E)$$

(Definition 7-6).

$i'^{\mathscr{U}}$ is induced by the inclusion $h: K_{\mathscr{U}'} \to K_{\mathscr{U}}$. But (cf. proof of Theorem 6-10) there is a commutative diagram

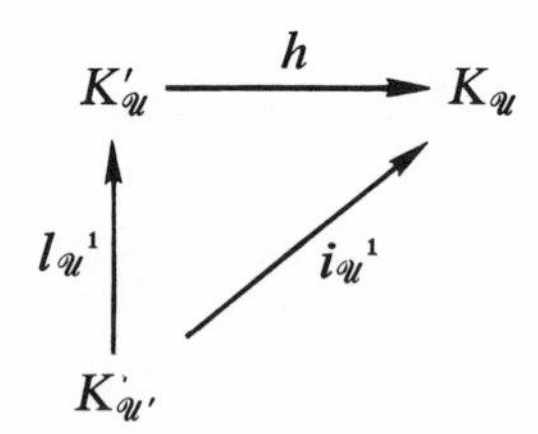

where $i^1_{\mathscr{U}}$; is induced by the inclusion $i: F \to E$, $\mathscr{U}' = i^{-1}(\mathscr{U})$ and $l^1_{\mathscr{U}}$; induces a map on simplicial chains with an algebraic homotopy inverse. So a commutative diagram can be constructed as follows:

$$\begin{array}{ccccccc}
\mathscr{H}^p(K'_{\mathscr{U}}) & \xleftarrow{\;i'^{\mathscr{U}}\;} & \mathscr{H}^p(K_{\mathscr{U}}) & \xleftarrow{\;j^{\mathscr{U}}\;} & \mathscr{H}^p(K_{\mathscr{U}}, K'_{\mathscr{U}}) & \xleftarrow{\;\partial'^{\mathscr{U}}\;} & \mathscr{H}^p(K'_{\mathscr{U}}) \\
{\scriptstyle l^{\mathscr{U}}}\downarrow & \swarrow {\scriptstyle i^{\mathscr{U}}} & & & & & \downarrow{\scriptstyle l^{\mathscr{U}}} \\
\mathscr{H}^p(K_{\mathscr{U}'}) & & & & & & \mathscr{H}^{p-1}(K_{\mathscr{U}'})
\end{array}$$

(7-1)

Here $l^{\mathscr{U}}$ is an isomorphism onto. As in the proof of Theorem 6-10, it must now be checked that all the homomorphisms in (7-1) satisfy the commutativity conditions so that direct limits can be taken. The pattern is so close to that of Theorem 6-10 that this point is left as an exercise. Take direct limits of all the groups and homomorphisms in (7-1) over the set of finite open coverings of E and use the notation:

$$j^* = \lim j^{\mathscr{U}},\ i^* = \lim i^{\mathscr{U}},\ i'^* = \lim i'^{\mathscr{U}}$$

$$l = \lim l^{\mathscr{U}},\ \partial' = \lim \mathscr{J}'^{\mathscr{U}},\ \check{H}'^p(F) = \lim \mathscr{H}^p(K'_{\mathscr{U}})$$

In this way obtain the following commutative diagram:

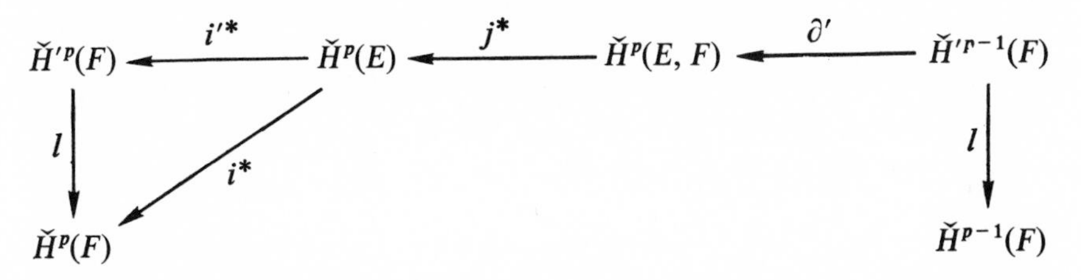

The commutativity follows from Exercise 7-5. Here the sequence along the top is exact (Exercise 7-6), and since the l are isomorphisms onto, an exact cohomology sequence

$$. \leftarrow \check{H}^p(F) \overset{i^*}{\leftarrow} \check{H}^p(E) \overset{j^*}{\leftarrow} \check{H}^p(E, F) \overset{\partial}{\leftarrow} \check{H}^{p-1}(F) \tag{7-2}$$

is obtained by writing $\partial = \partial' l^{-1}$.

Exercise. 7-9. Imitate the procedure of the proof of Theorem 6-10 to show that ∂ in (7-2) commutes with the homomorphisms induced by continuous maps.

Summing up, the following exactness theorem has been obtained.

Theorem 7-6. *There is a homomorphism ∂ commuting with the homomorphisms induced by continuous maps such that* (7-2) *is exact.*

7-7. THE CONTINUITY THEOREM

A continuity theorem holds for Čech cohomology, as for homology, with the appropriate changes from inverse to direct limits. Let $\{E_\alpha, \pi_{\alpha\beta}, A\}$ be an inverse system of compact topological spaces and let $E = \lim E_\alpha$ be its inverse limit. The map $\pi_\alpha : E \to E_\alpha$ (cf. Definition 6-4) induces a homomorphism

$$\pi_\alpha^* : \check{H}^p(E_\alpha) \to \check{H}^p(E)$$

At the same time, each $\pi_{\alpha\beta}$ (for $\beta > \alpha$) induces a homomorphism

$$\pi^{\beta\alpha} = \pi_{\alpha\beta}^* : \check{H}^p(E_\alpha) \to \check{H}^p(E_\beta)$$

and it is clear that $\{H^p(E_\alpha), \pi^{\alpha\beta}, A\}$ is a direct system. The diagram

$$\begin{array}{ccc} \check{H}^p(E_\alpha) & \xrightarrow{\pi_\alpha^*} & H^p(E) \\ {\scriptstyle \pi^{\beta\alpha}} \downarrow & & \downarrow {\scriptstyle identity} \\ H^p(E_\beta) & \xrightarrow{\pi_\beta^*} & H^p(E) \end{array}$$

is commutative for $\beta > \alpha$ (Theorem 7-3). Therefore, the direct limit of the π_α^* (Definition 7-4) can be defined: call it π.

Theorem 7-7. $\pi\colon \lim \check{H}^p(E_\alpha) \to \check{H}^p(\lim E_\alpha) = \check{H}^p(E)$ *is an isomorphism onto.*

Note here that the limit on the left is direct while that on the right is inverse.

Proof. The lemmas used in the proof of Theorem 6-16 now will be used again. The actual proof here is simpler on account of the simpler structure of the direct limit.

First, it will be shown that π is onto. Take $\theta \in \check{H}^p(E)$. To show that there is a ϕ such that $\theta = \pi(\phi)$ it is sufficient to find a representative of ϕ in $\Sigma\check{H}^p(E_\alpha)$ and, by Exercise 7-1, such a representative can be assumed to be in some $\check{H}^p(E_\alpha)$. Thus, it must be shown that there is a ϕ_α in some $\check{H}^p(E_\alpha)$ such that $\theta = \pi_\alpha^*(\phi_\alpha)$.

Now θ itself is an element of a direct limit, so θ has a representative $\theta_{\mathscr{U}'}$ in $\mathscr{H}^p(K_{\mathscr{U}'})$ for some finite open covering $\mathscr{U}'$ of E. Lemma 6-14 says that $\mathscr{U}'$ has a refinement of the form $\pi_\alpha^{-1}(\mathscr{U})$ for some finite open covering $\mathscr{U}$ of some E_α and it may be assumed (replacing $\mathscr{U}$ by a refinement if necessary) that $\mathscr{U}$ has the property that if $U_0 \cap U_1 \cap \cdots \cap U_r \cap \pi_\alpha(E) = \varnothing$ with the U_i in $\mathscr{U}$ then $\overline{U}_0 \cap \overline{U}_1 \cap \cdots \cap \overline{U}_r \cap \pi_\alpha(E) = \varnothing$ (Lemma 6-13). As soon as θ is known to have a representative in $\mathscr{H}^p(K_{\mathscr{U}'})$ the same is true if $\mathscr{U}'$ is replaced by a refinement. So it may as well be assumed that $\mathscr{U}'$ is now taken to be $\pi_\alpha^{-1}(\mathscr{U})$.

Next, Lemma 6-15 says that there is a β such that, if $\mathscr{U}'' = \pi_{\alpha\beta}^{-1}(\mathscr{U})$, then the nerves $K_{\mathscr{U}''}$ and $K_{\mathscr{U}'}$ are homeomorphic under the simplicial map $\pi_{\beta\mathscr{U}''}^1$. Then the homomorphism $\pi_\beta^{\mathscr{U}''}\colon \mathscr{H}^p(K_{\mathscr{U}''}) \to \mathscr{H}^p(K_{\mathscr{U}'})$ induced by $\pi_{\beta\mathscr{U}''}^1$ is an isomorphism onto and $\phi_{\mathscr{U}''} \in \mathscr{H}^p(K_{\mathscr{U}''})$ can be defined by

$$\phi_{\mathscr{U}''} = (\pi_\beta^{\mathscr{U}''})^{-1}(\theta_{\mathscr{U}'})$$

Write $\phi_\beta \in \check{H}^p(E_\beta)$ for the element represented by $\phi_{\mathscr{U}''}$. Since $\pi_\beta^{\mathscr{U}''}(\phi_{\mathscr{U}''}) = \theta_{\mathscr{U}'}$ it follows that $\pi_\beta^*(\phi_\beta) = \theta$. This is what was to be shown. $\phi \in \lim \check{H}^p(E_\alpha)$ can now be taken as the element represented by $\phi_\beta \in \check{H}^p(E_\beta)$.

To show that the kernel of π is 0, let $\phi \in \lim \check{H}^p(E_\alpha)$ be such that $\pi(\phi) = 0$. ϕ has a representative ϕ_α in some $\check{H}^p(E_\alpha)$ and $\pi(\phi) = 0$ means that $\pi_\alpha^*(\phi_\alpha) = 0$. Let $\mathscr{U}$ be a finite open covering of E_α and write $\mathscr{U}' = \pi_\alpha^{-1}(\mathscr{U})$. $\pi_\alpha^*(\phi_\alpha) = 0$ means that if $\phi_{\alpha\mathscr{U}}$ represents ϕ_α in $\mathscr{H}^p(K_{\mathscr{U}})$ then the induced homomorphism $\pi_\alpha{}^{\mathscr{U}}$ maps $\phi_{\alpha\mathscr{U}}$ into the relation group in $\Sigma\mathscr{H}^p(K_{\mathscr{V}})$. But $\pi_\alpha{}^{\mathscr{U}}(\phi_{\alpha\mathscr{U}}) \in \mathscr{H}^p(K_{\mathscr{U}'})$, so (Exercise 7-2) there is a refinement $\mathscr{V}'$ of $\mathscr{U}'$ such that

$$f^{\mathscr{V}'\mathscr{U}'}\pi_\alpha{}^{\mathscr{U}}(\phi_{\alpha\mathscr{U}}) = 0 \tag{7-3}$$

By Lemma 6-14 there is a β such that $\mathscr{V}'$ has a refinement of the form $\pi_\beta^{-1}(\mathscr{V})$ where $\mathscr{V}$ is a finite open covering of E_β. Since (7-3) remains true if $\mathscr{V}'$ is replaced by a refinement, it can be assumed that $\mathscr{V}' = \pi_\beta^{-1}(\mathscr{V})$ and also that $\mathscr{V}$ is taken as a refinement of $\mathscr{U}'' = \pi_{\alpha\beta}^{-1}(\mathscr{U})$. The relation $\pi_\alpha = \pi_{\alpha\beta}\pi_\beta$ implies that $\pi_\alpha{}^{\mathscr{U}} = \pi_\beta^{\mathscr{U}''}\pi_{\alpha\beta}^{\mathscr{U}}$ and (7-3) becomes

$$f^{\mathscr{V}'\mathscr{U}'}\pi_\beta^{\mathscr{U}''}\pi_{\alpha\beta}^{\mathscr{U}}(\phi_{\alpha\mathscr{U}}) = 0 \tag{7-4}$$

Now if $\phi_\beta = \pi_{\alpha\beta}^*(\phi_\alpha)$ is a representative of ϕ in $\check{H}^p(E_\beta)$, then the coordinate of ϕ_β in $\mathscr{H}^p(K_{\mathscr{U}''})$ is $\pi_{\alpha\beta}^{\mathscr{U}}(\phi_{\alpha\mathscr{U}})$. Write this as $\phi_{\beta\mathscr{U}''}$. Then (7-4) becomes

$$f^{\mathscr{V}'\mathscr{U}'}\pi_\beta^{\mathscr{U}''}(\phi_{\beta\mathscr{U}''}) = 0 \tag{7-5}$$

Note that this is essentially the same as (7-3) except that now both $\mathscr{V}'$ and $\mathscr{U}'$ are inverse images of finite open coverings of E_β. Now the $\pi_\beta{}^{\mathscr{V}}$ satisfy a commutativity relation $\pi_\beta{}^{\mathscr{V}}f^{\mathscr{V}'\mathscr{U}''} = f^{\mathscr{V}'\mathscr{U}'}\pi_\beta^{\mathscr{U}''}$ (cf. Section 7-4), so it follows that

$$\pi_\beta^{\mathscr{V}}f^{\mathscr{V}\mathscr{U}''}(\phi_{\beta\mathscr{U}''}) = 0$$

But now, replacing $\mathscr{V}$ if necessary by a refinement, assume that it satisfies the intersection property in the hypothesis of Lemma 6-15 and that β is chosen so that $\pi_\beta{}^{\mathscr{V}}$ is an isomorphism onto. Hence (7-5) implies that $f^{\mathscr{V}\mathscr{U}''}(\phi_{\beta\mathscr{U}''}) = 0$. It follows from this that $\phi_\beta = 0$ and, finally, $\phi = 0$. Thus the kernel of π is zero, and the proof is complete. ∎

Exercises. 7-10. Reformulate and prove the continuity theorem for pairs of compact spaces.

7-11. Let $f\colon (E, F) \to (E', F')$ be a relative homeomorphism, with E, F, E', F' all compact. Following the pattern of the proof of Theorem 6-17, show that $f^*\colon \check{H}(E', F') \to \check{H}^p(E, F)$ is an isomorphism onto for each p.

7-12. Prove that the uniqueness theorem, Theorem 6-22, holds for cohomology as well as homology.

Appendix A

The Fundamental Group

The study of the one-dimensional homology group of a space was motivated geometrically as the study of classes of closed paths on the space. There was an algebraic operation, commutative and written as addition, and equivalence was defined by making a path equivalent to 0 if it formed a boundary. On the other hand, a closed path, or any path for that matter, can be thought of as having a direction, as being in fact the path of a moving point traveling from the initial point toward the final point. (The path is closed if the initial and final points happen to coincide.) So it is natural to try to compose two closed paths f and g by an operation that takes this direction into account, obtaining a new path by traveling first along f and then along g. This time there would be no reason to believe that this operation would be commutative. With a suitable definition of equivalence, however, this operation induces a group structure on the set of equivalence classes. Some of the properties of this group will now be studied.

A-1. PATHS, THEIR COMPOSITIONS AND HOMOTOPIES

Definition A-1. A *path on topological space E* is a continuous map $f: I \to E$, where I is the unit interval of real numbers $0 \leqq t \leqq 1$. If $f(0) = x$ and $f(1) = y$, the path is said to *join x and y*, and x and y are called the *initial* and *final points*, respectively. If $x = y$ the path is said to be *closed.*

Note here that the path is the map, not a set of points (cf. the definition of a singular simplex, Definition 1-7).

The idea of the next definition is to express precisely the notion of joining two paths end to end, traveling first along one and then along the other.

Definition A-2. Let f be a path from x to y in a topological space E and let g be a path from y to z. Define a map $h: I \to E$ by the formulas

$$h(t) = f(2t) \qquad 0 \leqq t \leqq \tfrac{1}{2}$$

$$h(t) = g(2t - 1) \qquad \tfrac{1}{2} \leqq t \leqq 1$$

h is called the *composition of f and g* and is written as fg.

The continuity of h is easy to check, so h is a path on E. Clearly, h joins x to z.

Note that h is defined by dividing the interval I in half and then essentially applying f and g to the two subintervals. Thus the definition seems to depend on the choice of a ratio of division of I. However, in the intuitive notion of joining two paths together there seems to be no natural reason for such a dependence. It should, in fact, be possible to set up a whole family of joins of f and g, depending continuously on the ratio of division of I, and in some sense, the resulting paths should all be equivalent. This remark suggests the appropriate definition for equivalence of paths, namely, homotopy.

Definition A-3. Let f and g be two paths on a space E, both joining x to y. Let $\dot{I}$ be the subspace of I consisting of the two end points and let F be the subspace of E consisting of the two points x and y. Then if f and g are homotopic as maps of the pair $(I, \dot{I})$ into the pair (E, F), they are said to be *homotopic with respect to fixed end points*. This relation can be written $f \simeq g$. In particular, if $x = y$, so that f and g are closed paths, f and g are said to be *based on* x, and the homotopy relation just defined is called *homotopy with respect to the base point* x.

Of immediate interest is showing that the relation really is an equivalence relation, that is to say, that it is symmetric, reflexive, and transitive. The proof of each of these properties depends on the construction of a homotopy, in other words, a map of the square $I \times I$ into the space E. The condition of fixed end points x and y means that the two vertical sides $\{0\} \times I$ and $\{1\} \times I$ are mapped, respectively, into x and y. While it is important to write the appropriate formulas out explicitly, this is left as an exercise and the explanations given here will serve to show the geometric meanings of the proofs.

In the first place, if f is a path joining x and y, a map H of $I \times I$ into E can be defined by $H(t, s) = f(t)$. This shows at once that $f \simeq f$.

Next, if $f \simeq g$, a homotopy with respect to fixed end points, then a map H of $I \times I$ into E is already given. To show that $g \simeq f$, construct a new map by first turning the square upside down and then applying H.

Finally, to prove transitivity, let $f \simeq g$ and $g \simeq h$; both are homotopies with respect to fixed end points. Thus maps H and K of $I \times I$ into E are already given. Both carry the vertical sides into the fixed end points, H coincides with f on the bottom of $I \times I$ and with g on the top, and K coincides with g on the bottom and with h on the top. These maps are put together to define a map L on $I \times I$. The square $I \times I$ is divided in half by a horizontal line and L is defined to be H on the lower half and K on the upper, with a doubling of the vertical coordinate similar to the doubling of the parameter in the definition of composition of paths. L then clearly defines a homotopy between f and h with respect to fixed end points.

So, summing up, the relation $\simeq$ is an equivalence relation.

Definition A-4. The corresponding equivalence classes are called *homotopy classes*, with respect to fixed end points or a base point, as the case may be.

The next step is to show that the compositions of homotopic paths are homotopic, so that products of homotopy classes can be defined.

Lemma A-1. *If $f_1 \simeq f_2$ and $g_1 \simeq g_2$, where f_1 and f_2 join x and y and g_1 and g_2 join y and z, and the homotopies are with respect to fixed end points, then $f_1 g_1 \simeq f_2 g_2$ with respect to fixed end points x and z.*

Proof. Here maps F and G of $I \times I$ into E are given to define the homotopies $f_1 \simeq f_2$ and $g_1 \simeq g_2$, respectively. On the other hand, $f_1 f_2$ is defined by applying f_1 and f_2 to the halves of I, with the appropriate change in scale, and similarly for $g_1 g_2$. Now if $I \times I$ is divided in half by a vertical line and F and G are applied to the left and right halves, respectively, with the appropriate doubling of the horizontal coordinate, it is easy to verify that a map is obtained that defines the homotopy $f_1 f_2 \simeq g_1 g_2$, as required. ∎

Exercise. A-1. Write out explicitly the map on $I \times I$ constructed in Lemma A-1.

The next lemma treats formally the remark preceding Definition A-3.

Definition A-5. Let f and g be two paths in E, f joining x and y and g joining y and z. Let a be an interior point of the interval I and let h_a be a path on E defined as follows:

On the subinternal $[0, a]$ h is the composition of the linear map of $[0, a]$ onto $[0, 1]$ and the map f;

On $[a, 1]$ h is the composition of the linear map of $[a, 1]$ onto $[0, 1]$ and the map g.

It is easy to see that h is well defined at a and that it is continuous. Note that, if $a = \frac{1}{2}$, $h_a = fg$ is defined in Definition A-2.

Lemma A-2. *The homotopy class of* h_a $(0 < a < 1)$ *with respect to the fixed end points* x *and* z *is independent of* a.

Proof. Take $0 < a < b < 1$ and compare h_a and h_b. Let $a(t) = (1 - t)a + tb$. As t varies from 0 to 1, $h_{a(t)}$ varies continuously from h_a to h_b, thus yielding a homotopy between these paths. In detail, the required homotopy is defined by a map $H: I \times I \to E$ where

$$H(s, t) = h_{a(t)}(s)$$

It should be checked as an exercise that H has all the necessary properties. ▌

In particular, note that this lemma means that fg can be defined, up to homotopy class, as h_a, in the above notation, for any a. When this remark is applied twice over it follows that, if f, g, and h are paths on E joining x to y, y to z, and z to w, respectively, then $(fg)h$ is homotopic with respect to the fixed end points x and w to a path ϕ that is defined as follows. Divide the interval I into three equal subintervals I_1, I_2, and I_3. Then ϕ restricted to I_1 is the composition of f with the linear map of I_1 onto I; restricted to I_2 it is the composition of g with the linear map of I_2 onto I; and restricted to I_3 it is the composition of h with the linear map of I_3 onto I. But exactly the same thing could clearly be said of $f(gh)$. The following result is thus obtained.

Lemma A-3. *If* f, g, *and* h *are as just described, then* $(fg)h \simeq (f(gh)$ *with respect to the fixed end points* x *and* w.

Note that this means that, up to homotopy class, the composition of any number of paths can be written without brackets.

Since the ultimate object here is to define a group, the concepts of identity element and inverse of an element must be constructed, in the context of homotopy classes of closed paths. In the meantime, if f is a path from x to y in E, consider the possibility of constructing a path g such that $fg \simeq f$ with respect to the fixed end points x and y. Intuitively the path g, which maps the whole of I onto y, may be expected to satisfy this condition; for if the parameter on I is thought of as time, fg is in effect the same path as f but traveled at double speed for the first half of I, the moving point then remaining stationary for the second half. It should be possible to change this path continuously to f simply by reducing continuously the amount of stationary time.

This motivates the following definition and lemma.

Definition A-6. *Let* $e_x: I \to E$ map the whole of I on the point x of E. e_x is called the *constant path at* x.

Lemma A-4. *Let f be a path in E from x to y. Then $e_x f$ and fe_y are homotopic to f with respect to fixed end points x and y.*

Proof. Let f_t be a map of I into E defined as follows. On the interval $[0, \frac{1}{2} + \frac{1}{2}t]$ f_t is the composition of f with the linear map of $[0, \frac{1}{2}, \frac{1}{2}t]$ on I, and f_t maps $[\frac{1}{2} + \frac{1}{2}t, 1]$ entirely on y. The idea is that f_t is a path that changes continuously from fe_y (when $t = 0$) to f (when $t = 1$). The appropriate homotopy is a map $H: I \times I \to E$ defined by

$$H(s, t) = f_t(s)$$

It must be checked (as an exercise!) that H is continuous and satisfies the correct conditions at the end points. This shows that $fe_y \simeq f$. Similarly, it can be shown that $e_x f \simeq f$. ∎

Now let f be a path in E from x to y. When f is traced, followed by f in reverse, a doubled loop lying along f, results which begins and ends at x, and it should be possible to pull back this loop, along f itself, to the point x. This motivates the idea that gives the inverse of a homotopy class under composition of paths.

Definition A-7. Let f be a path on E from x to y. Then the *reversed path* f' is defined by

$$f'(s) = f(1 - s)$$

Lemma A-5. *In the notation just introduced $ff' \simeq e_x$ and $f'f \simeq e_y$.*

Proof. A homotopy is to be constructed which will reproduce the intuitive idea of pulling back the loop ff' along the path f. Thus, at stage t of the deformation, the loop should only reach as far as the point $1 - t$ on f. A suitable homotopy is H defined as follows:

$$H(s, t) = f(2s(1 - t)) \, (0 \leqq x \leqq \tfrac{1}{2})$$
$$H(s, t) = f(2(1 - s)(1 - t)) \, (\tfrac{1}{2} \leqq s \leqq 1)$$

It is easy to check that H is continuous and that it defines the homotopy $ff' \simeq e_x$ with respect to fixed end points, both of which are equal to x.
Similarly, $f'f \simeq e_y$. ∎

Note that this lemma and Lemma A-4 imply that if a composition of paths contains consecutive factors ff', then these factors can be cancelled without affecting the homotopy class (with respect to the initial and final end points).
The special case of closed paths will now be considered. Some additional terminology is needed in this case.

Definition A-8. Let E be a space and x a point. The set of homotopy classes of closed paths on E based on x are denoted by $\pi(E, x)$.

Let $\bar{f}$ and $\bar{g}$ be two elements of $\pi(E, x)$, that is, two homotopy classes, and let f and g be closed paths representing them. Lemma A-1 implies that the homotopy class of $\bar{f}\bar{g}$ with respect to the base point x is independent of the choice of representatives for $\bar{f}$ and $\bar{g}$ and, in fact, depends only on the classes. The homotopy class of fg is denoted by $\bar{f}\bar{g}$ and is called the product of $\bar{f}$ and $\bar{g}$. Thus a product operation has been defined on the set $\pi(E, x)$.

Theorem A-6. *$\pi(E, x)$ with the product operation just defined is a group.*

Proof. In the first place, the operation of multiplication is associative, by Lemma A-3. The homotopy class of the constant map e_x is the identity element by Lemma A-4. Finally, if $\bar{f}$ is an element of $\pi(E, x)$ and f is a path representing it, Lemma A-5 implies that the homotopy class of the reversed path f' is the inverse of $\bar{f}$. All the group axioms thus are satisfied. ∎

Definition A-9. The group $\pi(E, x)$ is called the *fundamental group of E with respect to the base point x.*

It is natural to ask to what extent the fundamental group depends on the base point x. The following theorem answers this question.

Theorem A-7. *If x and y are points of E that can be joined by a path in E then $\pi(E, x)$ and $\pi(E, y)$ are isomorphic.*

Proof. Let h be a path from x to y and let h' be the same path in reverse; that is, $h'(t) = h(1 - t)$. Let $\bar{f}$ be an element of $\pi(E, x)$ and let f be a representative closed path based on x. Let g be the path $h'fh$, a closed path based on y. If $f_0 \simeq f$, with respect to the base point x, then Lemma A-1 implies that $h'f_0 h \simeq h'fh$, with respect to the base point y, so the homotopy class $\bar{g}$ of g with respect to y depends only on $\bar{f}$ and not on the representative path

Write $\bar{g} = \phi(\bar{f})$. Thus ϕ is a map from $\pi(E, x)$ into $\pi(E, y)$. Similarly, if the roles of h and h' are interchanged a map ψ can be defined from $\pi(E, y)$ into $\pi(E, x)$. It will now be checked that ϕ and ψ are homomorphisms and are inverse to each other. Suppose that $\bar{g} = \phi(\bar{f})$ and that $\bar{f}$ and $\bar{g}$ are represented by f and $h'fh$, respectively. Then $\psi(\bar{g})$ is represented by $hh'fhh'$. By Lemmas A-4 and A-5, this is homotopic to f. Thus $\psi\phi(\bar{f}) = \bar{f}$ and so $\psi\phi$ is the identity. Similarly, $\phi\psi$ is the identity; that is, ϕ and ψ are inverse to each other. To check that they are homomorphisms (and so isomorphisms) let $\bar{f}_1$ and $\bar{f}_2$ be two elements of $\pi(E, x)$ with representative paths f_1 and f_2. Then $\phi(\bar{f}_1\bar{f}_2)$ is represented by $h'f_1 f_2 h$, which is homotopic to $h'f_1hh'f_2 h$ by Lemmas A-4 and A-5, and this is a representative of $\phi(\bar{f}_1)\phi(\bar{f}_2)$. Thus ϕ, and similarly ψ, is a homomorphism and the proof is complete. ∎

Note. If $y = x$, so that h is a closed path, the operation described in the proof of this theorem gives an inner automorphism of $\pi(E, x)$.

The proof of Theorem A-7 depends on the fact that x and y can be joined by a path. So a special case arises if this is true for any pair of points.

Definition A-10. If E is a space such that every pair of points can be joined by a path in E, then E is said to be *arcwise connected.*

Then Theorem A-7 says that, if E is arcwise connected, $\pi(E, x)$ is independent of x and so can be denoted unambiguously by $\pi(E)$.

Exercises. A-2. Show that the following spaces are arcwise connected:

(1) Euclidean space of any dimension;

(2) a Euclidean simplex of any dimension;

(3) the n-dimensional sphere for $n \geqq 1$.

A-3. Let E and F be spaces, x a point of E, and $\phi: E \to F$ a continuous map. Write $y = \phi(x)$. Let f be a closed path on E based on x. Then the composition $g = \phi f$ is a path on F based on y. Prove that the homotopy class $\bar{g}$ of g depends only on the homotopy class $\bar{f}$ of f. Writing $\bar{g} = \phi_*(\bar{f})$ show that ϕ_* is a homomorphism of $\pi(E, x)$ into $\pi(F, y)$. If $\phi: E \to F$ and $\psi: F \to G$ are continuous maps and $y = \phi(x)$, $z = \psi(y)$, show that the homomorphisms constructed in this way between $\pi(E, x)$, $\pi(F, y)$, $\pi(G, z)$ satisfy the relation $(\psi\phi)_* = \psi_* \phi_*$. Also prove that if ϕ is the identity then ϕ_* is the identity. (Note the analogy with Theorem 1-5.)

A-4. Deduce from Exercise A-3 that the fundamental group is a topological invariant.

A-5. Let ϕ and ψ be continuous maps from E to F and let $\phi(x) = \psi(x) = y$. If ϕ and ψ, as maps of the pair (E, x) into $F(, y)$, are homotopic, prove that $\phi_* = \psi_*$, in the rotation of Exercise 1-3. (Cf. Theorem 1-11.)

A-6. Deduce from the last exercise that if $\phi: E \to F$ has a homotopy inverse (that is, a map $\psi: F \to E$ such that $\phi\psi$ and $\psi\phi$ are both homotopic to the identity) then $\pi(E, X)$ and $\pi(F, y)$ are isomorphic. In particular, show that if E is a closed or open cell than $\pi(E)$ reduces to the identity element.

A-7. An arcwise connected space E for which $\pi(E)$ consists of the identity element only is called simply connected. Prove that if E is simply connected and if f and g are two paths in E joining x and y then $f \cong g$ with respect to the fixed end points x and y.

A-8. The object of this and the next two exercises is to work out a method for computing the fundamental group of the one-dimensional sphere S^1, in other words, the circumference of a circle. Let S^1 be identified with the set of complex numbers of absolute value 1, let R be the real numbers and let $p\colon R \to S^1$ be the exponential map defined by $p(x) = e^{2\pi ix}$. Let f be a closed path in S^1 based on the point 1. Prove that there is a unique path $\hat{f}$ in R with initial point 0 such that $p\hat{f} = f$.

(*Hint*: Roughly speaking, define $\hat{f}$ by taking logarithms, but first divide I into subintervals $I_1, I_2, \ldots$ such that, on each $f(I_i)$, the logarithm is well defined by its initial value.)

A-9. Remember that R is simply connected, so $\hat{f}$ in the last exercise is homotopic, with respect to fixed end points, to a linear map. Hence show that f is homotopic, with respect to the base point 1, to a map $f_n\colon I \to S^1$, for some n, where $f_n(t) = e^{2\pi int}$.

A-10. In the notation of the last exercise, show that $f_m f_n \simeq f_{m+n}$ with respect to the base point 1. Hence show that $\pi(S^1)$ is isomorphic to the additive group of integers.

A-11. Let E and F be arcwise connected spaces. Prove that $\pi(E \times F)$ is isomorphic to the direct product $\pi(E) \times \pi(F)$. Here if x and y are the base points for paths in E and F, respectively, (x, y) should be taken as the base point in $E \times F$.

(*Hint:* First note that, if p and q are the projection maps of $E \times F$ on E and F and if f is a path on $E \times F$, then pf and qf are paths on E and F whose homotopy classes depend only on that of f. Hence obtain a map $\phi\colon \pi(E \times F) \to \pi(E) \times \pi(F)$. Then check that ϕ is a homomorphism, that it is onto, and that it has kernel equal to the identity element.)

A-12. Deduce from the last exercise that the fundamental group of the torus is free Abelian with two generators.

A-2. GROUP SPACES

It frequently happens that a topological space has some additional structure, in particular, that of a group. It will now be shown that the fundamental group of such a space is commutative, the point being that here there are two multiplications with which to work, namely, group multiplication and path multiplication.

So let G be a topological group, that is, a group that is at the same time a topological space with the properties that the product $x \cdot y$ is a continuous map of $G \times G$ into G and the map $x \to x^{-1}$ is a continuous map of G onto itself. Let G be arcwise connected and let u denote its identity element. The fundamental group of G will be studied with u as base point.

If f and g are two closed paths in G based on u, define $f \cdot g$ by the equation

$$(f \cdot g)(t) = f(t) \cdot g(t)$$

The continuity of the group product implies that $f \cdot g$ is continuous, so it is a path on G. It is easy to see that it is a closed path based on u. This path must not, for the moment, be confused with fg, defined as in Definition A-2. On the other hand, $f \cdot g$ does have a property similar to one already obtained for fg (cf. Lemma A-1).

Lemma A-8. *If $f \simeq f'$ and $g \simeq g'$ then $f \cdot g \simeq f' \cdot g'$; all homotopies are with respect to the base point u.*

Proof. Let H and K be the maps of $I \times I$ into G which define the homotopies $f \simeq f'$ and $g \simeq g'$, respectively. Define the map L of $I \times I$ into G by the equation

$$L(s, t) = H(s, t) \cdot K(s, t)$$

It can easily be checked as an exercise that L satisfies all the conditions needed to define a homotopy between $f \cdot g$ and $f' \cdot g'$, with respect to the base point u, as required. ∎

This lemma means, in particular, that, without changing the homotopy class of $f \cdot g$, one can replace f and g by fe or ef and ge or eg, respectively, where e stands for the constant path, mapping all of I on u. Note now that fe is the path f traveled at double speed during the first half of I (thought of as a time interval) while eg is g traveled at double speed during the second half of I. It follows at once that the $\cdot$ product $fe \cdot eg$ is the same as fg. This is more formally stated in the next lemma.

Lemma A-9. *$f \cdot g \simeq fg$, with respect to the base point u.*

Proof. As just remarked, Lemma A-8 implies that $f \cdot g \simeq fe \cdot eg$. But $(fe \cdot eg)(t) = (fe)(t) \cdot (eg)(t)$, where

$$(fe)(t) = f(2t) \qquad 0 \leqq t \leqq \tfrac{1}{2}$$
$$(fe)(t) = u \qquad \tfrac{1}{2} \leqq t \leqq 1$$
$$(eg)(t) = u \qquad 0 \leqq t \leqq \tfrac{1}{2}$$
$$(eg)(t) = g(2t - 1) \qquad \tfrac{1}{2} \leqq t \leqq 1$$

Hence

$$(fe)(t) \cdot (eg)(t) = f(2t) \cdot u = f(2t) \qquad 0 \leqq t \leqq \tfrac{1}{2}$$
$$(fe)(t) \cdot (eg)(t) = u \cdot g(2t - 1) = g(2t - 1) \qquad \tfrac{1}{2} \leqq t \leqq 1$$

This is exactly the definition of $(fg)(t)$. ∎

Exactly the same proof applied to $(ef) \cdot (ge)$ shows that $f \cdot g \simeq gf$. So in terms now of homotopy classes, the following result has been proved.

Theorem A-10. *If G is an arcwise connected topological group then $\pi(G)$ is Abelian.*

Example

A-1. This theorem has already been illustrated in the case in which G is the circle S^1 or the torus $S^1 \times S^1$.

Exercises. The following exercises contain a number of results for the computation of fundamental groups.

A-13. Let E be a simplicial complex and let $f: I \to E$ be a path on E (not necessarily closed) joining two vertexes of E. Prove that f is homotopic, with respect to fixed end points, to a path in the 1 skeleton of E.

(*Hint*: The continuity of f implies that I can be subdivided into subintervals, each of which is mapped by f into the star of some vertex of E. But the star of a vertex is simply connected, so f can be replaced by a homotopic path f', which is a product of paths along straight line segments in E. In particular, the part of f' in an n simplex S of $E(n > 1)$ will fail to pass through some interior point and can be deformed, by radial projection from that point, into a path on the boundary of S. Repeated application of this process gives the result.)

A-14. Use Exercise A-13 to show that a sphere of any dimension greater than 1 is simply connected.

A-15. Let E be a sphere with p handles. This can be represented as a polygon with $4p$ sides with identifications as shown in Fig. 11, Section 2-9. The vertexes of the polygon are all identified; call the common point x. Show that any closed path on E that is based on x is homotopic, with respect to the base point x, to a path on the perimeter of the polygon. Thus, $\pi(E)$ is generated by elements $\alpha_i, \beta_i, i = 1, 2, \ldots, p$, where the α_i and β_i are represented by linear maps of I onto the sides of the polygon.

A-16. Exercise A-13 shows that any closed path on a simplicial complex E based on a vertex, is homotopic to a path in the 1 skeleton. Use a method similar to that of Exercise A-13 to show that if a path in the 1 skeleton is homotopic to a constant path in E then it is already homotopic to the constant path in the 2 skeleton of E; that is, the fundamental group of E is the fundamental group of the 2 skeleton of E.

A-17. Note that the last exercise essentially gives a method for expressing $\pi(E)$ by means of generators and relations. To make this more explicit, take the vertex x as base point and for each vertex y of E fix a path along the edges of E from x to y. Call this path $\lambda(y)$. Then, if a is the linear map of I onto an edge of E with end points y and z, let a' be the closed path $\lambda(y)a\lambda^{-1}(z)$. Verify that the set of a', for all edges of E, represents a set of generators of $\pi(E)$. If a', b', c' are as just described, corresponding to the three sides of a triangle in E, and their directions are suitably chosen, show that $a'b'c'$ is homotopic to a constant. Show that a basic set of relations for $\pi(E)$ can be obtained in this way.

A-18. Use the last exercise to show that, in exercise A-15, the only relation between the α_i and β_i is $\Pi\ \alpha_i \beta_i \alpha_i^{-1} \beta_i^{-1} = 1$.

A-19. For any topological space there is an important relation between the fundamental group and the one-dimensional singular homology group over the integers. Let the space E be arcwise connected and pick a point x as base point. Let $\bar{\alpha}$ be an element of $\pi(E, x)$ and let α be a representative closed path. Note that I can be identified with Δ_1, so α is a one-dimensional singular simplex on E. Prove that α is a cycle and that its homology class depends only on the homotopy class $\bar{\alpha}$, not on the representative path. Denote this homology class by $\phi(\bar{\alpha})$. Thus ϕ is a map of $\pi(E)$ into $H_1(E)$ (homology over the integers). Prove that ϕ is a homomorphism onto and that its kernel is the commutator subgroup of $\pi(E)$.

(*Hint*: The hard part is finding the kernel. Start by showing that $H_1(E)$ can be computed by using singular simplexes in E all of whose vertexes are at x. In effect, such simplexes generate a subcomplex of $C(E)$ and the point is to show that the inclusion has an algebraic homotopy inverse. Next, if α represents an element of ker ϕ, then the cycle α is a boundary $\Sigma\ a_i\ d\sigma_i$. If $d\sigma_i = \tau_{i1} + \tau_{i2} + \tau_{i3}$, the τ_{ij} are all closed paths, and if their product with the exponents a_i is taken as representing an element of $\pi(E)$ the point is to see that the result is homotopic to a constant and that in the product of this element with α^{-1} the exponents of each different factor add up to 0, thus showing that α is a representative of a commutator.)

Appendix B

General Topology

B-1. SET THEORY

The following notations are used for set theoretic concepts:

$A \cup B$	union of sets A and B
$A \cap B$	intersection of sets A and B
$A \subset B$	the set A is contained in the set B (or A is a subset of B)
$A \supset B$	the set A contains the set B
$x \in A$	the element x belongs to the set A
$A - B$	the set of elements in A and not in B
$\bigcup A_i$	the union of the sets A_i
$\bigcap A_i$	the intersection of the sets A_i
$\complement A$	the complement of A
$A \times B$	the Cartesian product of the sets A and B
ΠA_i	the Cartesian product of the sets A_i
$\varnothing$	the empty set
$f\colon A \to B$	f is a map from A to B
$f(C)$	the image of C under f
$f^{-1}(D)$	the inverse image of D under f

B-2. GEOMETRIC NOTIONS

Euclidean space R^n of dimension n is the set of n tuples $(x_1, x_2, \ldots, x_n)$ of real numbers.

The set in R^n satisfying the equation $\Sigma x_i^2 \leqq 1$ is the $(n-1)$-dimensional sphere S^{n-1}, with center the origin and radius unity.

The set of points in R^n satisfying the inequality $\Sigma x_i^2 \leqq 1$ is the solid sphere E^n of dimension n, with center the origin and radius 1. The open solid sphere is the set satisfying $\Sigma x_i^2 < 1$. Solid spheres are also called cells. (Note that both cells and spheres can be constructed with other centers and radii simply by translation and change of scale.)

B-3. TOPOLOGICAL SPACES

A topological space is a set E with a collection of subsets, called open sets, satisfying the following conditions:

(1) The union of any collection of open sets is open.
(2) The intersection of a finite collection of open sets is open.
(3) $\varnothing$ and E are open sets.

The elements of E are called points. A neighborhood of a point $x \in E$ is any set containing an open set that contains x.

A basis for the topology of a topological space is a collection $\mathcal{U}$ of sets such that the open sets are all the unions of members of $\mathcal{U}$.

The standard topology for Euclidean n space is that which has as a basis the collection of all open solid n spheres.

A subspace F of a topological space E is a subset of E, a set U in F being defined as open if and only if $U = F \cap U'$, where U' is open in E.

The interior of a set A in a topological space is the largest open set contained in A. The interior of A is denoted by $\mathring{A}$.

A covering of a topological space E is a collection of sets whose union is E. If all the sets of a covering $\mathcal{U}$ are open, $\mathcal{U}$ is called an open covering. A finite covering is one consisting of finitely many sets.

If A is a set in a topological space then $\bar{A}$, the closure of A, is defined as the set of all points x such that every neighborhood of x meets A. If $\bar{A} = A$, the set A is called closed. For any A, $\bar{A}$ is a closed set. Also, $\bar{A}$ is the smallest closed set containing A. Closure is related to interior by the operation of taking complements: $\complement\mathring{A} = \overline{\complement A}$.

B-4. CONTINUOUS MAPS

Let $f\colon E \to F$ be a map from the topological space E to the topological space F. f is continuous if, whenever A is an open set in F, $f^{-1}(A)$ is open in E.

If f is one-to-one and both f and f^{-1} are continuous, f is called a homeomorphism and E and F are said to be homeomorphic.

Topological properties are those that, when they belong to one topological space, belong to all homeomorphic spaces.

B-5. PRODUCT AND QUOTIENT SPACES

Let E_i be a family of topological spaces with i running through some set A of indexes. The topological product of the E_i is the set ΠE_i, with a basis for the topology taken as the collection of all products $\Pi\, U_i$ where U_i is open in E_i and $U_i = E_i$ for all but a finite number of indexes.

A point $x \in \Pi E_i$ is specified by its set of coordinates x_i, $x_i \in E_i$. Let $p_i(x) = x_i$. Then p_i is a continuous map of ΠE_j onto E_i, called the projection onto E_i.

Let E be a topological space and R an equivalence relation on E. Let E/R be the set of equivalence classes and let $f\colon E \to E/R$ be the map taking x in E onto its equivalence class. E/R is made into a topological space by making U open if and only if $f^{-1}(U)$ is open in E. The space E/R is called the quotient space of E by the relation R. Note that f is automatically continuous.

B-6. COMPACT SPACES

Here the definition of compactness will include the condition of being Hausdorff (this condition is not always included). A Hausdorff space E is a space such that, whenever x and y are distinct points, there are neighborhoods U and V of x and y, respectively, such that $U \cap V = \varnothing$. This condition itself is of some interest and importance. For example, a sequence of points x_n in a space E is said to have a limit x if, given a neighborhood U of x, x_n is in U for all sufficiently large values of n. If E is Hausdorff then the limit x is unique. As another example, let f and g be continuous maps of E into F. If the space F is Hausdorff then the set of points in E such that $f(x) = g(x)$ is closed.

A topological space is compact if it is Hausdorff and if every open covering contains a finite subcovering. A set in a space is compact if, regarded as a subspace, it is compact.

A closed subset of a compact is compact and a compact subset of a Hausdorff space is closed.

A set in Euclidean space is compact if and only if it is closed and bounded (Heine–Borel–Lebesgue theorem).

If E is compact and $f\colon E \to F$ is a continuous map then $f(E)$ is compact. In particular, this means that if f is a real valued function on a compact space then it attains a maximum and minimum value.

If $f: E \to F$ is a continuous map and if E is compact and F is Hausdorff, then if f is one-to-one it is automatically a homeomorphism into F. Note that, in general, a one-to-one continuous map need not have a continuous inverse.

If the spaces E_i are compact then the product space ΠE_i is compact (Tychonoff theorem). In particular, if I is a closed line interval, the product of any number of copies of I is compact. Any compact space E can be shown to be a subspace of a product of closed intervals.

If E is compact and A and B are two closed sets such that $A \cap B = \varnothing$, then there are open sets U and V such that $A \subset U$, $B \subset V$, and $U \cap V = \varnothing$. A space with this separation property is called normal; thus a compact space is normal. An alternative statement of this result is as follows: If E is compact and A is a closed set and U an open set containing A, then there is an open set V such that $A \subset V \subset \overline{V} \subset U$.

B-7. CONNECTED SPACES

Let E be a topological space. E is called connected if it is not the disjoint union of two nonempty open sets. A set in a space is connected if it is connected when regarded as a subspace. On the real line, for example, the only connected sets are the intervals, finite or infinite, with or without end points.

If E is connected and $f: E \to F$ is a continuous map then $f(E)$ is a connected set.

If $f: I \to E$ is a continuous map of the unit interval I of real numbers, $0 \leqq t \leqq 1$, into E with $f(0) = x$ and $f(1) = y$, f is called a path joining x and y. If there is a path joining every pair of points of E, E is said to be arcwise connected.

An arcwise connected space is connected.

If x is a point of a topological space E, the largest connected set of E containing x is called the connected component of x. Similarly, the arcwise connected component of a point x can be defined.

Bibliography

The following titles are those actually referred to in the text.

[1] S. Eilenberg and N. Steenrod, *Foundations of Algebraic Topology.* Princeton University Press, Princeton, New Jersey, 1952.

[2] J. L. Kelley, *General Topology.* Van Nostrand, Princeton, New Jersey, 1955.

[3] A. H. Wallace, *Introduction to Algebraic Topology.* Pergamon Press, New York, 1957.

The following titles are suggestions for complementary reading, showing different approaches to the subject of algebraic topology with different emphases on the various topics.

S. S. Cairns, *Introductory Topology.* Ronald Press, New York, 1961.

M. Greenberg, *Lectures on Algebraic Topology.* Benjamin, New York, 1966.

S. Lefschetz, *Introduction to Topology.* Princeton University Press, Princeton, New Jersey, 1949.

H. Seifert and W. Threlfall, *Lehrbuch der Topologie.* Chelsea, New York.

The following are suggestions for further reading, at a more advanced level.

P. J. Hilton and S. Wylie, *Homology Theory.* Cambridge University Press, Cambridge, 1960.

S. MacLane, *Homology.* Springer-Verlag, New York, 1963.

E. Spanier, *Algebraic Topology.* McGraw-Hill, New York, 1966.

Finally, while most of this text is concerned with homology and cohomology groups, there is another class of algebraic invariants that can be attached to topological spaces, the simplest example being the fundamental group. The following are suggested as introductions to the theory of these invariants.

P. J. Hilton, *Introduction to Homotopy Theory.* Cambridge University Press, Cambridge, 1953.

S. T. Hu, *Homotopy Theory.* Academic Press, New York, 1959.

Index